**Progress in Scientific Computing**
**Vol. 2**

**Edited by**
**S. Abarbanel**
**R. Glowinski**
**G. Golub**
**H.-O. Kreiss**

Birkhäuser
Boston · Basel · Stuttgart

# Numerical Treatment of Inverse Problems in Differential and Integral Equations

## Proceedings of an International Workshop, Heidelberg, Fed. Rep. of Germany August 30 - September 3, 1982

P. Deuflhard
E. Hairer, editors

1983

Birkhäuser
Boston • Basel • Stuttgart

Editors:

Peter Deuflhard
Ernst Hairer
Universität Heidelberg
Institut für Angewandte Mathematik
Abt. Numerische Mathematik
6900 Heidelberg 1
Fed.Rep. of Germany

Library of Congress Cataloging in Publication Data
Main entry under title

Numerical treatment of inverse problems in differential
and integral equations.

(Progress in scientific computing ; v. 2)
1. Inverse problems (Differential equations)--
Numerical solutions--Congresses. 2. Differential
equations--Numerical solutions--Congresses.
3. Integral equations--Numerical solutions--Congresses.
I. Deuflhard, P. (Peter) II. Hairer, E. (Ernst)
III. Series.

QA370.N85 1983 515.3'5 83-3815

CIP-Kurztitelaufnahme der Deutschen Bibliothek

Numerical treatment of inverse problems in differential and integral equations : proceedings of an internat. workshop, Heidelberg, Fed. Rep. of Germany, August 30 - September 3, 1982 / P. Deuflhard ; E. Hairer, ed. - Boston ; Basel ; Stuttgart : Birkhäuser, 1983.
(Progress in scientific computing ; Vol. 2)

ISBN-13:978-0-8176-3125-3 e-ISBN-13:978-1-4684-7324-7
DOI: 10.1007/978-1-4684-7324-7

NE: Deuflhard, Peter (Hrsg.); GT


Softcover reprint of the hardcover 1st edition 1983

CONTENTS

PREFACE

In many scientific or engineering applications, where ordinary differential equation (ODE),partial differential equation (PDE), or integral equation (IE) models are involved, numerical *simulation* is in common use for prediction, monitoring, or control purposes. In many cases, however, successful simulation of a process must be preceded by the solution of the so-called *inverse problem*, which is usually more complex: given measured data and an associated theoretical model, determine unknown parameters in that model (or unknown functions to be parametrized) in such a way that some measure of the "discrepancy" between data and model is minimal. The present volume deals with the numerical treatment of such inverse probelms in fields of application like chemistry (Chap. 2,3,4, 7,9), molecular biology (Chap. 22), physics (Chap. 8,11,20), geophysics (Chap. 10,19), astronomy (Chap. 5), reservoir simulation (Chap. 15,16), elctrocardiology (Chap. 14), computer tomography (Chap. 21), and control system design (Chap. 12,13).

In the actual computational solution of inverse problems in these fields, the following typical difficulties arise: (1) The evaluation of the *sensitivity coefficients* for the model may be rather time and storage consuming. Nevertheless these coefficients are needed (a) to ensure (local) uniqueness of the solution, (b) to estimate the accuracy of the obtained approximation of the solution, (c) to speed up the iterative solution of nonlinear problems. (2) Often the inverse problems are *ill-posed*. To cope with this fact in the presence of noisy or incomplete data or inevitable discretization errors, *regularization* techniques are necessary. (3) *Inappropriate modelling* may lead to unsuccesful or, at least, unsatisfactory computer runs: this phenomenon may occur either, if the models are too small (so that important sub-processes are missing), or, if the model is too large (so that the available experimental data determine only a parameter subspace in the model).

In recent years rapid developments in computing methods for ordinary, and partial differential equations as well as for integral equations have taken place. Therefore, a workshop on the numerical treatment of the important class of inverse problems in these fields seemed to be both timely and needed. Such a workshop was held from Aug. 30- Sept. 3, 1982, at the University of Heidelberg under the auspices of the Deutsche Forschungsgemeinschaft (Sonderforschungsbereich 123). The workshop brought together numerical mathematicians, who work on the theory and development of algorithms for inverse problems, and scientists, who have to actually solve challenging real life inverse problems. The present proceedings combine the invited papers and a selection of the contributed papers.

Part I covers the *initial value problem* approach to parameter identification in ODEs. Chap. I (Gear, Vu) contains a basic study, warning users of ODE software about the non-smooth parameter dependence of most presently available numerical integrators. The next three chapters deal with inverse problems in chemical mass action kinetics. In Chap. 2 (Nowak, Deuflhard) two user-oriented software packages for parameter identification in complex reaction systems are presented. One of them is based on a new theoretical concept ($L_2$-minimization in lieu of the standard $l_2$-minimization). In Chap. 3 (Schlöder, Bock) a special treatment for reaction systems exhibiting hysteresis behavior (such as the well-known Zhabotinski-Belousov reaction) is proposed: here the parameters can be determined from measurements of the limit points by solving a constrained, highly nonlinear least squares problem of special structure. The modelling aspect for small reaction systems is essentially focussed in Chap. 4 (Lachmann). Finally, in Chap. 5 (Zadunaisky) the recovery of small perturbations of the right-hand side of ODEs is treated. Modelling problems of this kind are familiar in astronomy, but may become increasingly important also in further disciplines.

Part II covers the *boundary value problem* approach to parameter identification in ODEs (Chap. 6-9), inverse Sturm-Liouville problems (Chap. 10, 11), and the modern field of control system design, which basically reduces to new interesting numerical linear algebra questions (Chap. 12, 13). In Chap. 6 (Deuflhard, Bader) a unified theoretical perturbation analysis of analytic and discrete boundary value problems and multiple shooting techniques is given. On this basis, a new, more efficient multiple shooting code is developed. In Chap. 7 (Bock) the multiple shooting approach and a collocation approach for parameter identification are compared. Moreover, experimental evidence for the superiority of generalized Gauss-Newton techniques over the (formerly used) random optimization technique is demonstrated. In Chap. 8 (England) a series of real life nonlinear eigenvalue problems is shown to be tractable by a special multiple shooting code. In Chap. 9 (Seelig, Füllemann) certain small reaction-diffusion systems are analytically reduced to a system of periodic nonlinear boundary value problems, which are treated by means of truncated Fourier series. The next two chapters deal with inverse Sturm-Liouville problems: given measurements for a number of eigenvalues, what can be stated about the potential? Chap. 10 (Hald) contains a short contribution about the type of problem encountered in models for the Earth mantle, while Chap. 11 (Fiedeldey, Lipperheide, Sofianos) treats a typical application in quantum mechanics. The last two chapters of this part deal with the rapidly developing field of control system design. Chap. 12 (Fletcher) gives a survey of some generalized matrix eigenvalue problems that arise in this context. In Chap. 13 (Kautsky, Nichols, van Dooren, Fletcher) two algorithms for robust eigenstructure assignment are suggested to apply to state feedback ODE systems.

Part III contains the treatment of various inverse problems in *partial differential equations*, which may, in some cases, also include the reformulation in terms of integral equations. In Chap. 14 (Colli Franzone) some inverse problems from electrocardiology are analyzed and solved numerically. Computer simulations in this field are used in hospitals to monitor the infarct size in the intensive care units. The basic inverse problem is to determine the potential distribution on the heart surface from measurements on the body surface. The associated ill-posed elliptic Cauchy problem is discretized using appropriate finite elements, and a special regularization for the discrete equations is imposed. In Chap. 15 (Ewing) model problems for use in reservoir simulation for primary and secondary recovery of petroleum are presented. The governing porous media equations (assuming Darcy's law) are known to be parabolic for compressible fluids and elliptic for incompressible fluids. Porosity and permeability (pointwise parametrized) appear as unknowns in a highly ill-conditioned, extremely large constrained nonlinear least square problem. The article gives an extensive survey on the various methods and the different sources of errors arising in actual field computations. In Chap. 16 (Hornung) the uniqueness question for the inverse problem in a special infiltration experiment is studied. Additional measurements are suggested to guarantee uniqueness of the solution in certain parameter domains. In Chap. 17 (Eriksson, Dahlquist) the class of inverse diffusion problems is considered in which the diffusion coefficient depends nonlinearly on the solution. For this class a working algorithm is suggested. In Chap. 18 (Eldên) a special ill-posed parabolic problem is transformed into a Volterra integral equation of the first kind, which, in turn, is solved using regularization. In Chap. 19 (Mundry) geoelectric field measurements are used to identify unknown resistivities and thicknesses of ground layers. A semi-analytic preparation leads to an ill-conditioned system of nonlinear equations, which is solved by a Marquardt algorithm. Finally, Chap.20 (Hermans) sketches an approach to identify unknown small perturbations of the refraction index in a two-dimensional Helmholtz equation including the case of incomplete data (which is related to the Radon transform discussed in Chap. 21).

Part IV covers *Fredholm integral equations of the first kind*. Chap. 21 (Natterer) treats the limited angle problem in computer tomography. On a theoretical basis, algorithmic considerations about the possible exploitation of the structure of the ranges of Radon transforms are presented. In Chap. 22 (Provencher, Vogel) numerical techniques for several typical inverse probelms arising in molecular biology are discussed. Among these, numerical deconvolution using a special B-spline representation and three-dimensional image reconstruction from electron microscopy are described in detail. The next two chapters deal with numerical deconvolution. Chap. 23 (Davies, Iqbal, Maleknejad, Redshaw) gives a survey and comparison of several ways to choose the regularization parameters (a challenging practical problem!). In Chap. 24 (Philip) a comparative theoretical study of the deconvolution problem for different non-negative kernels is given. Finally, Chap. 25 (Engl) contains a detailed functional analytic investigation of the discretization error in collocation methods for linear Fredholm equations of the first kind.

As the organizers of the workshop, we would like to acknowledge financial support by the Deutsche Forschungsgemeinschaft for the Sonderforschungsbereich 123. Moreover, we wish to thank I. Heitz and G. Gienger for their careful reading and unifying the final form of the manuscripts.

We hope that our meeting in Heidelberg has stimulated further work in the field, and that the present volume will be helpful in the hands of numerical analysts, applied mathematicians, and scientists who work on practical inverse problems in differential and integral equations.

Heidelberg, December 1982

*P. Deuflhard, E. Hairer*

# PART I

# INVERSE INITIAL VALUE PROBLEMS IN ORDINARY DIFFERENTIAL EQUATIONS

# SMOOTH NUMERICAL SOLUTIONS OF ORDINARY DIFFERENTIAL EQUATIONS

C.W. Gear and Thu Vu


## Abstract

When the function evaluation in parameter estimation by error norm minimization or in a range of related optimization problems involves the numerical solution of differential equations, the smoothness of that solution with respect to parameter changes is critical to the behavior of the minimization code (which might be trying to estimate partial derivatives numerically). If fixed stepsize, fixed order methods are used for integration, it is easy to see that the integrands have the desired smoothness, but such methods are not always efficient or even possible. Modern automatic integrators have very poor smoothness properties-sometimes not even continuous. If the solution of $y' = f(y,t,p)$, $y(0) = y_0(p)$ is denoted by $y(t,p)$ where p is a parameter, and the numerical solution from a code using tolerance $\varepsilon$ is $y(t,p;\varepsilon)$, we hope that

$$\| y(t,p) - y(t,p;\varepsilon) \| = O(\varepsilon) .$$

In a "smooth" method we also want

$$\left\| \frac{\partial^s y(t,p)}{\partial p^s} - \frac{\partial^s y(t,p;\varepsilon)}{\partial p^s} \right\| \leq O(\delta)$$

In practice, the second term is computed numerically. If the integrator is not smooth, we are forced to use $\varepsilon = O(\delta(\Delta p)^s)$ which can be very expensive. In this paper we examine methods for which the above inequality can hold with $\varepsilon = \delta$. Preliminary experiments with Runge-Kutta like codes show some promise.


## 1. Introduction

Figure 1 shows the output from a typical, modern, automatic integrator as an input parameter to the program is varied. In this case, the integrator is RKF45, the parameter is the tolerance $\varepsilon$ and the problem is

$$y' = -(y-F) + (y-F)^2 + F'$$

$$F = \sin t, \quad y(0) = 1 ,$$

integrated to a particular value of t. (Note: the figures in this paper are sketches from actual computer output and are qualitatively correct.) Although some users may vary the tolerance to get an idea of the global accuracy, the usual reason for varying a parameter is to compute Jacobians of a function defined by an initial value problem with respect to parameters in order to do parameter estimation. The behavior of the output of an automatic integrator as any parameter is varied behaves in the same fashion as that shown in Figure 1, namely the output is a piecewise-smooth function of the parameter with frequent jump discontinuities. These jumps are due to the stepsize and order selection mechanisms and it is easier to see exactly what is happening when the parameter closest to the control mechanism, $\varepsilon$, is varied directly. Hence, our examples in this paper will generally consider $\varepsilon$ as the parameter, although our remarks apply to the case of more general interest. The discontinuities arise when a change of a parameter causes the code to employ a different path of execution.

As we can see from Figure 1, numerical estimation of the derivatives can introduce serious errors due to the jumps in the solution of the order of the integrator tolerance. If a parameter is changed by $O(\Delta)$ to get an $O(\Delta)$ accurate estimate of the derivative, the integration error could cause an additional derivative estimate errror of $O(\varepsilon/\Delta)$ so we must use $\varepsilon=O(\Delta^2)$ for optimality. This leads to a very expensive numerical integration. We would like to have an integrator that yields smooth solutions in the sense that

$$\left\|\frac{\partial^s \hat{y}}{\partial p^s} - \frac{\partial^s y}{\partial p^s}\right\| \le k_s \varepsilon \tag{1}$$

where $y = y(t_e,p)$ is the solution of the problem

$$y' = f(y,t,p) \ , \quad y(0) = y_0(p) \tag{2}$$

at the output point $t_e = t_e(p)$, p is a parameter, and $\hat{y}$is the corresponding computed solution. It is of theoretical interest to try to achieve this for all s such that $\partial^s y/\partial p^s$ is continuous, but in practice we will be happy with s = 1 and possibly s = 2. (Of course, in practice this is impossible because in the presence of finite precision, $\hat{y}$ is constant almost everywhere, but we assume that we are working well above roundoff error levels so that its effect can be ignored.)

If the original problem can be differentiated symbolically, we can solve the variational equations

$$\left(\frac{\partial y}{\partial p}\right)' = \frac{\partial f}{\partial p} + \frac{\partial f}{\partial y}\frac{\partial y}{\partial p} \tag{3}$$

directly, and when these equations are not unduly complicated, this is probably the best strategy. However, if $\partial f/\partial y$ and/or $\partial f/\partial p$ are expensive to evaluate compared to f, or if f is given in a form in which symbolic integration is impractical (e.g., it involves tabular data or "black box" code), we are forced into numerical differentiation.

In the next section we will discuss a simple well known scheme briefly. This scheme is based on disabling the automatic control so that it cannot cause the difficulties seen in Figure 1. When this scheme is applicable but symbolic differentiation is not, it is probably the best technique for computing derivatives, but there are situations in which it cannot be used, so in the third section we will discuss automatic integrators which can yield solutions with the desired smoothness.

## 2. Fixed-mesh Techniques

The most obvious technique is to use a fixed-order, fixed-stepsize technique. One can see immediately that under reasonable conditions this yields as much differentiability as that of the original problem. Consider, for example, the backward Euler method:

$$y^{n+1} = y^n + hf(y^{n+1},p,t^{n+1}) \tag{4}$$

where $y^n$ is the computed approximation to $y(t^n)$ and $t^n = nh$. Writing $d = \partial y/\partial p$ and $d^n = \partial y^n/\partial p$ we find that

$$d^{n+1} = d^n + hf_y d^{n+1} + hf_p \tag{5}$$

where subscripts denote partial differentiation. Equation (5) is the result of applying the same backward Euler method to the defining equation for d, equation (3). If we are in a region in which $\|(I - hf_y)^{-1}\| \le k$, the solution of equations (4) and (5) will both lead to O(h) approximations to the solutions of equations (2) and (3) and we get the desired behavior.

Equal stepsizes do not lead to an efficient integrator for most problems, but it is not essential that the stepsizes be equal, only that they be the same in the two numerical solutions differenced to estimate a derivative. Note that we can replace h by $h^n$ in equations (4) and (5), where $t^{n+1} = t^n + h^n$ to get the variable stepsize method, and the same result follows. In the case of a general variable-order, variable-step-size code it is necessary only to follow exactly the same stategy in two program executions to ensure that the derivative of the numerical scheme is the result of applying the same numerical method to equation (3).This can be seen by noting that once we have fixed the sequence of stepsizes, orders, number of iterations, etc., the code gives an explicit computational scheme for determining the output from the input. This scheme can be differentiated with respect to p. This, differentiation causes y terms to be replaced by d terms and f(y,p,t) terms to be replaced by $f_y d + f_p$ terms. The resulting scheme is the result of applying the same numerical method applied to equation (3).

In practice, the easiest way to estimate the derivative numerically is to perform both integrations simultaneously, computing $\hat{y}(t^n,p)$ and $\hat{y}(t^n,p+\Delta)$ while selecting the stepsize and order sequence to control the sum of the errors in the two components. This is called "internal differentiation" by Bock (these proceedings).

Bock has pointed out that if the value t at which the derivative is needed depends on p, say $t_E(p)$, we must calculate the derivative

$$\frac{d}{dp} y(t_E(p),p) = y_p(t_E(p),p) + y'\frac{\partial t_E}{\partial p} \tag{6}$$

The first term can be estimated numerically while the second term can be handled directly (although $\partial t_E/\partial p$ may have to be handled numerically).

There is another way to handle differentiation at a p-dependent point which is worth examining because it indicates the source of a condition we will need in the next section. Suppose we compute the solution at a sequence of mesh points $t^n(p)$ which depends (smoothly) on p. Let us introduce another independent variable, $\tau$, and a relation $t^n(q)=t(\tau^n,q)$; $p = q$. Since $\tau$ is largely undetermined, set $\tau = t(\tau,p_0)$ at the parameter value, $p_0$, of interest. Now we can view all functions as functions of t and p or $\tau$ and q. Define

$$D(t,p) = \frac{\partial y}{\partial q} = \frac{\partial y}{\partial p} + y'\frac{\partial t}{\partial q} \tag{7}$$

Here, prime still denotes differentiation with respect to t. As before, let $y^n = \hat{y}(t^n,p)$ and define

$$D^n = \frac{\partial y^n}{\partial q} \tag{8}$$

Note that this is the value we will compute if we differentiate the code with respect to p. Considering the backward Euler method equation (4) as an example with $h = h^n$ we get by differentiating (4) with respect to q

$$D^{n+1} = D^n + h^n[f_y D^{n+1} + f_p + f_t t_q^{n+1}] + fh_q^n \tag{9}$$

Differentiate equation (7) with respect to t to get

$$D' = \frac{d}{dt}(y_p) + y''\frac{\partial t}{\partial q} + y'\frac{d}{dt}\left(\frac{\partial t}{\partial q}\right)$$

$$= \frac{\partial}{\partial p}(y') + y''\frac{\partial t}{\partial q} + y'\frac{d}{dt}\left(\frac{\partial t}{\partial q}\right)$$

$$= f_y y_p + f_p + (f_y y' + f_t)\frac{\partial t}{\partial q} + y'\frac{d}{dt}\left(\frac{\partial t}{\partial q}\right)$$

or

$$D' = f_y D + f_p + f_t\frac{\partial t}{\partial q} + y'\frac{d}{dt}\left(\frac{\partial t}{\partial q}\right) \tag{10}$$

We note that equation (9) is the result of applying the backward Euler method to equation (10) if

$$h^n \frac{d}{dt}\left(\frac{\partial t}{\partial q}\right) = \frac{\partial h^n}{\partial q} \tag{11}$$

If this holds, then the derivatives $D^n$ of the numerical solution will have the same order of approximation to D as $y^n$ does to y.

What about condition (11)? One situation in which this holds is that of linear stretching. If the final point $t_E(q)$ depends on q, we can apply a uniform stretching as q changes by using

$$t(\tau,q) = \frac{t_E(q)}{t_E(p_0)}\, t(\tau,p_0) = \frac{t_E(q)}{t_E(p_0)}\,\tau \tag{12}$$

For this $\partial t/\partial q = R(\tau)\ \partial t_E/\partial q$ where $R(\tau) = \tau/t_E(p_0)$. Hence, along $p_0$

$$\frac{d}{dt}\left(\frac{\partial t}{\partial q}\right) = \frac{\partial t_E}{\partial q}\frac{1}{t_E(p_0)} \tag{13}$$

while

$$\frac{\partial h^n}{\partial q} = \frac{\partial(t^{n+1}-t^n)}{\partial q} = h^n \frac{\partial t_E}{\partial q}\frac{1}{t_E(p_0)} \tag{14}$$

Equations (13) and (14) can be seen to satisfy equation (11). A more careful analysis, mentioned in the next section, shows that equation (14) is stronger than needed, and we only require that $\partial h^n/\partial q = O(h_n)$ to get the result we want, namely that $\partial y^n/\partial q$ is an approximation to $D(t^n,p)$ of the same order as $y^n$ is an approximation to $y(t_n)$.

## 3. Smooth, Variable-stepsize Integrators

A property of an integrator that virtually guarantees the smoothness of its output is that it follows exactly the same computational path at the arithmetic level for all inputs. This means that there can be no MAX, MIN, INT, or ABS functions, and no IF statements, causing the computation to be explicit and of fixed length. Clearly this is impossible in an automatic integrator since, at the very least, there must be an IF statement (or equivalent) to decide when to stop, and the stopping point will occur after a number of steps that depend on the input. However, IF statements are permissible if the action taken in the THEN clause is smoothly connected to the action taken in the ELSE clause. This will be exploited in our code. The type of action that is not allowed is a test for acceptable error estimates in an integration step followed by outright rejection and choice of a different stepsize if the estimate is too large. This is the cause of the discontinuities in Figure 1. Similar problems can occur in a variable-order code when the order is changed because, with our current understanding and implementation of order, order changing is a discrete process rather than a smooth one.

We have considered fixed-order codes to avoid the latter problems, and have avoided all IF statements except ones concerned with determining when the output point has been reached. This has been possible by using a variant of the code described in [2] by Vu. This code was itself an extension of the Runge-Kutta starter described in [1]. We call the method behind the new code a Runge-Kutta-Taylor (RKT) method. The idea is to first evaluate a set of intermediate Runge-Kutta values based on a stepsize $h_R$, namely :

$$k_i = h_R f(y + \textstyle\sum_j \beta_{ij} k_j)$$

for $1 \leq i,j \leq q$ and $j < i$ (for explicitness). From these the scaled derivative estimates at $t^n$

$$s_m = \frac{(h_R)^m y^{(m)}}{m!} \cong \textstyle\sum_j \gamma_{mj} k_j \qquad (15)$$

are calculated for $1 \leq m \leq p$. It is possible to find coefficients such that these have errors of $O((h_R)^{p+1})$. We are using $p = 4$. Gear [1] shows that six is the smallest value of q for which this is possible. The values in equation (15) can be used to calculate $\hat{y}$ at $t^{n+1}$ via

$$\hat{y}^{n+1} = T^n(h^n) = \hat{y}^n + \textstyle\sum_{m=1}^{p} (\frac{h^n}{h_R})^m s_m \qquad (16)$$

This is called the Taylor part of the step. The Taylor stepsize $h^n$ is chosen from an error estimate calculated from the $k_i$ values. (This is described by Vu [2]. One extra stage is used to get a useful estimate.) Consequently, a step is never rejected; instead, a large error estimate E in the RK part of the step causes a small value of $h^n$ to be chosen while a small estimate causes a large value to be chosen via a smooth formula of the form

$$h^n = h_R \, \phi \, (\frac{E}{\varepsilon}) \qquad (17)$$

where $\varepsilon$ is the integrator tolerance parameter. Finally, the value of $h_R$ for the next step is chosen based on the most recent $h^n$ and possibly $E/\varepsilon$. As long as these functions are smooth, it is possible to get the desired smoothness in the result, but the choice of these functions is still the subject of experiment and will be described in forthcoming reports.

At this stage we are able to prove the following:

<u>Theorem</u>

If $\partial^j h^n / \partial p^j = O(h^n)$, $1 \leq j \leq s$ and $f(y,t,p)$ and the functions described above have continuous derivatives of order s, then

$$\left\| \frac{\partial^j}{\partial p^j} y(t^n(p),p) - \frac{\partial^j y^n}{\partial p^j} \right\| = O(\varepsilon) \qquad \text{for } 1 \le j \le s \tag{18}$$

Note that the theorem refers to derivatives along $t^n$ trajectories because $t^n$ changes as p changes. Typically we want the output at some specified $t_E(p)$ and would like $\hat{y}(t_E(p),p)$ to be a smooth function of p. A program must compute $\hat{y}(t,p)$ by means of some interpolation formula, say a

$$\hat{y}(t,p) = I(y^{n-j}, y^{n-j+1},\ldots,y^{n+k};t) \quad \text{for } t^n \le t \le t^{n+1} \tag{19}$$

Computing $\hat{y}(t,p)$ involves an IF statement to determine which interval of integration contains t, but it is easy to get the desired derivative continuity, for example, by using splines. However, derivative continuity is not the complete answer as Figure 2 illustrates. It shows the output of the RKT integrator using a $C^1$ piecewise cubic for interpolation. The size of the oscillations is unacceptable and would cause relatively large errors in derivative estimates.

The problem is to fit a smooth interpolant whose errors have small oscillations. This has proved to be one of the more difficult parts of the task. The first approach we tried, apparently the obvious one for the RKT method, was to blend the Taylor series of equation (16) forward from $t^n$ with the Taylor series backward from $t^{n+1}$ in the form

$$\hat{y}(t) = \alpha(s)\, T^n(t-t^n) + (1-\alpha(s))\, T^{n+1}(t-t^{n+1}) \tag{20}$$

where $s = (t-t^n)/(t^{n+1}-t^n)$, $\alpha(0) = 1$, $\alpha(1) = 0$, and the derivatives of $\alpha$ are chosen to get the desired derivative continuity. This was not very successful because intermediate values of $T^n(t)$ change fairly rapidly. We next tried a formula similar to equation (20) where $T^n(t-t^n)$ is defined to be the result of integrating by one step of the RK formula from $t^n$ to t. Thus, for $t^n < t < t^{n+1}$, $T^n(t-t^n)$ is the forward integral from $t^n$, while $T^{n+1}(t-t^n)$ is the backward integral from $t^{n+1}$. This gave a result of the form shown in Figure 3: a pronounced wiggle that can be explained as follows. For very small $t-t^n$, $T^n(t-t_n)$ provides a good approximation to the integral curve of the differential equation through $(y^n,t^n)$. Hence, the blended value is tangent to the integral curves for all n. Since the integral curve through $(y^n,t^n)$ does not, in general, pass through $y^{n+1}$ or $y^{n-1}$, as shown in Figure 3, the blended solution has an oscillatory second derivative. The difficulty in this approach arises because we are insisting that $\hat{y}(t)$ match the shape of the integral curves at $t^n$.

At first sight it could appear that an approach based on fitting a piecewise polynomial to just the solution values $y^n$ would be the best approach. However, our experiments indicate that the error in the interpolation greatly exceeds the error in the integration (see Figure 2) for sensible degrees of polynomials (e.g., quintics). A satisfactory solution to this problem still remains to be found. In some situations it is possible to use the approach in equation (6) to compute the derivative. If we write

$$\hat{y}(t) = y^n + I(s)$$

where $s = (t-t^n)/(t^{n+1}-t^n)$ and I is the interpolating function, we can write

$$\frac{\partial \hat{y}}{\partial p}(t) \cong \frac{\partial y^n}{\partial p} + \left.\frac{\partial I(s)}{\partial p}\right|_{s\text{ fixed}} - y'\frac{\partial t(s)}{\partial p} \tag{21}$$

In other words, to differentiate numerically we integrate with one value of p to get $\hat{y}(t)$, then integrate for the second value of p and evaluate the interpolant at the same point relative to the $t^n$ values as in the first integration. A correction factor $y'\partial t(s)/\partial p$ can then be subtracted from the estimated derivative. However, if this is possible, the internal differentiation of Bock discussed in the last section is probably possible and would be more efficient. Work is continuing on this problem.

## 4. Conclusion

We have found ways to compute values at the mesh points of an automatic integrator that are smooth functions of the parameters. Further work remains to be done on the interpolation formulas.

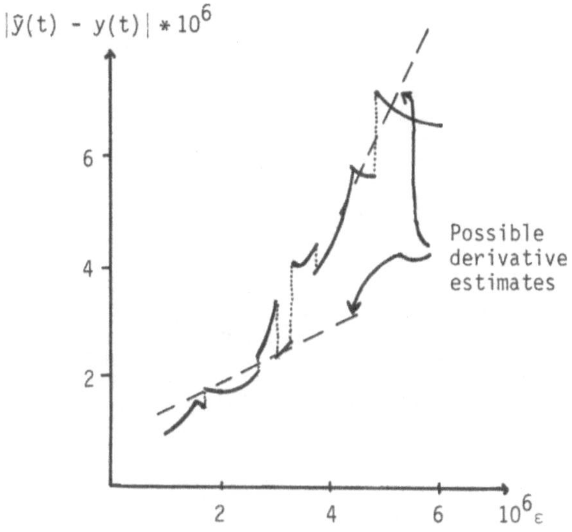


Figure 1. Output from RKF45

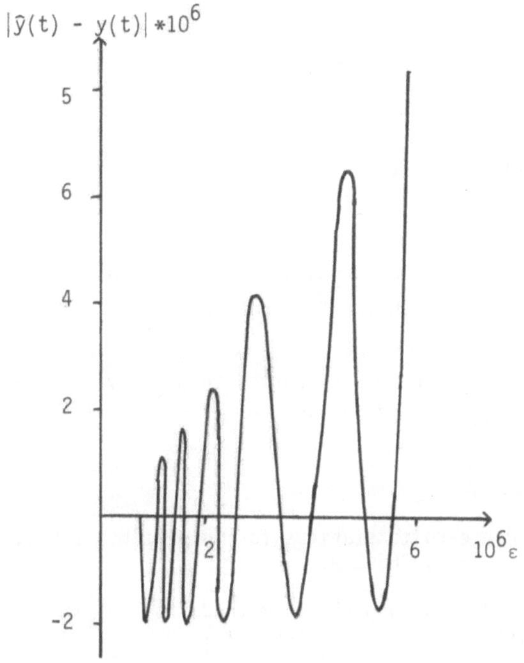


Figure 2. Output from "Smooth" Integrator

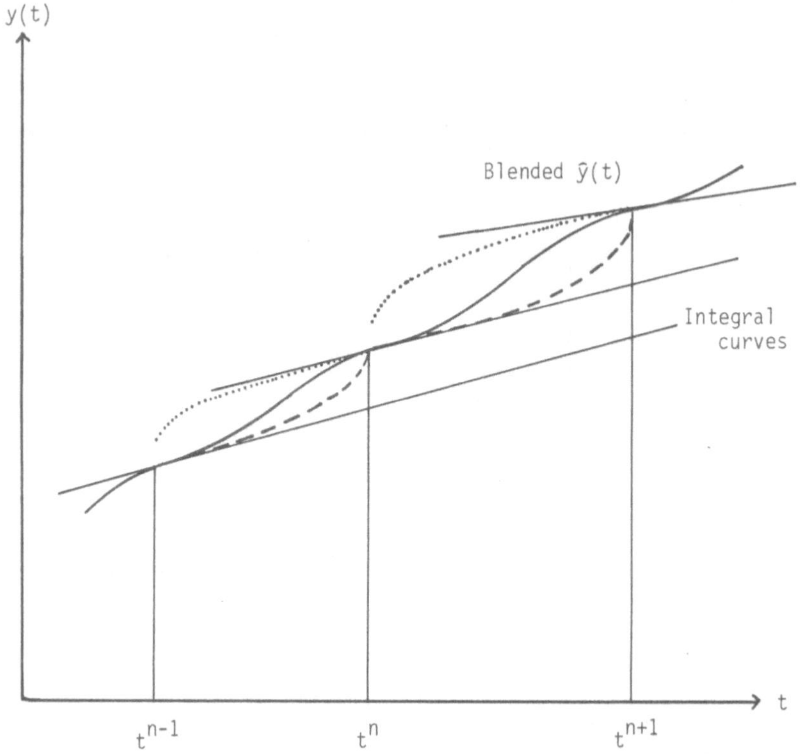


Figure 3. Blending of RK Solutions

Legend

......... RK solution back from next point

--------- RK solution forward from last point

Acknowledgement: This work was supported in part by the U.S. Department of Energy, Grant DOE DEAC0276ER02383.

References

[1] Gear, C.W., Runge-Kutta starters for multistep methods, ACM TOMS 6 (3), September 1980, 263-279.

[2] Vu, Thu V., Modified Runge-Kutta methods for solving odes, Dept.Rpt. UIUCDCS-R-81-1069, Dept. Computer Sci., Univ. Illinois, 1981.

# TOWARDS PARAMETER IDENTIFICATION FOR LARGE CHEMICAL REACTION SYSTEMS

U. Nowak, P. Deuflhard

## 0. Introduction

A standard task in the modelling of chemical reaction systems (CRS) is the *identification of rate constants* in the kinetic equations from given experimental data - which is the often so-called *inverse problem* (IP) of chemical kinetics (as opposed to simulation, the direct problem). For sufficiently complex CRS, the modelling problem itself is already rather intricate. So there is a need for user-oriented software that allows the chemist to concentrate on the chemistry of his process under investigation. As a first step in this direction, *simulation packages* have been developed - such as FACSIMILE [5], CHEMKIN [15] or LARKIN [10,3].

One possibility to attack the IP is to combine any such simulation package with any nonlinear least squares(NLSQ) fit routine. External combinations by the user, however, tend to lead to rather inefficient algorithms, which may nevertheless be of same value in the case of *small* CRS. For *large* systems, the special structure of the IP of chemical kinetics needs to be exploited in more detail.

In section 1, the result of telescoping the simulation package LARKIN due to [10,3] and the NLSQ program NLSQA due to [7] is described. This NLSQ technique was selected, since it is especially well-suited for numerically sensitive NLSQ problems and includes an efficient treatment of rank-deficiency of the Jacobian. The resulting code PARKIN1 realizes a Gauss-Newton iteration for the *discrete* NLSQ problem ($l_2$-minimization). In section 2, the IP is formulated as a special $L_2$-minimization problem - an approach, which seems to be new, at least in the field of chemical kinetics. The resulting code PARKIN2 realizes a Gauss-Newton iteration for a naturally constructed *continuous* NLSQ problem. In section 3, numerical comparisons of PARKIN1 and PARKIN2 are presented.

## 1. Discrete Functional Approach

The usual IP of chemical mass action kinetics arises in the following mathematical form: given a system of n in general nonlinear ordinary differential equations (ODEs)

$$y' = f(y;p)\ ,\quad y(0)\ \text{given},\ p \in \mathbb{R}^q\ , \tag{1.1.a}$$

determine the parameter vector p such that the solution y(t,p) "fits" to given experimental data

$$(t_1,z_1),\dots,(t_m,z_m) \qquad z_j \in \mathbb{R}^n \quad . \tag{1.1.b}$$

The deviations at the measurement points are

$$dy(t_j) := y(t_j;p) - z_j .$$

In the standard approach to be treated in this section, one determines p from minimizing some weighted $l_2$-norm of the errors (see Fig. 1)

$$\| F(p) \|^2 := \sum_{j=1}^{m} dy(t_j)^T D_j\ dy(t_j) = \min. \tag{1.2}$$

$D_j$ : positive-definite weighting (n,n)-matrix (usually diagonal)

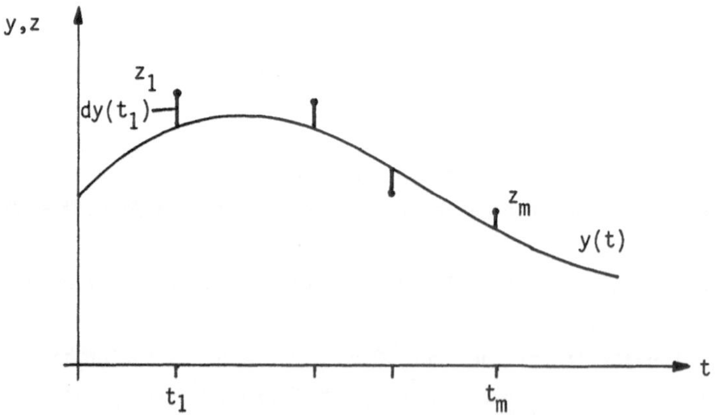


Fig. 1. Schematic representation of the discrete functional approach

*Remark.* In applications, often only *some* of the components of $z_j$ can be measured. This case is formally included here by setting the associated components of $dy(t_j)$ to zero. To avoid notational inconvenience, the presentation throughout the paper assumes that *all* components of $z_j$ are measured.

*Gauss-Newton method.* The determination of p requires an iterative procedure starting from some given initial guess $p^0$. For reasons not to be discussed here, the authors favor the Gauss-Newton iterative procedure in the special form suggested in [7]: this algorithm requires the actual computation of

$$\begin{aligned} \Delta p^k &:= -F'(p^k)^+F(p^k) && \text{ordinary GN correction} \\ \Delta \bar{p}^{k+1} &:= -F'(p^k)^+F(p^{k+1}) && \text{simplified GN correction} \end{aligned} \tag{1.3}$$

where F(p) is the above defined mn-vector, and F'(p) the associated Jacobian (mn,q)-matrix. A new iterate

$$p^{k+1} := p^k + \lambda_k \Delta p^k \ , \quad 0 < \lambda_k \leq 1 \tag{1.4.a}$$

is accepted, if the so-called natural monotonicity test

$$\| \Delta \bar{p}^{k+1} \|_2^2 \leq \| \Delta p^k \|_2^2 \tag{1.4.b}$$

holds (in a scaled norm). In order to expand the convergence domain, the theoretically backed damping strategy (determining $\lambda_k$) due to [6] is applied. Moreover, a successful treatment of rank-deficiency is included on the basis of the detailed rounding error analysis in [9].

*Approximation of the Jacobian.* In LARKIN, the polynomial right-hand side f in (1.1.a) is automatically generated (evaluation of a formal subroutine) from input in terms of chemical reactions - for details see [3], section 2. Using the same type of data structure, one may also generate the right-hand sides of the q *variational equations*

$$y_p' = f_y(y;p)y_p + f_p(y) \ , \quad y_p(0) = 0 \ , \tag{1.5}$$

where $y_p(t)$ is the *sensitivity* (n,q)-*matrix*. With this notation, the Jacobian reads

$$F'(p) = \begin{bmatrix} y_p(t_1) \\ \vdots \\ y_p(t_m) \end{bmatrix}$$

Thus, the sufficiently accurate numerical integration of (1.5) combined with (1.1.a) directly yields a sufficiently accurate approximation of the Jacobian. Note that, in contrast to some of the other types of Jacobian approximation techniques, the above version permits a reliable control of the approximation error and, at the same time, economic exploitation of the sparsity pattern of $f$, $f_y$, $f_p$. If $q$ is smaller than $n$ (standard case treated here), then the presently required additional storage (beyond that required from LARKIN) is roughly $10nq$. (For $q \geqq n$, a different implementation would be preferable, which has not yet been realized).

*Positivity constraints.* From the chemist's point of view, only positive parameters $p$ are of interest. Therefore, in view of the Arrhenius law, the transformation

$$p_i = e^{u_i} \qquad i = 1,\dots,q \tag{1.6}$$

may be used. This implies the changes

$$F(p) \to \hat{F}(u) = F(e^u)$$

$$F'(p) \to \hat{F}'(u) = \begin{bmatrix} y_u(t_1) \\ \vdots \\ y_u(t_m) \end{bmatrix}$$

Let $\bar{D} := \text{diag}(p_1,\dots,p_q)$, then $y_p \to y_u = y_p\,\bar{D}$, $f_p \to f_u = f_p\,\bar{D}$.

Note that, since $p$ arises linearly in $f$, the entries of $f_u$ are just those terms that anyway appear in $f$ - a fact, which permits computational simplifications. Now, Gauss-Newton iteration in the variable $u$ ensures that all iterates $p^k$ are positive, once $p^0$ is positive. For the termination of the u-iteration, an *absolute error criterion* is sufficient to guarantee final relative accuracy of the p-iterates, since

$$du = \frac{dp}{p} \ . \tag{1.6'}$$

*Incompatibility factor.* After a successful run, PARKIN1 supplies an estimate of the incompatibility factor $\kappa(p^*)$, which was introduced in

[7, 8]. In the full rank case, this factor indicates whether the underlying statistical problem was well-posed.

## 2. Continuous Functional Approach

Obviously, the discrete functional approach described in section 1 above just uses *pointwise* information. Thus, it does not reflect the fact that every observation $(t_j, z_j)$ defines a *sub-trajectory*

$$z_j(t;p),\quad t \in [t_j, t_{j+1}]$$

by virtue of the initial value problem

$$z_j' = f(z_j;p) \quad , \quad z_j(t_j) = z_j \quad , \quad t \in [t_j, t_{j+1}] \qquad (2.1)$$

The associated deviation along the trajectory is

$$dy(t) := y(t;p) - z(t;p)$$

where (with $t_{m+1}$ formally included):

$$z(t) := \{z_j(t) \quad \text{for } t \in [t_j, t_{j+1}]\}$$

In order to take the continuous structure of the IP into account, one may determine p from minimizing some weighted $L_2$-norm of the deviations:

$$I[p] := \sum_{j=1}^{m} \int_{t_j}^{t_{j+1}} dy(t)^T D(t) dy(t) dt = \min. \qquad (2.2)$$

$D(t)$ : positive-definite weighting matrix

For an illustration of this approach see Fig. 2. Such an approach (which is new to the knowlegde of the authors) might be promising for the following reasons:

- the nature of the underlying time-dependent process is reflected,
- the dependence on the node distribution is reduced,
- more information about the parameter influence on the process is included.

*Remark.* The definition of the $z_j$ above involves a bias in the positive time direction. Such a bias, however, seems to be appropriate in view of the fact that the underlying chemical reactions proceed in just this direction. Moreover, the usual stiffness of the ODE system exhibits the same kind of bias.

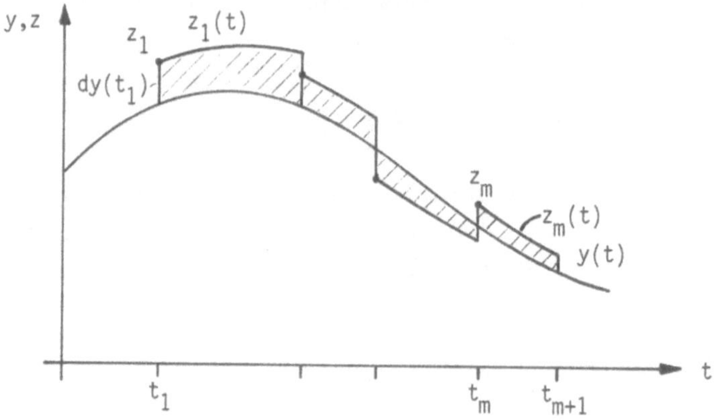


Fig. 2. Schematic representation of the continuous functional approach

*Ordinary Gauss-Newton (GN) method.* To simplify the subsequent presentation, the above functional I may be written as an inner product in the Hilbert space $L_2$:

$$I[p] \equiv \langle dy, dy \rangle \ . \qquad (2.2')$$

The derivation of the GN method requires

$$\begin{aligned} I'[p] &= \langle dy_p, dy \rangle & \text{q-vector} \qquad (2.3) \\ I''[p] \to \bar{I}''[p] &:= \langle dy_p, dy_p \rangle & \text{(q,q)-matrix} \end{aligned}$$

Then the ordinary GN correction satisfies

$$\langle dy_p^k, dy_p^k \rangle \, \Delta p^k = - \langle dy_p^k, dy^k \rangle \ , \quad k = 0,1,\ldots \ . \qquad (2.4)$$

Throughout this section, the left-hand (q,q)-matrix is assumed to be strictly positive definite - an assumption, which is actually monitored in the algorithm (see below).
The relation (2.4) defines an operator $\Gamma$ by

$$\begin{aligned} &\Delta p = -\Gamma(p) dy \\ &\Gamma : \mathbb{R}^q \to \mathcal{L}(L_2, \mathbb{R}^q) \ . \end{aligned} \qquad (2.4')$$

Under the above full rank assumption, one easily verifies the *projection property*

$$\Gamma dy_p = I \quad . \tag{2.5}$$

Then, under mild assumptions, Theorem 4 of [8] guarantees *local convergence* of the ordinary GN iteration for a class of $L_2$-minimization problems - to be called *adequate* as a direct extension of the definition in [7].

*Damping strategy.* In order to expand the convergence domain, some type of damping strategy should be applied. Following the lines of section 1, (1.3)/(1.4), would require the evaluation of

$$\Delta\bar{p}^{k+1} := -\langle dy_p^k, dy_p^k\rangle^{-1} \langle dy_p^k, dy^{k+1}\rangle \ ,$$

which requires essentially the same amount of computing costs as the evaluation of $\Delta p^k$ - in contrast to the discrete case of section 1. Therefore, the damping strategy due to [6] would be too expensive. As an alternative, one may realize a *Newton-like method* for the solution of $I'[p] = 0$, which would require the evaluation of

$$\Delta\hat{p}^{k+1} := -\langle dy_p^k, dy_p^k\rangle^{-1} \langle dy_p^{k+1}, dy^{k+1}\rangle \ .$$

Note that $\langle dy_p^{k+1}, dy^{k+1}\rangle$ is anyway needed for the evaluation of $\Delta p^{k+1}$. However, such a strategy leads to extremely small values of the damping factor $\lambda_k$. Therefore, a damping strategy in the spirit of the one suggested by [1] was implemented: one selects the greatest $\lambda_k \in \{1, \frac{1}{2}, \dots, \frac{1}{64}\}$, for which

$$I[p^k + \lambda_k \Delta p^k] \leq I[p^k] + \frac{1}{2}\lambda_k I'[p^k]^T \Delta p^k \quad . \tag{2.6}$$

This strategy turned out to be cheap and quite efficient.

*Computation of functionals.* The evaluation of the inner products is performed by introducing functions $\alpha(t)$ (scalar), $\beta(t)$(q-vector), and $\gamma(t)$((q,q)-matrix, reduced using symmetry):

$$\begin{aligned} \alpha(t_{m+1}) &= I \ , & \alpha(t_1) &= 0 \ , \\ \beta(t_{m+1}) &= I' , & \beta(t_1) &= 0 \ , \\ \gamma(t_{m+1}) &= \bar{I}'' , & \gamma(t_1) &= 0 \ . \end{aligned} \tag{2.7}$$

This means the numerical integration of the following large ODE system:

$$\begin{aligned}
&\text{a)} \quad y' = f(y;p) \\
&\qquad z' = f(z;p) \\
&\text{b)} \quad y_p' = f_y(y;p)y_p + f_p(y) \\
&\qquad z_p' = f_z(z;p)z_p + f_p(z) \\
&\text{c)} \quad \alpha' = dy^T D dy , \\
&\text{d)} \quad \beta' = dy_p^T D dy , \\
&\qquad \gamma' = dy_p^T D dy_p .
\end{aligned} \tag{2.8}$$

Initial values for $y, y_p, \alpha, \beta, \gamma$ at $t_j$ are obtained from continuity and (1.1.a),(1.5),(2.7), while the initial values for $z, z_p$ are

$$z(t_j) = z_j \quad , \; z_p(t_j) = 0 \quad .$$

The whole system is integrated by means of an extrapolation method: for the stiff part (2.8.a,b), the semi-implicit mid-point rule (SIMPR) due to [2] is applied, for the mere quadrature part (2.8.c,d), the explicit mid-point rule (EMPR) due to [14] is used. The existence of an asymptotic $h^2$-expansion for the total scheme is easily verified by applying Theorem 1 in [2] with a suitably partitioned matrix A. Note that evaluating I for the damping strategy (2.6) just requires the solution of (2.8.a,c). The similarities of the (2q+2) ODE systems (2.8.a,b) permit a rather efficient realization of the SIMPR in combination with the sparse matrix technique as worked out in [11].

*Linear system solution.* In contrast to the situation of section 1, the continuous approach of the present section directly yields a kind of *normal equations* - see (2.4). The symmetric, poisitive semi-definite matrix is treated by means of a *rational Cholesky decomposition* in the actual form published in [16]. The above full rank assumption is then easily checked by monitoring the positivity of the arising diagonal elements.

For the sake of completeness, it may be mentioned that *positivity constraints* and *incompatibility factor* can be realized as in section 1.

## 3. Numerical Comparisons

In this section, the two approaches described above are compared. These two approaches are implemented in two packages with the following features:

PARKIN1: based on standard *discrete* functional minimization (section 1), Gauss-Newton method with damping strategy due to [6], treatment of rank-deficient case included,

PARKIN2: based on new *continuous* functional minimization (section 2), Gauss-Newton method with damping strategy following [1], no treatment of rank-deficient case included (yet).

Both codes are in a user-oriented form and apply sparse matrix techniques as in the simulation package LARKIN [10,3]. Compared with PARKIN1, the algorithm in PARKIN2 requires about *double* the amount of computing time *per iteration* (just compare (2.8') with (1.1.a)/(1.5)) and considerably more storage. In smaller problems the PARKIN codes can be compared with the multiple shooting code PARFIT due to [4]. In its present form, however, that code is far from being user-oriented. Moreover, PARFIT requires drastically more storage in large systems.

All PARKIN runs were performed on the IBM 370/168 of the Computing Center of the University of Heidelberg. (in FORTRAN double precision using the optimizing compiler).

### 3.1. Denitrogenation of Pyridine (due to [4])

This rather small example comprises 11 chemical reactions, $n = 7$ ODEs, and $q = 11$ parameters to be identified from 77 measurements. As initial guess, one takes $p^0 = (1,\dots,1)^T$ from [4]. A rough comparison of PARKIN1, PARKIN2 and PARFIT is given in Table 1. In this example, the code PARKIN1 performs slightly faster than the code PARFIT with $m = 7$ multiple shooting nodes. Note, however, that finding this *optimal* set of nodes required several trials. Since for *large* systems PARFIT

Table 1. Comparative results in example 3.1 (PARFIT runs due to [17])

| | Time (sec) | Iteration number | Iterations with $\lambda < 1$ |
|---|---|---|---|
| PARFIT (m=2) | 42 | 18 | 13 |
| PARFIT (m=7) | 16 | 8 | 2 |
| PARKIN1 | 13 | 11 | 6 |
| PARKIN2 | 37 | 10 | 4 |

with $m > 2$ is too storage consuming, the single shooting run ($m = 2$) is also included. A detailed interpretation of the above numbers is difficult for the following reasons: (a) all codes use the same type of stiff integrator, but with different scaling, (b) in view of large systems, the PARKIN codes evaluate formal subroutines for $f$, $f_y$, $f_p$ (which is slower), while PARFIT requires a user given subroutine $f$, (c) PARKIN integrates the variational equations numerically, while PARFIT employs internal numerical differentiation, (d) the damping strategies in the Gauss-Newton methods differ. Finally, compared with PARKIN1, the code PARKIN2 seems to have a larger domain of local convergence (for $\lambda = 1$).

### 3.2. Glycolytic Pathway ( due to [13])

This medium size example comprises 89 chemical reactions, $n = 6$ ODEs, and a selection of q=6 parameters to be identified. For illustration purposes, the typical PARKIN input is presented in Table 2.

Table 2. Part of PARKIN input for example 3.2.

```
*REACTION SYSTEM
GLU  +  HKS        <=>  HKG                 (3.7D6)    (1.5D3)
HKG  +  1TP        <=>  HKP                 (4.D6)     (6.5D6)
HKP                <=>  HKU + ADP           (3.D-3)    (2.D6)*
HKU                <=>  HKS + GLP           (1.D3)*     (4.D5)*
ADP + HKG           =>  HKI                 (2.D6)
HKI                 =>  ADP + HKG           (1.D3)
GLP + ISM          <=>  ISG                 (8.7D7)*   (1.57D3)*
ISG                <=>  FRP + ISM           (1.84D3)*   (3.2D8)
FRP + FRP + PFK     =>  PFF                 (1.91D11)
FRP + FRP + PFK + ADP =>  PFF                (2.4D14)
PFF                 =>  PFK + FRP           (1.4D3)
PFF + 1TP           =>  ADP + FPP + PFK     (1.1D4)
PFF + FPP          <=>  PFI                 (9.D6)     (8.)
PFI + ADP           =>  PFF + FPP           (9.D5)
FPP + ALD          <=>  ALF                 (3.11D7)   (4.02D2)
ALF                <=>  ALA + GAP           (5.24D3)   (3.3D7)
ALA                <=>  DHA + ALD           (2.D2)     (8.33D6)
DHA + TIM          <=>  TID                 (1.29D5)   (2.5D2)
TID                <=>  GAP + TIM           (1.14D1)   (2.01D6)
ALA + TBM          <=>  ALD + TBD           (1.62D10)  (1.5D10)
TBD                <=>  TBM + DHA           (1.36D1)   (1.67D5)
TBD                <=>  TBM + GAP           (6.6D2)    (2.36D5)
TBM                 =>  TIM                 (1.5D1)
TIM + DHA           =>  TBM                 (2.D5)
GAP + MOD          <=>  MOB + DPH           (3.96D7)   (3.14D9)
```

| | | | | |
|---|---|---|---|---|
| MOB + PI* | <=> | MOX + DGA | (5.9D6) | (1.45D8) |
| MOX + DPN | <=> | MOD | (2.06D8) | (7.85D3) |
| DGA + PGK | <=> | PGG | (6.66D8) | (3.35D2) |
| PGG + ADP | <=> | 1TP + 3GA + PGK | (1.33D7) | (2.2D9) |
| 3GA + PGA | <=> | PGP | (7.75D7) | (1.89D3) |
| PGP | <=> | 2GA + PGA | (4.96D3) | (6.8D7) |
| PGA | <=> | DGA + PGM | (8.4D3) | (1.07D9) |
| 2GA + ENL | <=> | ENP | (3.56D8) | (6.D3) |
| PEP + ENL | <=> | ENP | (2.D8) | (6.83D3) |
| 3GA + ENL | <=> | ENI | (4.D6) | (2.5D2) |
| PYK + ADP | <=> | PYD | (9.7D6) | (3.5D2) |
| PYD + PEP | <=> | PYP | (4.35D6) | (1.17D1) |
| PYP | => | PYT + PYR | (1.59D2) | |
| PYT | <=> | 1TP + PYK | (4.D3) | (2.92D7) |
| PYR + LDD | <=> | LDL | (4.72D7) | (3.D4) |
| LDL | <=> | LAC + LDN | (1.2D4) | (1.6D5) |
| LDH + DPN | <=> | LDN | (4.5D8) | (1.88D4) |
| LDH + DPH | <=> | LDD | (1.11D10) | (6.5D3) |
| PYR + DIN | => | DIH | (9.D3) | |
| DIH + X.I + OXY | => | DIN + XSI | (2.5D8) | |
| XSI + PI* | => | XSP | (6.8D3) | |
| XSP + ADP + ADP | => | 2TP + X.I | (3.3D5) | |
| XSI + DBP | => | X.I | (4.D3) | |
| 2TP + DBP | => | 1TP | (10) | |
| X.I + 2TP | => | XSI + ADP + PI* | (6.D2) | |
| CON + 2TP | => | ADP + PI* | (2.) | |
| 1TP + PUE | => | PPP | (6.D3) | |
| PPP | => | PUE + ADP + PI* | (60) | |
| DHA + DPH | <=> | AGP + DPN | (1.67D4) | (60) |
| CON | => | GLP | (3.D-3) | |

To generate artificial measurements , the ODE system was solved with a relative integrator accuracy of $10^{-2}$ for the parameter

$$\bar{p}=(0.2D7,0.1D4,0.4D6,0.87D8,0.157D4,0.184D4)$$

- see Table 2. In this way, 260 measurements at the nodes {0.5,1.,2.,4.} were produced. Two sets of initial parameter guesses $p^0$ were selected:

I: $p^0$ := (0.24D7, 0.12D4, 0.36D6, 0.8D8, 0.1D4, 0.2D4)

II: $p^0$ := (0.3D7, 0.2D4, 0.2D6, 0.5D8, 0.8D3, 0.29D4)

The two codes (with relative integrator accuracy of $10^{-3}$) yield the solutions (rounded to 3 decimal digits)

PARKIN1 : p = (0.200D7, 0.942D3, 0.391D6, 0.891D8, 0.161D4, 0.184D4)

PARKIN2 : p = (0.200D7, 0.954D3, 0.393D6, 0.895D8, 0.161D4 0.184D4)

Of course, the two different approaches may lead to two different solutions. Comparison runs are presented in Table 3. The termination of PARKIN1 for II was caused by the occurrence of successive increase of $\|\Delta p^k\|$ - compare the associated considerations in [7]. As in the preceding smaller example, PARKIN2 requires less iterations and has a significantly larger convergence domain. As a consequence,PARKIN2 may be expected to be more robust.

Table 3. Comparative results in example 3.2 for two initial guesses of the parameters.

| | | Time (sec) | Iterations | Iterations with $\lambda<1$ |
|---|---|---|---|---|
| I | PARKIN1 | 55 | 6 | 2 |
| | PARKIN2 | 68 | 5 | 0 |
| II | PARKIN1 | Fail | (4) | (4) |
| | PARKIN2 | 111 | 8 | 0 |

The array storage requirements were

PARKIN1 : about 75 K,
PARKIN2 : about 110 K .

For comparison, an estimate of the array storage requirements for PARFIT in the form as in [4] would be

PARFIT (m=2) : about 390 K
PARFIT (m=5) : about 500 K .

If sparse matrix techniques were additionally implemented, these estimates would reduce to

PARFIT (m=2) : about 240 K ,
PARFIT (m=5) : about 350 K .

In view of the large storage requirements and of the fact that PARFIT requires the user to establish the subroutine f, comparison runs have not been made in this example.

## Conclusion

Both user-oriented packages PARKIN1 and PARKIN2 seem to be a first step towards the parameter identification of *large* chemical reaction systems. Both codes check whether the free parameters in a given system can be identified at all. The slightly faster code PARKIN1 based on the standard formulation seems to be essentially developed. The more robust code PARKIN2 based on the new formulation suggested herein seems to justify further efforts. Generally speaking, the higher complexity of the identification problem (compared with the simulation problem) is reflected in the fact that - within given restrictions on computing time and storage - the PARKIN codes permit the treatment of chemical reaction systems smaller than those, which can be simulated, even though much larger than those, which could be identified before.

*Acknowledgement:* This work has been supported by the Deutsche Forschungsgemeinschaft. The authors wish to thank Mrs. I. Heitz for her quick and careful typing of the article.

## References

[1] L. Armijo:
Minimization of functions having Lipschitz-continuous first partial derivatives.
Pacific J.Math. 16, 1-3 (1966).

[2] G.Bader, P. Deuflhard:
A Semi-Implicit Midpoint Rule for Stiff Systems of Ordinary Differential Equations.
Numer.Math., to appear (1983).

[3] G. Bader, U. Nowak, P. Deuflhard:
An Advanced Simulation Package for Large Chemical Reaction Systems.
In: Aiken (ed.): Stiff Computation. Oxford University Press (1983).

[4] H.G. Bock:
Numerical Treatment of Inverse Problems in Chemical Reaction Kinetics.
In [12], p.102-125 (1981).

[5] E.M. Chance, A.R. Curtis, I.P. Jones, C.R. Kirby:
FACSIMILE : a computer program for flow and chemistry simulation, and general initial value problems.
Harwell, AERE Tech.Rep.R. 8775 (Dec.1977).

[6] P. Deuflhard:
A Relaxation Strategy for the Modified Newton Method.
In: Bulirsch/Oettli/Stoer (ed.): Optimization and Optimal Control.
Springer Lecture Notes 477, 59-73 (1975).

[7] P. Deuflhard, V. Apostolescu:
A Study of the Gauss-Newton Method for the Solution of Nonlinear Least Squares Problems.
In: Frehse/Pallaschke/Trottenberg (ed.): Special Topics of Applied Mathematics.
Amsterdam: North-Holland Publ., p. 129-150 (1980).

[8] P. Deuflhard, G. Heindl:
Affine Invariant Convergence Theorems for Newton's Method and Extensions to Related Methods.
SIAM J.Numer.Anal. 16, 1-10 (1979).

[9] P. Deuflhard, W. Sautter:
On Rank-Deficient Pseudo-Inverses.
J.Lin.Alg.Appl. 29, 91-111 (1980).

[10] P. Deuflhard, G. Bader, U. Nowak:
LARKIN - a software package for the numerical simulation of LARge systems arising in chemical reaction KINetics.
In [12] , p.38-55 (1981).

[11] I.S. Duff, U. Nowak:
On sparse matrix techniques in a stiff integrator of extrapolation type.
Univ. Heidelberg, SFB 123: Tech.Rep. (1982).

[12] K.H. Ebert, P. Deuflhard, W. Jäger (ed.):
Modelling of Chemical Reaction Systems.
Springer Series Chem.Phys. 18 (1981).

[13] D. Garfinkel, B. Hess:
Metabolic Control Mechanisms VII. A detailed computer model of the glycolytic pathway in ascites cells.
J.Bio.Chem. 239, 971-983 (1964).

[14] W.B. Gragg:
On Extrapolation Algorithms for Ordinary Initial Value Problems.
SIAM J. Numer. Anal. 2, 384-404 (1965).

[15] R.J. Kee, J.A. Miller, T.H. Jefferson:
CHEMKIN: A General-Purpose, Problem-Independent, Transportable, Fortran Chemical Kinetics Code Package.
Sandia National Laboratories, Livermore: Tech.Rep. SAND80-8003 (1980)

[16] R.S. Martin, G. Peters, J.H. Wilkinson:
Symmetric Decomposition of a Positive Definite Matrix
Numer.Math. 7, 362-383 (1965).

[17] H.G. Bock:
Recent Advances in Parameter Identification Techniques for ODEs.
These proceedings, Chap. 7 (1983)

# IDENTIFICATION OF RATE CONSTANTS IN BISTABLE CHEMICAL REACTIONS

Johannes Schlöder and Hans Georg Bock

## *0. Introduction*

BELOUSOV [ 3 ] was the first to report on oscillating bromate oxidative reactions in 1958. Although isolated examples of oscillatory chemical reaction systems were known earlier, they were often ignored because such phenomena were considered to be ruled out by the second law of thermodynamics. In 1964, ZHABOTINSKII [24] exploited BELOUSOV's investigations and discovered additional temporal and spatial effects.

Not until 1968, an international audience became aware of these exciting facts, when ZHABOTINSKII reported on his work at a conference in Prague. His lecture initiated an ever growing wave of interest in such systems, since except for an importance in their own right, they serve as models for far more complicated large-scale biochemical systems. Eventually intensive efforts were made to elucidate the underlying kinetical mechanism.

In the early seventies, a detailed mechanism of the BELOUSOV-ZHABOTINSKII-reaction (BZR) in terms of ordinary differential equations was devised by FIELD, KÖRÖS, and NOYES [16]. Parallel to these theoretical investigations many thorough laboratory studies were conducted. While in early experimental studies the complete BZR in batch reactors was investigated, later also continuous stirred tank reactors (CSTR) were used in order to eliminate thermodynamical and spatial effects.

Experiments of this type were conducted by GEISELER and FÖLLNER [18], who studied the oxidation of cerous ions in sulfuric acid solution, which forms an important subsystem of the BZR. They observed bistability, i.e. the existence of two different stable steady states. BAR-ELI and NOYES [ 1 , 2 ] show that the theoretical mechanism devised by NOYES, FIELD, and THOMPSON (NFT) [20] provides an adequate model for the observed phenomena.

The present study treats the *inverse problem* of identifying the rate constants in the o.d.e. given by the NFT-mechanism from the experimental data in [17]. Since these data were obtained from a large number of experiments, the corresponding large-scale constrained least squares estimation problem exhibits a certain structure. This appears to be a typical feature of a broad class of problems where several different experiments are conducted in order to identify parameters, which are common to each of the experiments.

A general approach for the solution of such *multiexperiment* problems using as core a Gauss-Newton type method is given. On the basis of these considerations the software package MULTEX for general parameteridentification purposes is designed and applied to the identification of the NFT-mechanism and numerical results are discussed.

## *1. The Chemical System*

In order to gain some insight into the mathematical problem, first the experimental situation is briefly described. The oxidation of cerous ions by bromate in sulfuric acid solution in the presence of bromide in a CSTR - held at constant temperature - is investigated. The situation of each experiment is completely described by five experimental parameters, four inflow-concentrations ($[Br^-]_E$, $[H^+]_E$, $[Ce^{3+}]_E$, $[BrO_3^-]_E$) and the flow rate $k_E$. By varying them, different experiments can be conducted. The progress of the reaction is followed by measurement of bromide (Fig. 1.1).

The flow experiments are initiated as batch experiments. Two different steady states (SSI,SSII) can be reached for certain experimental parameters, depending on the moment when the continuous operation is started (Fig. 1.2).

A curve of values for one steady state is traced by slowly varying one of the experimental parameters (e.g. $[Br^-]_E$) (Fig. 1.3). The curve terminates with a sudden change to the other steady state. This is called a bistability point. Changing the value of the experimental parameter in the reverse direction, a whole hysteresis loop can be followed.

Now varying a second experimental parameter (e.g. $[BrO_3^-]_E$), curves of bistability points can be established (Fig. 1.4). Coordinates of 44 such bistability points were measured by GEISELER and BAR-ELI [17].

These 88 data form the basis for the actual identification problem.

Figure 1.1 Experimental device

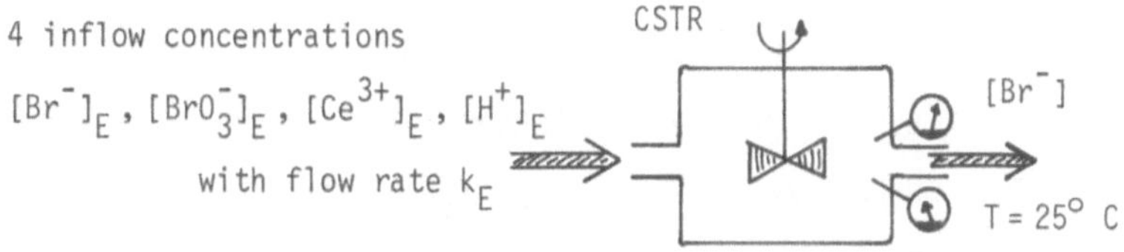


Figure 1.2 Time history of reaction

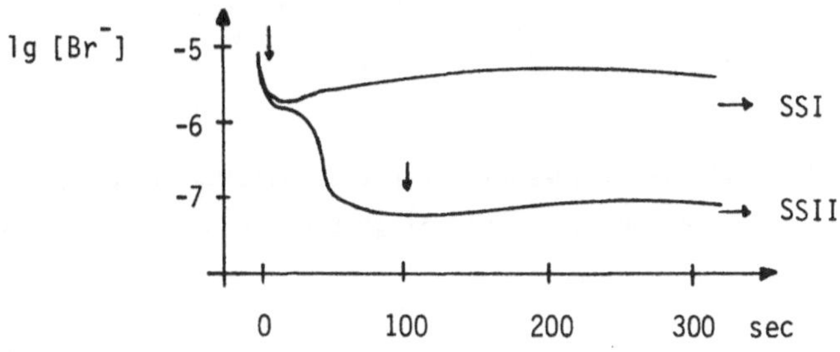


Figure 1.3 Steady state curves

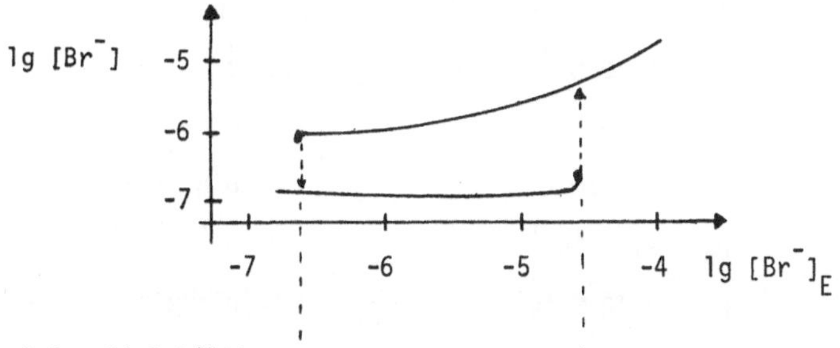


Figure 1.4 Bistability curves

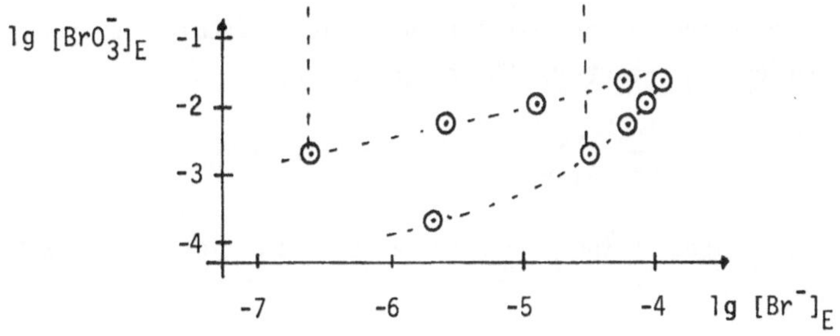

The NFT-mechanism is a system of seven reversible reactions which models the kinetics of the system under consideration.

Figure 1.5 NOYES-FIELD-THOMPSON mechanism

$$BrO_3^- + Br^- + 2H^+ \rightleftharpoons HBrO_2 + HOBr$$

$$HBrO_2 + Br^- + H^+ \rightleftharpoons 2HOBr$$

$$HOBr + Br^- + H^+ \rightleftharpoons Br_2 + H_2O$$

$$BrO_3^- + HBrO_2 + H^+ \rightleftharpoons 2BrO_2 + H_2O$$

$$Ce^{3+} + BrO_2 + H^+ \rightleftharpoons Ce^{4+} + HBrO_2$$

$$Ce^{4+} + BrO_2 + H_2O \rightleftharpoons Ce^{3+} + BrO_3^- + 2H^+$$

$$2HBrO_2 \rightleftharpoons BrO_3^- + HOBr + H^+$$

In [17] the following values of the rate constants $p_i$ , $p_{-i}$ , $i=1(1)7$ , are suggested. They serve as initial guesses in the present study.

Table 1.1 Rate constants

| | |
|---|---|
| $p_1 = 2.1M^{-3}s^{-1}$ | $p_{-1} = 1\times10^4M^{-1}s^{-1}$ |
| $p_2 = 2\times10^9M^{-2}s^{-1}$ | $p_{-2} = 5\times10^{-5}M^{-1}s^{-1}$ |
| $p_3 = 8\times10^9M^{-2}s^{-1}$ | $p_{-3} = 110s^{-1}$ |
| $p_4 = 1\times10^4M^{-2}s^{-1}$ | $p_{-4} = 2\times10^7M^{-1}s^{-1}$ |
| $p_5 = 6.5\times10^5M^{-2}s^{-1}$ | $p_{-5} = 2.4\times10^7M^{-1}s^{-1}$ |
| $p_6 = 9.6M^{-1}s^{-1}$ | $p_{-6} = 1.3\times10^{-4}M^{-3}s^{-1}$ |
| $p_7 = 4\times10^7M^{-1}s^{-1}$ | $p_{-7} = 2.1\times10^{-10}M^{-2}s^{-1}$ |

According to the law of mass-action, a closed chemical system, where NS species $C_j$ react in NR reversible reactions

$$\sum_{j=1}^{NS} l_{ij}\, C_j \rightleftharpoons \sum_{j=1}^{NS} r_{ij}\, C_j \qquad i=1(1)NR$$

can be described by the following set of o.d.e. for the concentrations $c_j$

$$\dot{c}_j = \sum_{i=1}^{NR} (r_{ij} - l_{ij}) R_i(c,p)$$

with the reaction rates

$$R_i(c,p) = p_i \Pi_{j=1}^{NS} c_j^{l_{ij}} - p_{-i} \Pi_{j=1}^{NS} c_j^{r_{ij}}$$

where $p_i$ , $p_{-i}$ are the rate constants of the i'th reaction (generally depending on temperature).

In the CSTR-case these equations are slightly modified to

$$\dot{c}_j = \sum_{i=1}^{NR} (r_{ij} - l_{ij}) R_i(c,p) + k_E(c_{Ei} - c_i) .$$

Here $k_E$ denotes the flow rate of the reactor and $c_{Ei}$ stands for the concentration of $C_i$ in the inflow.

Formulae for the first and second order derivatives of the right hand side with respect to concentrations and rate constants are easily derived analytically from these polynomial expressions. Explicit programming of the o.d.e. and corresponding derivatives involves time consuming and cumbersome labour, that is heavily prone to human error. Therefore in connection with MULTEX an automatic translation procedure (CREATE) for the evaluation of right hand side and first and second order derivatives was designed. The invaluable importance of such procedures is especially evident, if large chemical systems are treated as in the highly sophisticated simulation package LARKIN [14].

Application of the translation rules to the NFT-mechanism yields a system of nine o.d.e., whose right hand side consists of fourth order polynomials in the concentrations. Concentration of water is regarded as constant.

$$\dot{c} = f(c,p) + k_E(c_E - c) .$$

The differential equation of bromate e.g. reads

$$\begin{aligned}\frac{d}{dt} BrO_3^- = & -p_1 \cdot BrO_3^- \cdot Br^- \cdot H^+ \cdot H^+ + p_{-1} \cdot HBrO_2 \cdot HOBr - p_4 \cdot HBrO_2 \cdot BrO_3^- \cdot H^+ + \\ & + p_{-4} \cdot BrO_2 \cdot BrO_2 + p_6 \cdot BrO_2 \cdot Ce^{4+} - p_{-6} \cdot BrO_3^- \cdot Ce^{3+} \cdot H^+ \cdot H^+ + \\ & + p_7 \cdot HBrO_2 \cdot HBrO_2 - p_{-7} \cdot BrO_3^- \cdot HOBr \cdot H^+ + k_E(BrO_{3E}^- - BrO_3^-) .\end{aligned}$$

With these formulae the measured bistability points are characterized by two criteria.

(1) Steady state condition

$$f(c,p) + k_E(c_E - c) = 0$$

(2) Bistability condition

$$\frac{\partial}{\partial c}\,[f(c,p) + k_E(c_E - c)] =: A \qquad \text{singular} .$$

Assuming $\dim(\ker A) = 1$ the singularity can be described in a computational convenient way by

$$(2') \qquad \exists!\, h \quad \|h\|_2 = 1 \qquad Ah = 0 .$$

For each of the 44 experiments, we thus have nine concentrations $c_i$, two experimental parameters and nine auxiliary variables $h_i$ as experiment specific variables which correspond to nineteen equality conditions from (1) and (2') and two least squares conditions given by the two measured bistability-coordinates. The fourteen rate constants $p$ are the same for all experiments. They are additionally restricted by two thermodynamical conditions

$$p_4 \cdot p_5 \cdot p_6 = p_{-4} \cdot p_{-5} \cdot p_{-6}$$
$$p_1 \cdot p_{-2} \cdot p_7 = p_{-1} \cdot p_2 \cdot p_{-7} \quad .$$

In total the identification of $p$ requires the solution of a constrained least squares system with 926 conditions for 894 variables.

## 2. *Multiexperiment Problems*

The problem outlined above is quite typical of a class of problems which frequently arise in applied sciences such as e.g. chemistry [19], biology [23] or geophysics [21,22]. These multiexperiment problems have the general form:

Problem M1. Find $x_1,\ldots,x_m,x_G$ such that

$$\begin{aligned} \textstyle\sum_{j=1}^{m} \|F_{1j}(x_j,x_G)\|_2^2 &= \min \\ F_{2j}(x_j,x_G) &= 0 \qquad\qquad j=1(1)m \\ F_{3j}(x_j,x_G) &\geq 0 \end{aligned}$$

The functions $F_{ij}$ are (in general) nonlinear in the *local variables* $x_j$ which describe the special situation of the j-th experiment, and the *global variables* $x_G$ which are common to all $m$ experiments.

Note that each of the experiments may exhibit considerable complexity, e.g. may be a multipoint boundary value problem [ 7 ]. In the following an enhanced version of a generalized Gauss-Newton method is described for the efficient treatment of multiexperiment problems. The method and the corresponding code MULTEX take advantage of the special structure of these problems. Thus both storage requirements and computing time are considerably reduced.

*Remark:* Consider measurements which are functions of local and global variables

$$\eta_{ij} = g_{ij}(x_j,x_G) + \varepsilon_{ij} .$$

Under the standard statistical assumptions that the errors $\varepsilon_{ij}$ are independent and normally distributed with zero mean and known variances $\sigma_{ij}$ $(N(0,\sigma_{ij}))$ , the least squares functional

$$\textstyle\sum_{j=1}^{m} \|F_{1j}(x_j,x_G)\|_2^2 := \sum_{j=1}^{m} \sum_i (\sigma_{ij}^{-2}(\eta_{ij} - g_{ij}(x_j,x_G))^2)$$

is known to be a maximum likelihood estimator.

## 3. *Generalized Gauss-Newton Method*

Generalizations of Gauss-Newton methods to equality and inequality constrained least squares problems have proven to be vigorous and reliable tools in the treatment of complicated real life problems [4,5,12]. They show both good local and global convergence properties. Since an approximation of the covariance matrix can easily be computed, a strong aid for the statistical analysis of the solution is given.

In the following some essential properties of the generalized

Gauss-Newton method are briefly reviewed.

Consider the problem of finding $x$, such that

$$\begin{aligned} \|F_1(x)\|_2 &= \min \\ F_2(x) &= 0 \\ F_3(x) &\geq 0 \,. \end{aligned}$$

The generalized Gauss-Newton method iteratively improves a given estimate $x_k$ by

$$x_{k+1} = x_k + \Delta x_k$$

where the increment $\Delta x_k$ solves the linear problem

$$\begin{aligned} \|F_1(x_k) + J_1(x_k)\Delta x_k\|_2 &= \min \\ F_2(x_k) + J_2(x_k)\Delta x_k &= 0 \qquad , \; J_i(x_k) := \frac{\partial F_i(x_k)}{\partial x_k} \\ F_3(x_k) + J_3(x_k)\Delta x_k &\geq 0 \end{aligned}$$

Under suitable conditions it can be guaranteed by a perturbation theorem [6] that near the solution the set of active constraints remains unchanged. Thus the problem of *local convergence* is the same as in the case of the equality constrained problem.

Consider the linearized problem

$$\begin{aligned} \|F_1(x_k) + J_1(x_k)\Delta x_k\|_2 &= \min \\ F_2(x_k) + J_2(x_k)\Delta x_k &= 0 \end{aligned}$$

and denote by $J = (J_1^T, J_2^T)^T$ the Jacobimatrix and by $F = (F_1^T, F_2^T)^T$ the corresponding right hand side. Then under weak assumptions a generalized inverse $J^+$ can be shown to exist, which yields a solution of the linear problem

$$\Delta x_k = J^+(x_k)F(x_k) \,. \tag{3.1}$$

If $J_2$ is not of full rank, the equality constraints are relaxed to be satisfied in a least squares sense.

The following affine invariant convergence theorem is valid.

Theorem. Let $D \subset \mathbb{R}^n$ a region, $F \in C^1(D)$ .
Assume that $\forall\, t \in [0,1]$ $\forall\, x,y \in D$ with $x-y = J^+(x)F(x)$

$$\frac{\| J(y)^+(J(x+t(y-x))-J(x))(y-x)\|}{t\| y-x\|^2} \le \omega < \infty$$

$$\frac{\|(J(y)^+-J(x)^+)(I-J(x)J(x)^+)F(x)\|}{\| y-x\|} \le \kappa(x) \le \kappa < 1$$

For all starting vectors $x_0 \in D$ with $\|J^+(x_0)F(x_0)\| =: \alpha_0$ ,
$\delta_0 := \frac{1}{2}\alpha_0\omega + \kappa < 1$ , $D_0 := K(x_0,\frac{\alpha_0}{1-\delta_0}) \subset D$ holds

(i) the iteration (3.1) is well-defined with $x_k \in D$

(ii) $x_k \to x^*$ $(k \to \infty)$ with $J(x^*)^+F(x^*) = 0$

(iii) the convergence rate can be estimated by

$$\|\Delta x_{k+1}\| \le \frac{\omega}{2}\|\Delta x_k\|^2 + \kappa\,\|\Delta x_k\| \le \delta\,\|\Delta x_k\|$$

(iv) $x^*$ is a strict local minimum if $J_2$, $J$ have full rank .

Proof. cf. [ 6],[15]

The nonlinearity of the problem is described by the curvature $\omega$ . The asymptotic convergence rate $\kappa$ describes the incompatibility of the data with the model. If $\kappa \ge 1$ the model can no longer be identified.

In complex real life problems such as the indicated multiexperiment problems one cannot hope to have a starting vector that fulfills the conditions of the local convergence theorem. Therefore it is necessary to extend the domain of convergence by a damped iteration

$$x_{k+1} = x_k + \lambda_k \Delta x_k \qquad , \; 1 \ge \lambda_k \ge \varepsilon > 0 \; .$$

The relaxation factor $\lambda_k$ is chosen in order to decrease an appropriate level function $T(x)$

$$T(x_{k+1}) < T(x_k) \; ,$$

which has to satisfy

$$\frac{d}{d\varepsilon+} T(x_k + \varepsilon \Delta x_k) < 0 \Leftrightarrow \Delta x_k \neq 0 .$$

This generalized underrelaxed Gauss-Newton method can be proven to be *globally convergent* for the inequality constrained least squares problem by means of the appropriate level function

$$T_1(x) = ||F_1(x)||_2^2 + \sum \mu_i |F_{2i}(x)| + \sum \nu_i |\min(0, F_{3i}(x))|$$

where $\mu_i$ and $\nu_i$ are upper bounds for the adjoint variables [ 5 ].

Practical disadvantages of this *low order* level function however are the lack of reliable a-priori-estimates for $\mu_i$ and $\nu_i$ . Moreover, extremely small stepsizes are suggested for (mildly) ill-conditioned problems, since then the steepest descent direction is nearly orthogonal to the Gauss-Newton increment.

As a remedy the locally defined level function

$$T_k(x) = ||J_k^+ F(x)||_2^2$$

is used, which is an adequate generalization of the natural level function due to DEUFLHARD [10].

## *4. Continuation Methods*

Despite the good convergence properties of the generalized Gauss-Newton method, it can be favourable to apply continuation methods in the case of difficult problems, especially if they depend *naturally* on a parameter, say $t$ , and the solution for $t = t_0$ is available while the solution for $t = t_{end} \neq t_0$ is desired.

Consider the parameter dependent problem.

Problem C1. Find $x$ , such that

$$||\hat{F}_1(x,t)||_2 = \min_x$$
$$\hat{F}_2(x,t) = 0$$
$$\hat{F}_3(x,t) \geq 0$$

Given a continuous parametrization of the homotopy path (existence assumed)

$$(x,t):\ [0,1] \to \mathbb{R}^{n+1}$$

$$z(s) = (x(s),t(s)) \text{ solves C1} \qquad t(0) = t_o\ ,\ t(1) = t_{end}$$

one can proceed as follows.
Divide

$$0 = s_o < s_1 < \ldots < s_n = 1$$

and solve problem C1 for $t = t(s_i)$, $i=1(1)n$ by a local iterative procedure, using an initial guess, which is predicted from results for $s_j$, $j < i$. This procedure is in principle well known and used with advantage e.g. in optimal control problems.

The quality of a continuation method depends crucially on the choice of the stepsize $s_{j+1} - s_j$ in connection with the generation of an initial guess for the next step, and on the local iterative procedure.

In a slight modification of the method of section 3, the active sets are determined as a function of the homotopy path, so that the local procedure only has to solve equality constrained problems. An effective *local iterator* is the following extension of the method described above, where in each step the *multistage system*

$$\|(\Delta x,\Delta t)\|_2 = \min$$

$$\text{subject to} \quad \|F_1 + J_1\Delta x + j_1\Delta t\|_2 = \min_{\Delta x} \quad , J_i = \frac{\partial}{\partial x_j} F_i\ ,\ j_i = \frac{\partial}{\partial t} F_i\ ,$$

$$\|F_2 + J_2\Delta x + j_2\Delta t\|_2 = 0 \qquad \mathcal{J} = \begin{pmatrix} J_1 & j_1 \\ J_2 & j_2 \end{pmatrix}$$

is easily solved by means of a generalized inverse $\mathcal{J}^+$. Note, that in the case of pure nonlinear equations $\mathcal{J}^+$ is replaced by a (weighted) Moore-Penrose pseudoinverse. When minimizing $\|\Delta t\|_2$ alone, the local iteration reduces to the method of section 3.

In the classical case $t(s) = s$ DEUFLHARD [11] has given an automatic *stepsize strategy* for the case of nonlinear equations in connection with Newton's method as local iterator. These results are generalized in BOCK [ 6 ] to generalized Gauss-Newton methods for constrained least squares problems.

The essential results are:
Given a predictor $\hat{z}(s)$ of order $p$ for $z(s+h)$

$$\|\hat{z}(s+h) - z(s+h)\| \leq \eta h^p$$

and a homotopy path with a neighbourhood

$$\mathcal{U} = \cup_{s\in[0,1]} K(z(s),\varepsilon(s)) \quad , \; \varepsilon(s) \geq \varepsilon$$

where the local convergence conditions are fulfilled:

$$\frac{\|\bar{J}(z_2)^+(\bar{J}(z_1+\lambda(z_2-z_1))-\bar{J}(z_1))(z_2-z_1)\|}{\lambda\,\|z_2-z_1\|^2} \leq \omega < \infty$$

$$\frac{\|\bar{J}(z_2)^+(I-\bar{J}(z_1)\bar{J}(z_1)^+)F(z_1)\|}{\|z_2-z_1\|} \leq \kappa(z_1) \leq \kappa < 1$$

If $h^p \leq \tau \frac{\sqrt{5-4\kappa}-1}{\omega\cdot\eta}$, $(\tau < 1)$ ; $\tilde{s} := s+h$ ; $\tilde{z} := \hat{z}(\tilde{s})$ ;
$\alpha_0 := \|\bar{J}(\tilde{z})^+\hat{F}(\tilde{z})\|$ ; $\delta := \frac{1}{2}\alpha_0\omega+\kappa$ ; $K(h) := K(\tilde{z}, \frac{\alpha_0}{1-\delta}) \subset K(\tilde{z},\varepsilon(\tilde{s}))$

then the full step converges in $K(h)$ to a solution of C1 at $\tilde{s}$ with the initial guess $z_0 = \tilde{z}$ .

Proof. cf. [ 6 ]

Note that good estimates for $\kappa$, $\omega$ and $\eta$ are computationally available from the previous step.

## 5. *The Multiexperiment Case*

In order to solve multiexperiment problems by a generalized Gauss-Newton method, problem M1 is successively linearized. Each iteration of the generalized Gauss-Newton method requires the solution of the linear problem.

Problem LM1. Find $\Delta x_1,\ldots,\Delta x_m,\Delta x_G$ such that

$$
\begin{aligned}
F_{L1}(\Delta x_1,\ldots,\Delta x_m,\Delta x_G) &:= \textstyle\sum_{i=1}^{m} \|F_{1i}(x_i,x_G) + J_{1i}^{L}\Delta x_i + J_{1i}^{G}\Delta x_G\|_2^2 &= \min \\
F_{L2i}(\Delta x_i,\Delta x_G) &:= F_{2i}(x_i,x_G) + J_{2i}^{L}\Delta x_i + J_{2i}^{G}\Delta x_G &= 0 \\
F_{L3i}(\Delta x_i,\Delta x_G) &:= F_{3i}(x_i,x_G) + J_{3i}^{L}\Delta x_i + J_{3i}^{G}\Delta x_G &\geq 0
\end{aligned}
$$

$$
i=1(1)m\ ,\ J_{ji}^{L} = \frac{\partial}{\partial x_i} F_{ji}(x_i,x_G)\ ,\ J_{ji}^{G} = \frac{\partial}{\partial x_G} F_{ji}(x_i,x_G)
$$

*5.1 Solution of the linear system*

This implies determination of a feasible point, followed by an active set strategy to determine the constraints which are active at a solution point.

The first step can be very conveniently accomplished by solving

$$
\textstyle\sum_{i=1}^{m} \left( \|\max(0,F_{L3i}(\Delta x_i,\Delta x_G))\|_2^2 + \|F_{L2i}(\Delta x_i,\Delta x_G)\|_2^2 \right) = \min .
$$

For a given actual set of active constraints the full linear system is suitably decomposed. It has the block-sparse structure:

$$
M_0 \Delta x = -F
$$

with

$$
M_0 = \begin{bmatrix}
E_{1L} & 0 & & \cdots & 0 & E_{1G} \\
S_{1L} & 0 & & & 0 & S_{1G} \\
0 & E_{2L} & 0 & \cdots & 0 & E_{2G} \\
0 & S_{2L} & 0 & & 0 & S_{2G} \\
\vdots & & \ddots & & & \vdots \\
0 & & \cdots & 0 & E_{mL} & E_{mG} \\
0 & & & 0 & S_{mL} & S_{mG}
\end{bmatrix} \quad ; \quad
\begin{bmatrix}
f_{1E} \\ f_{1S} \\ f_{2E} \\ f_{2S} \\ \vdots \\ f_{mE} \\ f_{mS}
\end{bmatrix} = F
$$

$M_0$ consists of $m$ blocks, each corresponding to one experiment, which are built up by the local part $(E_{iL}^{T},S_{iL}^{T})^{T}$ and the global part $(E_{iG}^{T},S_{iG}^{T})^{T}$. The E-matrices are associated with the active constraints and the S-matrices with the least squares conditions. For ease of notation it is assumed in the following, that $M_0$ is already ordered in such a way, that pivoting - which is crucial to ensure feasibility and stability of the triangulation transformations - causes no permutations.

By elementary equivalence transformations $T_\alpha$ the local parts of $M_0$ are reduced to triangular form. The right hand sides and the global parts undergo the same transformations.

$$\left[\begin{array}{c} \boxed{\begin{array}{c}E_{iL} \\ \hline S_{iL}\end{array}} \cdots \boxed{\begin{array}{c}E_{iG} \\ \hline S_{iG}\end{array}}\end{array}\right] = \left[\begin{array}{c}\boxed{\begin{array}{c|c}T_i & O \\ \hline L_i & Q_i\end{array}}\end{array}\right] \cdot \left[\begin{array}{c}\boxed{\begin{array}{c}\diagdown R_i \\ O\end{array}}\end{array}\right] \cdots \left[\begin{array}{c}\boxed{\begin{array}{c}\tilde{G}_i \\ \hline \tilde{\tilde{G}}_i\end{array}}\end{array}\right]$$

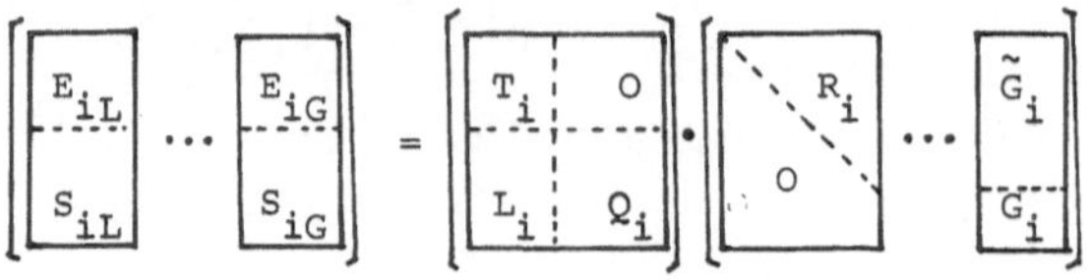

$T_i$ reduces the E-part to upper triangular form (e.g. by Gaussian elimination or orthogonal transformation [8]), $L_i$ eliminates the corresponding local variables. The S-part is handled by the orthogonal transformations $Q_i$ .

In the next step the remaining conditions $\tilde{\tilde{G}}_i$ of each block are assembled by a permutation $P^T$ to form the matrix $G_G$ , which is then triangulized in a similar way by $T_\beta$

$$\left[\boxed{\begin{array}{c}G_G\end{array}}\right] = \left[\boxed{\begin{array}{c|c}T_G & O \\ \hline L_G & Q_G\end{array}}\right]\left[\boxed{\begin{array}{cc}R_{G1} & R_{G2} \\ \diagdown & R_{G3} \\ O & O\end{array}}\right]$$

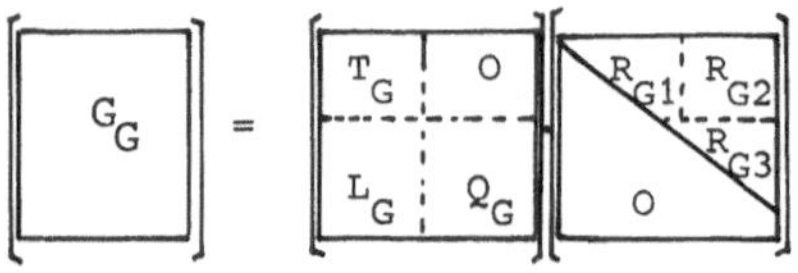

We thus have:

Lemma 1. With $T_\alpha$ , $P$ , $T_\beta$ as defined above

$$M_1 = T_\beta^{-1}P^T T_\alpha^{-1}M_0 = \left[\begin{array}{cccccc} R_1 & O & \cdots & O & \tilde{G}_1 \\ O & R_2 & \cdots & O & \tilde{G}_2 \\ & O & & & \vdots \\ & & \ddots & & \\ & & & R_m & \tilde{G}_m \\ & & & O & R_G \\ O & & \cdots & & O \end{array}\right] ; \left[\begin{array}{c}\tilde{u}_1 \\ \tilde{u}_2 \\ \vdots \\ \tilde{u}_m \\ \tilde{u}_G \\ u_G\end{array}\right] = \hat{u} = T_\beta^{-1}P^T T_\alpha^{-1}F$$

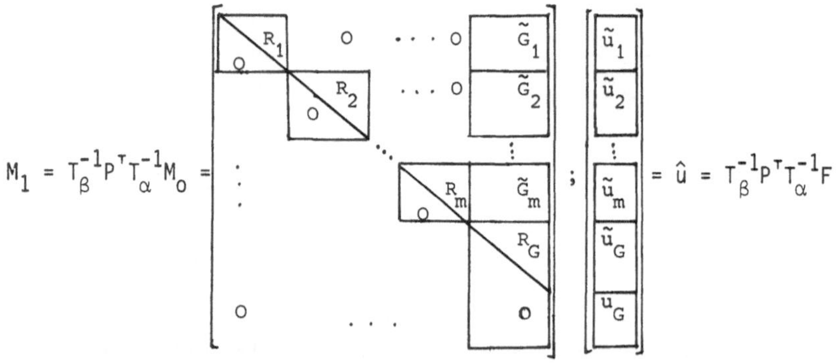

$M_1$ is of triangular form.

The unique solution is given by backward substitution

$$\Delta x_G = -R_G^{-1}\,\tilde{u}_G$$
$$\Delta x_i = -R_i^{-1}(\tilde{G}_i \Delta x_G + \tilde{u}_i) \ . \quad \square$$

*Remark 1:* The triangulation of the local blocks and the computation of the local variables can be performed by parallel processing.

*Remark 2:* If the local blocks are sparse or have a special structure, adequate techniques for their triangulation are to be used (e.g. the condensing algorithm in case of multiple shooting). If a hierarchy of the constraints is defined, multistage QR-decomposition is possible [ 6 ].

*Remark 3:* In case of rank-deficiency generalized inverses can be built up.

*Remark 4:* Since devices like pivoting depend heavily on the scale of the variables, internal scaling is performed in MULTEX to equilibrate the variables.

## *5.2 Evaluation of adjoint variables*

Using the same notation as in the previous section, we define

$\alpha_G := Q_G(0,u_G^T)^T$ the residual of the global matrix

$y_G := (T_G^{-1})^T L_G^T \alpha_G$ the adjoint variables of the global matrix

$(y_1^T,\alpha_1^T,y_2^T,\alpha_2^T,\ldots,y_m^T,\alpha_m^T)^T := P(0,0,\ldots,y_G^T,\alpha_G^T)^T$ .

With these definitions the following lemma holds:

Lemma 2. The adjoint variables $\lambda_i$ and the residuals $r_i$ of the local blocks are given by

$$r_i = Q_i\alpha_i$$
$$\lambda_i = (T_i^{-1})^T(y_i+L_i^T r_i) .$$

Proof. Verify $(\lambda_1^T,-r_1^T,\ldots,\lambda_m^T,-r_m^T)M_0 = 0$ by straightforward calculation.

*Remark 1:* Once the residuals and the adjoint variables for the global matrix are established, the residuals and adjoint variables for the local blocks can be evaluated parallely.

*Remark 2:* The case of upper and lower bounds is theoretically treated in the same way. However computationally substantial simplifications and storage reductions are possible and exploited in MULTEX.

## *6. Analysis of the Solution*

Statistical analysis of the solution can be conducted by means of the covariance matrix. In case of the generalized Gauss-Newton method a linear approximation is readily established by

$$\mathrm{cov}(x) = E(\delta x \delta x^T) = J^+ E(\delta\varepsilon\delta\varepsilon^T)(J^+)^T$$

with $J$ the Jacobian of the last iterate. Taking into account the triangular form derived in lemma 1 the covariance matrix in the multi-experiment case is easily derived. Under the standard statistical assumptions given above, the interesting part for the global variables is - up to permutation - given by

$$C = \mathrm{cov}(x_G) = \begin{pmatrix} ABA^T & AB \\ BA^T & B \end{pmatrix} \qquad \begin{aligned} A &= -R_{G1}^{-1} R_{G2} \\ B &= R_{G3}^{-1}(R_{G3}^{-1})^T \end{aligned} .$$

The standard error of the global variables is approximated by

$$\sigma(x_{Gi}) = \sqrt{c_{ii}} .$$

*Remark:* The full covariance matrix can be computed in a similar way. But since this $n_V \times n_V$ matrix is *non-sparse*, this is usually deliberately avoided. The variances however can be computed without additional storage requirement.

## *7. Numerical Investigations*

The identification of the NFT-mechanism from the measured bistability-points is treated by the generalized Gauss-Newton method for multiexperiment problems derived above.

### *7.1 Generation of initial guesses*

Initial guesses for the iteration process are obtained by simulating the chemical process as far as possible.

The simulation of the situation in Fig. 1.2 requires the solution of an initial value problem with discontinuous right hand side

$$\dot{c} = f(c,p) + k_E(c_E - c)$$
$$c(0) = c_o \;,\; k_E = 0 \;\text{ if }\; t < t_s \;,\; k_E \neq 0 \;\text{ if }\; t \geq t_s \;.$$

The stiff o.d.e. originating from the NFT-mechanism are numerically integrated using the routine METAN1 [13]. Depending on the value of $t_s$ (the moment of switching the model from batch to CSTR mode) two different steady states for the same set of experimental parameters $(c_E, k_E)$ are reached.

In the next step the curves of steady states are traced by applying the homotopy techniques described in section 5. One of the experimental parameters, say $c_{E\alpha}$ (e.g. $[Br^-]_E$ as in Fig. 1.3), is used as a natural homotopy parameter. The ends of the homotopy paths are indicated by the steplength strategy.

The conditions (1),(2') of section 1 characterize the exact end of the homotopy path. This nonlinear system in $(c,h,c_{E\alpha})$ is solved by Newton's method using initial guesses which are readily established from the variables and the nearly singular Jacobian of the last homotopy step. The solution determines a bistability point of the model for all coordinate-pairs $(c_{E\alpha}, c_{E\beta})$, with $c_{E\beta}$ an experimental parameter different from $c_{E\alpha}$.

Choose now $c_{E\beta}$ as homotopy parameter (cf. Fig. 1.4, $c_{E\beta} = [BrO_3^-]_E$ in the experiment) bistability curves - solutions of system (1),(2') - are followed.

The continuation process is stopped, as soon as $c_{E\beta}$ reaches one measured coordinate of a bistability point. This approximation of the local variables is slightly improved by detecting that point on the bistability curve, which is closest to the measured values of the bistability point in the least squares sense. Note, that this means just solving the local experiment least squares problem for fixed global parameters.

In this way approximations to all measured bistability points are obtained and initial guesses for all local variables are generated.

*7.2 Estimation of rate constants*

According to chemist's information four parameters of the model $(p_1, p_2, p_4, p_{-6})$ are expected to be sensitive with respect to the experimental data. Therefore, these parameters are chosen to vary, while the other ten are fixed to their initial guesses. Indeed, a drastic change of these values is observed.

Table 7.1 Estimation of four parameters

| | $p_1$ | $p_2$ | $p_4$ | $p_{-6}$ |
|---|---|---|---|---|
| initial guesses | 2.10 | 2.00 (09) | 1.00 (04) | 1.30 (-4) |
| estimated by MULTEX | 6.54 | 6.23 (09) | 1.15 (04) | 1.49 (-4) |

The bistability curves for these improved parameter values are established in the different subspaces of the five-dimensional parameter-space in much the same way as above by continuation. The typical situation is shown in Fig. 7.1 and Fig. 7.2.

Figure 7.1 $[BrO_3^-]_E$ - $[Br^-]_E$ subspace

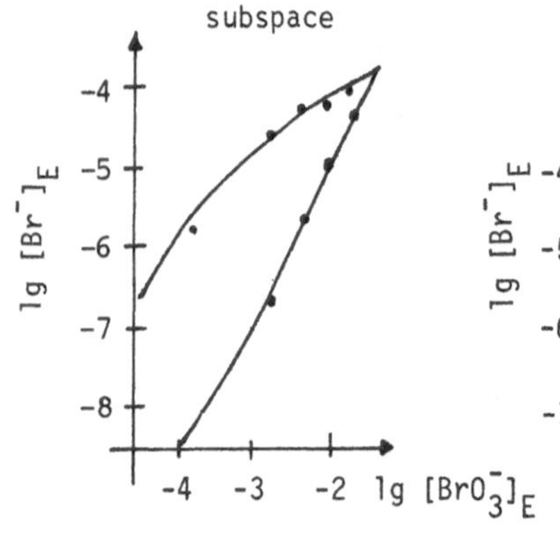


Figure 7.2 $[Ce^{3+}]_E$ - $[Br^-]_E$ subspace

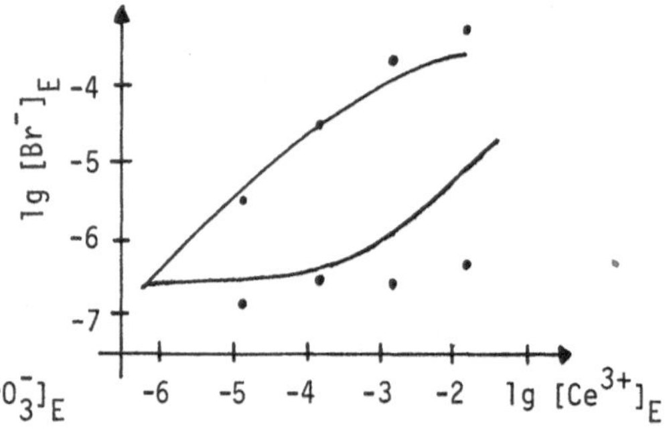


While in Fig. 7.1 the data are fitted satisfactorily, in Fig. 7.2 severe deviations are observable. The a-posteriori estimation of the data error yields a value approximately seven times greater than is expected from the experimental conditions. This and the high incompatibility factor of $\kappa = 0.6$ strongly indicate, that the four-parameter-model is incomplete.

Consequently, the question arises, whether this deficiency can be fixed up by the complete fourteen-parameter-model. However, preliminary numerical investigations into this direction have proven unsuccessful so far. Numerical results show, that the sensitivity matrices become nearly singular on the one hand and no significant improvement of the fit is obtained on the other hand. This behaviour suggests, that additional - and different - experimental data are necessary to identify the remaining ten parameters of the NFT-mechanism, but also, that further modelling is required in order to obtain a completely satisfactory explanation of the data already observed. Hopefully the methods outlined in this present paper will also contribute to the design of such experiments e.g. by the performance and optimization of controlled numerical experiments.

Thus after all this computational effort, the work on the identification of the BZR seems to have only just begun.

## *8. Conclusion*

The present paper treats the challenging inverse problem of identification of the Noyes-Field-Thompson mechanism from real life data, which are obtained from 44 different experiments. The mathematical formulation leads to a nonlinear constrained least squares problem of formidable size, which can be imbedded in the general framework of multiexperiment problems. The structure of this very broad class of problems is analyzed and adequate numerical methods,in particular a generalized Gauss-Newton method are designed. These analyses form the basis of the general purpose code MULTEX, which is applied to the NFT-identification problem. The numerical results exhibit features which seem to be typical of the treatment of real life inverse problems. With respect to the data the full model turns out to be both incomplete and non unique. The numerical analysis calls for the revision of the model and the conduction of further different experiments. These problems - which can only be successfully attacked in close cooperation of the chemist and the mathematician - will be the subject of further work.

*Acknowledgement.* The authors are indebted to K. Bar-Eli (Tel Aviv University, Israel) who brought the problem of identifying the NFT-mechanism to their attention and provided them with all necessary information.
This work was supported by the Deutsche Forschungsgemeinschaft, Sonderforschungsbereich 72, and the DFVLR, Köln-Porz.

*References*

[1] K. Bar-Eli, R.M. Noyes, J. Phys. Chem. 81, 1988 (1977)

[2] K. Bar-Eli, R.M. Noyes, J. Phys. Chem. 82, 1352 (1978)

[3] B.P. Belousov, Sborn. Referat. Radiats. Med., 1958, Medgiz, Moscow, 145 (1959)

[4] H.G. Bock: A Multiple Shooting Method for Parameter Identification in Nonlinear Differential Equations, GAMM Conference, Brussels (1978)

[5] H.G. Bock: Numerical Treatment of Inverse Problems in Chemical Reaction Kinetics, in [9]

[6] H.G. Bock: Randwertproblemmethoden zur Parameteridentifizierung in Systemen nichtlinearer Differentialgleichungen (in preparation)

[7] H.G. Bock: Recent Advances in Parameter Identification Techniques for O.D.E., these proceedings

[8] P. Businger, G.H. Golub: Linear Least Squares Solutions by Householder Transformations, Numer. Math. 7, 269 (1965)

[9] K.H. Ebert, P. Deuflhard, W. Jäger (eds.): Modelling of Chemical Reaction Systems, Springer Series in Chemical Physics 18, Heidelberg (1981)

[10] P. Deuflhard: A Modified Newton Method for the Solution of Ill-conditioned Systems of Nonlinear Equations with Applications to Multiple Shooting, Numer. Math. 22, 289 (1974)

[11] P. Deuflhard: A Stepsize Control for Continuation Methods and its Application to Multiple Shooting, Numer. Math. 33, 115 (1979)

[12] P. Deuflhard, V. Apostolescu: An Underrelaxed Gauss-Newton Method for Equality Constrained Nonlinear Least Squares Problems, in: Optimization Techniques, Proc. 8th IFiP Conf., Würzburg, Aug. 77, ed. by J. Stoer, Lecture Notes Control Inf. Sci., 7/2, 22 (1978)

[13] P. Deuflhard, G. Bader: A Semi-implicit Mid-point Rule for Stiff Systems in Ordinary Differential Equations, SFB 123, Techn. Rep. 114, Univ. Heidelberg (1981)

[14] P. Deuflhard, G. Bader, U. Nowak: LARKIN - A Software Package for the Numerical Simulation of Large Systems Arising in Chemical Reaction Kinetics, in [9]

[15] P. Deuflhard, G. Heindl: Affine Invariant Convergence Theorems for Newton's Method and Extensions to Related Methods, SIAM J. Numer. Anal. 16/1, (1979)

[16] R.J. Field, E. Körös, R.M. Noyes, J. Am. Chem. Soc. 94, 8649 (1972)

[17] W. Geiseler, K. Bar-Eli, J. Phys. Chem. 85, 908 (1981)

[18] W. Geiseler, H. Föllner, Biophys. Chem. 6, 107 (1977)

[19] J. Grievink, Koinklijke Shell-Laboratorium, Amsterdam, Private Communication (1981)

[20] R.M. Noyes, R.J. Field, R.C. Thompson, J. Am. Chem. Soc. 93, 7315 (1971)

[21] V. Pereyra, H.B. Keller, W.H.K. Lee: Computational Methods for Inverse Problems in Geophysics: Inversion of Travel Time Observations, Phys. Earth Planet. Inter. 21, 120 (1980)

[22] Th. Reiners, Diploma Thesis (in preparation)

[23] J. Swartz, J.H. Bremermann: Discussion of Parameter Estimation in Biological Modelling: Algorithms for Estimation and Evaluation of the Estimates, J. Math. Biol. 1, 241 (1975)

[24] A.M. Zhabotinskii, Dokl. Akad. Nauk. SSSR 157, 392 (1964)

# NEW METHODS OF PARAMETER IDENTIFICATION IN KINETICS OF CLOSED AND OPEN REACTION SYSTEMS

H. Lachmann

By means of multiwavelength spectrometry the inverse problem of chemical kinetics may be solved by analysing not the concentration matrix, but the absorbance matrix, directly. The number of linearly independent reaction steps can be determined by graphical matrix rank analysis. Closed and open reaction systems may be kinetically analysed with high precision and significance by simultaneous evaluation of absorbance-time curves at different wavelengths.

The so-called inverse problem of chemical reaction kinetics is defined by mathematicians in the following way:
It is assumed that the mathematical model, i.e. the system of ordinary differential equations (ODE) is known and that the concentration-time curves of all reaction components or most of them can be measured. The aim is then the identification and determination of all rate constants [1].

In real practice of experimental kinetics two further problems must be solved before:

1. to find the best ODE system for a given chemical system, that is to find out the "true" reaction mechanism, which contains all kinetically relevant reaction steps;
2. to calculate the concentration time curves from the measured signal-time curves, which are in many cases linearly dependent on one or more concentration variables. For this purpose a set of calibration constants must be determined in separate experiments.

For example in spectrophotometric measurements the molar absorptivities of all components of the reaction system should be known. Unfortunately the molar absorptivities of unstable intermediates of chemical reactions cannot be determined in direct experiments. Because of this lack of information it seems to be difficult to solve the inverse problem. Equation (1) and (2) summarize this difficulty.

In the general formulation (2) of Beer-Lambert's Law the optical path length is omitted ($d = 1$ cm).

$$\underline{\underline{A}} = \begin{pmatrix} A_{11} & A_{12} & \cdots & A_{1m} \\ A_{21} & A_{22} & \cdots & \vdots \\ \vdots & \vdots & & \vdots \\ A_{11} & A_{12} & \cdots & A_{1m} \end{pmatrix} \qquad (1),$$

$$\underline{\underline{A}} = \underline{\underline{\varepsilon}} \cdot \underline{\underline{c}}$$

$$= \begin{pmatrix} \varepsilon_{11} & \varepsilon_{12} & \cdots & \varepsilon_{1n} \\ \varepsilon_{21} & \varepsilon_{22} & \cdots & \vdots \\ \vdots & \vdots & & \vdots \\ \varepsilon_{11} & \varepsilon_{12} & \cdots & \varepsilon_{1n} \end{pmatrix} \cdot \begin{pmatrix} c_{11} & c_{12} & \cdots & c_{1m} \\ c_{21} & c_{22} & \cdots & \vdots \\ \vdots & \vdots & & \vdots \\ c_{n1} & c_{n2} & \cdots & c_{nm} \end{pmatrix} \qquad (2)$$

with: $1 \mathrel{\hat{=}}$ wavelength $\lambda$,
$m \mathrel{\hat{=}}$ time t,
$n \mathrel{\hat{=}}$ number of components,
$\varepsilon \mathrel{\hat{=}}$ molar absorptivity,
$c \mathrel{\hat{=}}$ concentration.

By spectrophotometric measurements the absorbance matrix $\underline{\underline{A}}$, i.e. absorbance-time curves at different wavelengths, may be determined. For solving the inverse problem in the conventional way the concentration matrix is needed, i.e. concentration-time curves of the reacting species. The matrix $\underline{\underline{A}}$ and $\underline{\underline{c}}$ are connected by the matrix $\underline{\underline{\varepsilon}}$ of all molar absorptivities. And a great lot of them cannot be determined in direct experiments.

To overcome this fundamental problem we have developed a new approach by analysing not the concentration-time curves, but the measured quantities, directly. This technique may be applied to any measured signal which is linearly dependent on one ore more concentration variables. Examples are VIS-UV-absorption, fluorescence, CD, ORD, IR, different chromatographies or combinations of them.

First of all, the whole measured information, for example all absorbance-time curves, are examined by means of graphical matrix rank analysis. The principles of this method shall be derived using as an example the most simple first order reaction A → B in a closed reaction system.

If the condition of mass conservation

$$a_o = a + b \tag{3}$$

is inserted into Beer-Lambert's Law for two species A and B, we get

$$\begin{aligned} A_\lambda &= d \cdot (\varepsilon_{\lambda A} \cdot a(t) + \varepsilon_{\lambda B} \cdot b(t)) \\ &= d \cdot (\varepsilon_{\lambda A} - \varepsilon_{\lambda B}) \cdot a(t) + d \cdot \varepsilon_{\lambda B} \cdot a_o \end{aligned} \tag{4}$$

with: $d \;\hat{=}$ sample path length,
$\varepsilon_\lambda \;\hat{=}$ molar absorptivity at the wavelength $\lambda$,
$a,b =$ concentrations of A and B,
$a_o =$ initial concentration of A,
$A_{\lambda\infty} = d \cdot \varepsilon_{\lambda B} \cdot a_o$.

Subtraction of the absorbance at infinite time $A_{\lambda\infty}$ leeds to

$$\Delta A_\lambda := A_\lambda - A_{\lambda\infty} = d \cdot (\varepsilon_{\lambda A} - \varepsilon_{\lambda B}) \cdot a(t) \tag{5}.$$

The consequences of this equation are: whenever the molar absorptivities $\varepsilon_{\lambda A}$ and $\varepsilon_{\lambda B}$ are equal, we have the condition of an isosbestic point. And when two equations of the type (5) are divided for any two wavelengths one by another, we get

$$\Delta A_{\lambda_1} = \frac{\varepsilon_{\lambda_1 A} - \varepsilon_{\lambda_1 B}}{\varepsilon_{\lambda_2 A} - \varepsilon_{\lambda_2 B}} \cdot \Delta A_{\lambda_2} \tag{6}.$$

That means, if the $\Delta A_\lambda$ values at different wavelengths but at equal time values are plotted one against the other, we get straight lines without an intercept (Absorbance-Difference diagrams or AD-diagrams, Fig. 1a). Two absorbance time curves at different wavelengths which are plotted one against the other give straight lines, too, but which an intercept (Absorbance diagrams or A-diagrams, Fig. 1b).

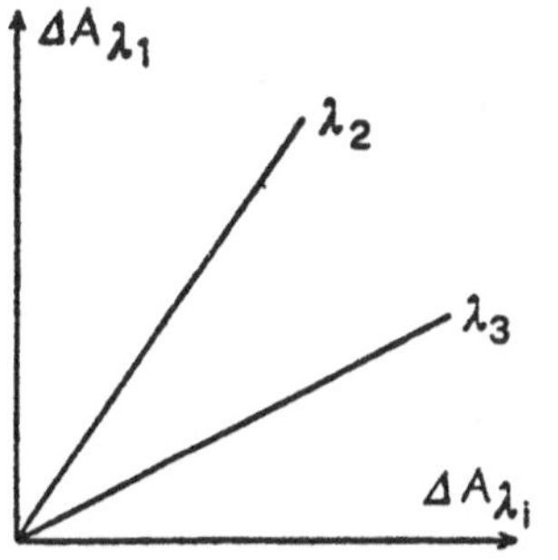


Figure 1a) AD-diagrams (schematically)

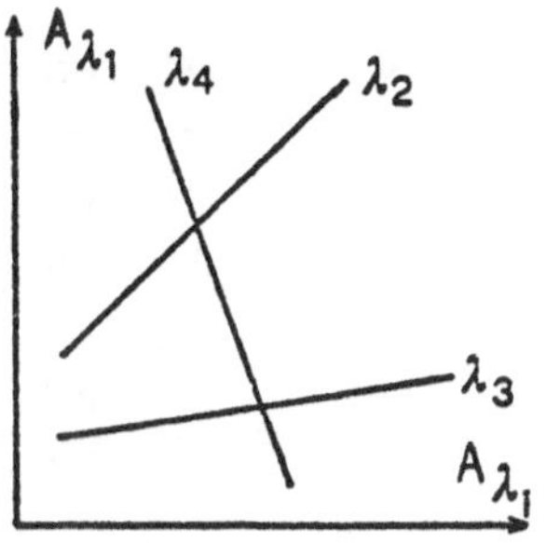


Figure 1b) A-diagrams (schematically)

The existence of isosbestic points, if the time dependent spectra are crossing, and of linear A- and AD-diagrams are necessary conditions for a simple reaction system A → B [2,3]. Some examples of such reactions which where extensively studied are the hydrolysis of p-nitrophenyl acetate at pH 8 and 25°C [4] and of 2-hydroxy-5-nitro-α-toluenesulfonic acid sultone at pH 7 and 25° C [5]. The time dependent spectra (so-called reaction spectra) of these reactions show up to 6 isosbestic points and the A- and AD-diagrams at any combination of different wavelengths are straight lines. For the calculation of the rate constant of such reactions A → B only one absorbance-time curve is needed, because multiwavelength information is redundant in this most simple case.

The methods of multiwavelength analysis for more complicated reaction systems are derived elsewhere [2,3]. The results are summarized in table 1.

Table 1: Graphical matrix rank analysis of closed reaction systems

| System | Rank | Reaction Spectra | A-Diagrams | ADQ-Diagrams |
|---|---|---|---|---|
| $A \rightarrow B$ | s=1 | Isosbestic Points | Linear | Point |
| $A + B \rightleftharpoons C$ | s=1 | Isosbestic Points | Linear | Point |
| $A \rightarrow C \rightarrow D$ | s=2 | No Isosbestic Points (generally) | Curved | Linear |
| $A+B \rightleftharpoons C \rightleftharpoons D$ | s=2 | No Isosbestic Points (generally) | Curved | Linear |
| $S+E \rightleftharpoons ES'+P1$, $\downarrow$ $E+P2$ | s=2 | No Isosbestic Points (generally) | Curved | Linear |

In contrast to the well-known methods of numerical matrix rank analysis we have called this technique "graphical matrix rank analysis" [6,7]. As derived above a reaction consisting of one linear independent concentration variable must show isosbestic points and linear A- and AD-diagrams. This is independent of the order of the reaction. If there are two linearly independent concentration variables, that means if the rank of the concentration matrix is s = 2, the reaction spectra show no isosbestic points, generally, and the A- and AD-diagrams are curved. We can make use now of another diagram, the so-called Absorbance-Difference-Quotient diagram or ADQ-diagram (2). In this diagram three absorbance-time curves are combined for evaluation and the quotients $\Delta A_{\lambda_1}/\Delta A_{\lambda_2}$ vs. $\Delta A_{\lambda_3}/\Delta A_{\lambda_2}$ are plotted one against the other at equal reaction times. It is important, that the same wavelength $\lambda_2$ is used in the denominator at both axes (compare Fig.4 and 10). If the rank of the system is s = 2, the ADQ-diagrams must be straight lines for any combination of wavelengths. In table 1 different examples of two linear independent reaction steps are compiled.

By means of these diagrams we are able to determine the number of linearly independent reaction steps as a very important criterion for the formulation of a reaction mechanism.

Figure 2 gives the reaction scheme of a more complicated reaction. Pyridoxal and histidine (as an excess component) at pH 8 and 25°C produce a Schiff's base as an intermediate, and the stable result of the reaction is a cyclisation product with a tetrahydropyridine ring [8].

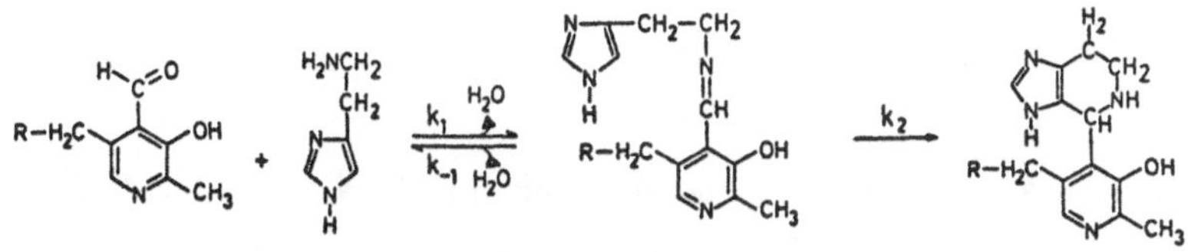

Figure 2 Reaction scheme of the cyclisation reaction of pyridoxal + histidine

The reaction spectra show intersecting regions, but no sharp isosbestic points [8]. The A-diagrams are curved (Fig.3), the ADQ-diagrams are straight lines (Fig.4), that is the rank of the system is s = 2.

How can we determine now the rate constants of such a reaction system by means of multiwavelength analysis? We can do it by using the so-called "Formal Integration method" [9].

For the simple consecutive reaction A → B → C the ODE system of two absorbance-time curves at different wavelengths $\lambda_1$ and $\lambda_2$ is given by

$$\begin{pmatrix} \dot{A}_{\lambda_1} \\ \dot{A}_{\lambda_2} \end{pmatrix} = \begin{pmatrix} z_{10} \\ z_{20} \end{pmatrix} + \begin{pmatrix} z_{11} & z_{12} \\ z_{21} & z_{22} \end{pmatrix} \cdot \begin{pmatrix} A_{\lambda_1} \\ A_{\lambda_2} \end{pmatrix} \quad (7).$$

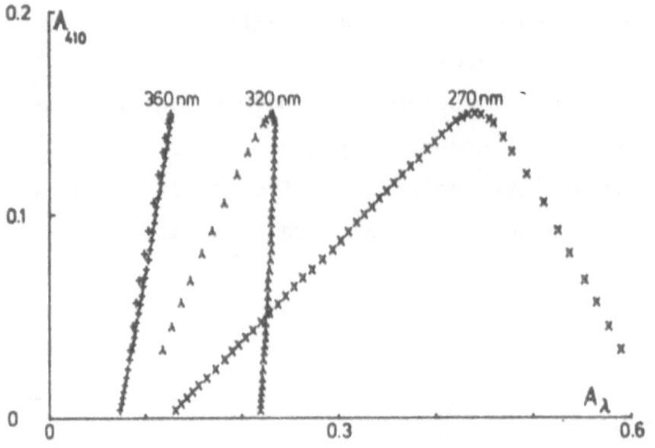


Figure 3 A-diagrams of the reaction (Fig.2) at pH 8.0 and 25°C in 0.15 M phosphate buffer; $a_o = 10^{-4}$ mol/l; $b_o = 8 \cdot 10^{-2}$ mol/l; wavelength λ (in nm)

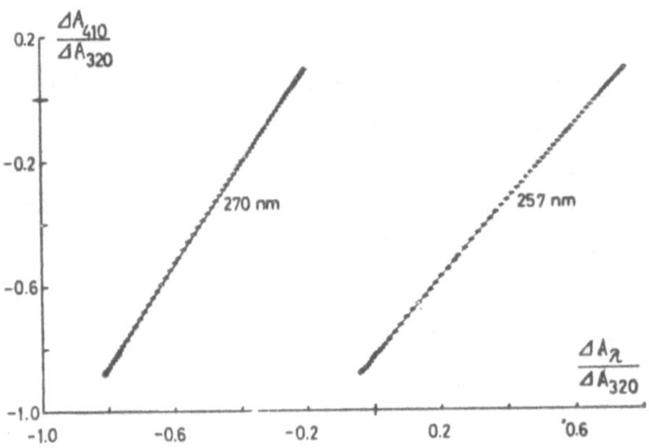


Figure 4 ADQ-diagrams (reaction conditions as in Fig.3)

By numerical integration of (7) we get

$$\Delta A_{\lambda_1} = z_{10} \cdot \Delta t + z_{11} \cdot \int A_{\lambda_1} + z_{12} \cdot \int A_{\lambda_2} dt$$
$$\Delta A_{\lambda_2} = z_{20} \cdot \Delta t + z_{21} \cdot \int A_{\lambda_1} + z_{22} \cdot \int A_{\lambda_2} dt \qquad (8).$$

Each line implies a combination of both wavelengths. By evaluation of this linear system all six coefficients z can be determined. From the trace T and the determinant D of the coefficient matrix both eigenvalues may be calculated [9].

$$T = z_{11} + z_{22}; \quad D = z_{11} \cdot z_{22} - z_{21} \cdot z_{12} \qquad (9),$$

$$k_{1,2} = 0.5 \cdot (T \pm \sqrt{T^2 - 4D}) \qquad (10).$$

In this most simple consecutive reaction the eigenvalues are identical with the rate constants. The absorbance at infinite time can be evaluated, too

$$E_{\lambda_{1\infty}} = \frac{z_{20} \cdot z_{12} - z_{10} \cdot z_{22}}{D}$$
$$E_{\lambda_{2\infty}} = \frac{z_{10} \cdot z_{21} - z_{20} \cdot z_{11}}{D} \qquad (11).$$

Fig. 5 shows absorbance-time curves of the cyclisation reaction of pyridoxal and histidine. The symbols are measured points at four different wavelengths. The solid lines are simulated by means of a Runge-Kutta-Merson program on the basis of the z-values as determined by the Formal Integration method. As to be seen the agreement is very good.

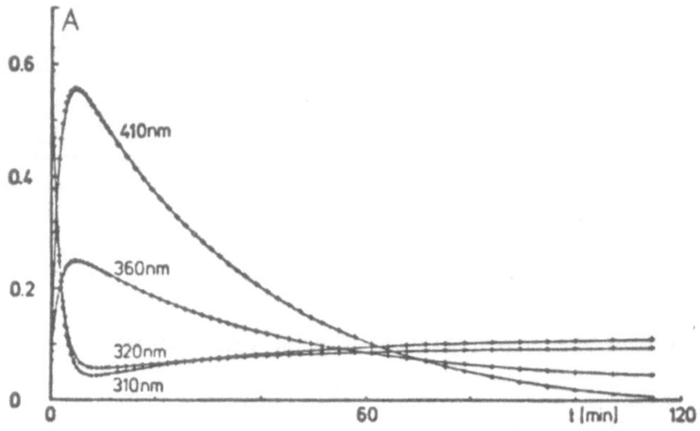


Figure 5 A(t)-diagrams and back simulations (reaction conditions as in Fig.3)

When a consecutive reaction A to B to C contains reversible steps, the information is not sufficient to calculate all three or four rate constants; we can only determine both eigenvalues.

For certain consecutive reactions this problem may be solved in a very elegant way: if one reaction step is of pseudo-first order, the concentration of the excess component can be varied in different experiments ($A + B \rightleftharpoons C \rightleftharpoons D$). In this case the concentration dependence of the trace and determinant of the kinetic matrix contain the lacking information, and it is possible to discriminate between different mechanisms and to calculate all rate constants [8].

Fig.6 shows this evaluation for the cyclisation reaction of pyridoxal and histidine. The trace T and determinant D are plotted as a function of the concentration of the excess component histidine. The result of the evaluation is: the first step is reversible, the second step is practically irreversible [8]. Measurable quantities of a carbinolamine as an intermediate do not exist. All three rate constants may be calculated from the slope and intercept of Fig.6a or b, respectively:

$k_1 = 1.36 \ l \cdot mol^{-1} \cdot s^{-1}$,
$k_{-1} = 3.74 \cdot 10^{-3} \ s^{-1}$,
$k_2 = 7.50 \cdot 10^{-4} \ s^{-1}$.

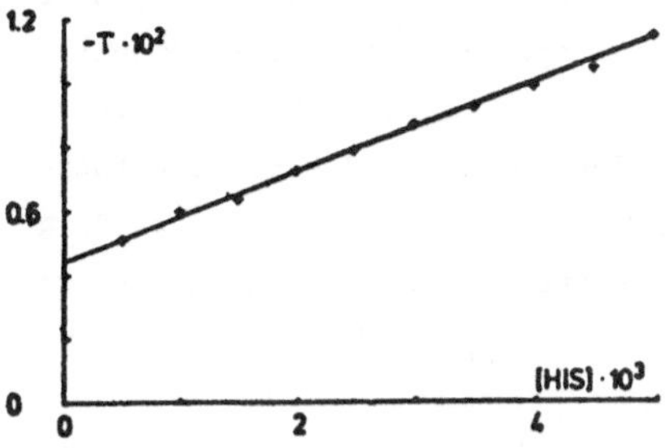


Figure 6a) Trace T vs. $b_o$

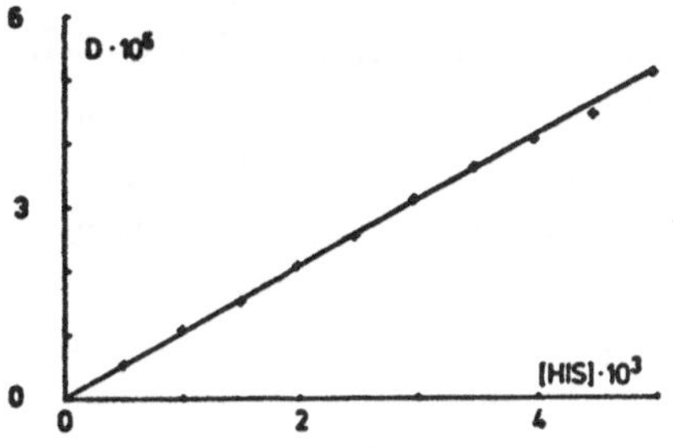


Figure 6b) Determinant D vs. $b_o$

For the evaluation of first order consecutive reactions with more than two exponentials an iterative curve-fitting procedure should be used, which is able to combine the whole multiwavelength information simultaneously in the curve-fitting procedure. We have already applied sucessfully a simultaneous curve-fitting program to spectrophotometric titrations [10,11] with up to six linearly independent titration steps.

All kinetic evaluation methods mentioned so far were originally derived for closed reactions. In the last part their application to open raction systems shall be considered.

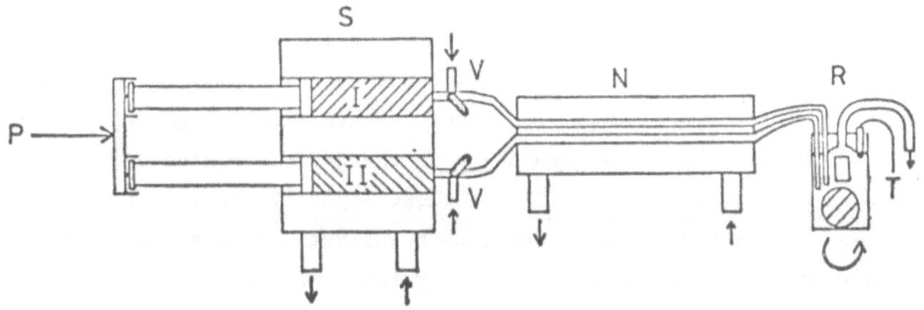


Figure 7 Construction of the CSTR system:
P = linear pump (Infors precidor);
S = syringes (Hamilton 20 ml) in thermostated block;
V = miniature valves (Hamilton); N = teflon needles (Hamilton) with temper jacket; R = reactor cuvet (Hellma) with thermistor T and stirring ball

Fig.7 shows the construction of a very little continuously stirred tank reactor especially designed for spectrometric measurements [11]. The reaction vessel is a modified sample cell with 1 cm path length which is magnetically stirred. The volume is 1.8 ml. Two thermostated syringes are driven by a linear pump of high precision. Both input tubes are thermostated as well as the sample cell in the spectrometer. $\tau_f = 1/k_f$ is the mean residence time, $k_f$ is the rate constant of the influx.

In table 2 some first order reactions in closed systems and the CSTR are compared, schematically:

Table 2: First order reactions in closed and open reaction systems

| Rank | s = 1 | s = 2 | s = 3 |
|---|---|---|---|
| Closed System | $A \xrightarrow{k_1} B$ | $A \xrightarrow{k_1} B \xrightarrow{k_2} C$ | $A \xrightarrow{k_1} B \xrightarrow{k_2} C \xrightarrow{k_3} D$ |
| Open System (CSTR) | $k_f\downarrow$ A $k_f\downarrow$ | $k_f\downarrow$ $A \xrightarrow{k_1} B$ $k_f\downarrow$ $k_f\downarrow$ | $k_f\downarrow$ $A \xrightarrow{k_1} B \xrightarrow{k_2} C$ $k_f\downarrow$ $k_f\downarrow$ $k_f\downarrow$ |

In the CSTR the rank of a chemical reaction system is increased by one compared with the corresponding closed system because the rate constant $k_f$ of input and output represents an additional eigenvalue. For that reason, a simple reaction $A \rightarrow B$ in the CSTR is equivalent to a consecutive reaction $A \rightarrow B \rightarrow C$ in a closed system. And it can be derived that the methods of multiwavelength analysis, as described previously, can be applied for the open system, too [11].

Fig.8 shows the reaction spectra of the hydrolysis of p-nitrophenyl acetate at pH 10 and 25$^{o}$C in the CSTR [11]. In contrast to a closed system, this simple reaction does not show isosbestic points in the CSTR. The A-diagrams(Fig.9) are curved at every combination of wavelengths. Fig.10 shows linear ADQ-diagrams, that is, the rank is s = 2. In Fig.11 absorbance-time measurements at three wavelengths are depicted. The curves are back-simulations on the basis of the z-values as determined by the Formal Integration method. Simulations and measured points agree very well.

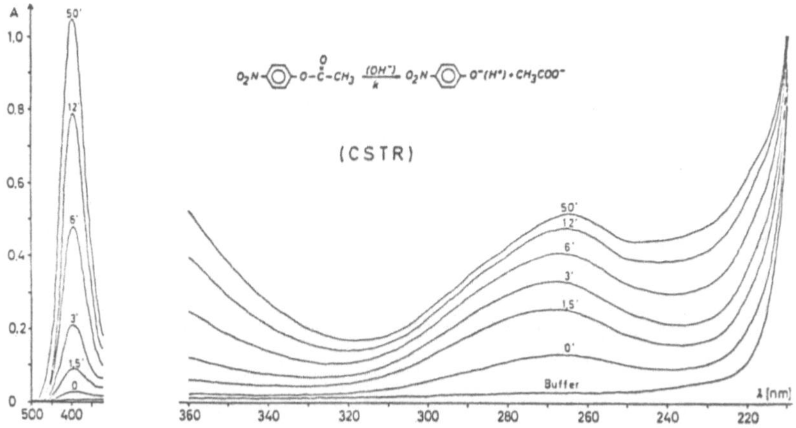


Figure 8 Reaction spectra of the hydrolysis of p-nitrophenyl acetate in the CSTR at pH 10.0 and 25°C in 0.15 M phosphate buffer; $a_o = 10^{-4}$ mol/l; reaction time (in minutes) as a parameter

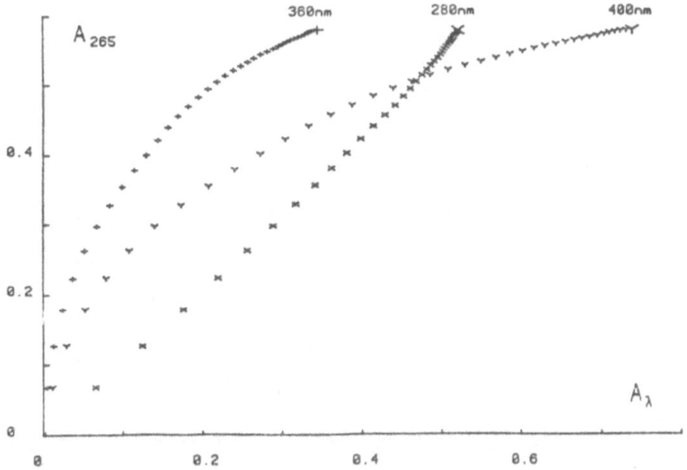


Figure 9 A-diagrams (reaction conditions as in Fig.8)

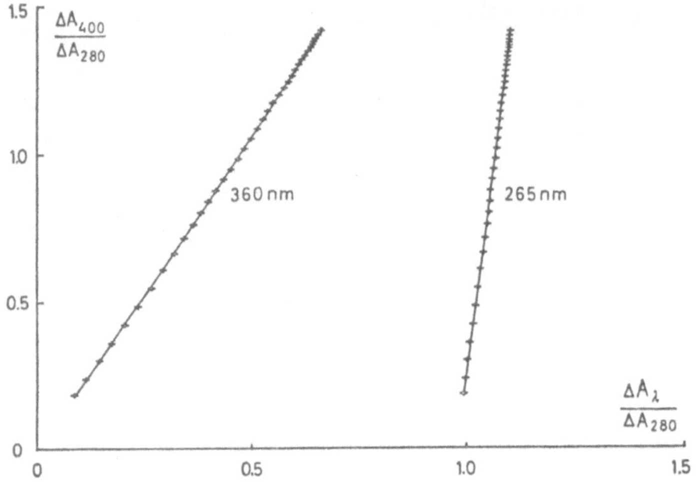


Figure 10 ADQ-diagrams (reaction conditions as in Fig. 8)

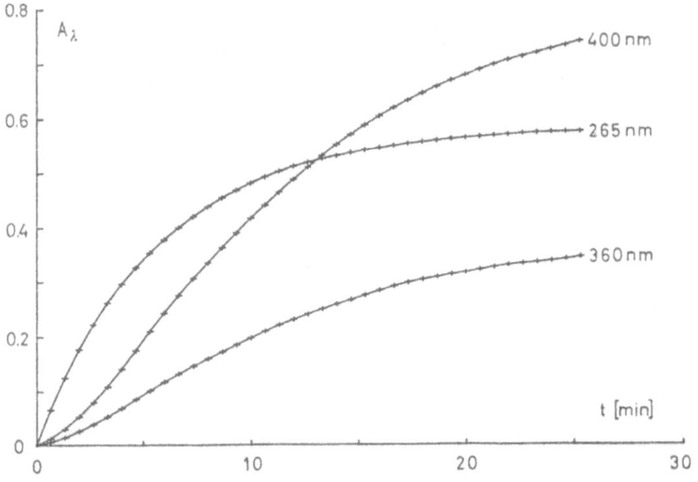


Figure 11 A(t)-diagrams and back simulations (reaction conditions as in Fig. 8)

The rate constants of the chemical reaction (= k) and the physical inflow (= $k_f$) may be calculated from the z-values according to equations (9,10). The results are $k = 1.50 \cdot 10^{-3}\ s^{-1}$; $k_f = 2.20 \cdot 10^{-3}\ s^{-1}$. These values are in good agreement with separate determinations of k in a closed system and of $k_f$ with other methods [11].

In the moment we are applying the methods of multiwavelength analysis to more complicated autocatalytic reaction systems in a CSTR, which show critical slowing down, multiple steady states and oscillations.

References:


[1] K.H.Ebert, P.Deuflhard, W.Jäger, (eds.), Modelling of Chemical Reaction Systems, Springer, Berlin (1981) p.VII, 92, 102,

[2] H.Mauser, Z.Naturforsch. 23b (1968) 1021, 1025,

[3] H.Mauser, Formale Kinetik, Bertelsmann Universitätsverlag, Düsseldorf (1974),

[4] H.Lachmann, H.Mauser, F.Schneider, H.Wenck, Z.Naturforsch. 26b (1971) 629,

[5] H.Lachmann, Z.Anal.Chem. 301 (1980) 148,

[6] H.Lachmann, Z.Anal.Chem. 290 (1978) 117,

[7] G.Lachmann, H.Lachmann, Z.Anal.Chem. 290 (1978) 118,

[8] G.Lachmann, H.Lachmann, H.Mauser, Z.Phys.Chem.NF 120 (1980) 19,

[9] G.Lachmann, H.Lachmann, H.Mauser, Z.Phys.Chem.NF 120 (1980) 9,

[10] F.Göbber, H.Lachmann, Hoppe-Seyler's Z.Physiol.Chem. 359 (1978) 269,

[11] H.Lachmann, Habilitationsschrift, Tübingen (1982).

# ON THE ESTIMATION OF SMALL PERTURBATIONS IN ORDINARY DIFFERENTIAL EQUATIONS

PEDRO E. ZADUNAISKY

## 1. INTRODUCTION

The essential subject of this work is concerned with those problems represented by a system of ordinary differential equations involving one or several small perturbing functions to be determined in order to obtain either a solution given in advance (control problems) or a solution that approximates a set of measurements that may be affected by random errors. The traditional solution of such problems consists of the parameter identification of a model of the perturbations by means of statistical methods. The inconvenience of these procedures is that some peculiarities in the residuals that are small but physically significant can be smoothed out either by insufficiency of the model or by the intrinsic nature of the statistical methods. We present here a direct method based only on the assumptions that the solutions of the differential equations can be expressed piecewise in short intervals by a Taylor convergent expansion and that the unknown perturbations can be approximated by elementary functions.

## 2. SOME THEORETICAL BACKGROUNDS

Although our method is applicable to a set of o.d.e. of any order we shall consider for conciseness a single equation of the form

$$\ddot{y}=f(t,y)+P\ ,y(t_o)=y_o\ ,\ \dot{y}(t_o)=\dot{y}_o \tag{2.1}$$

where P is an unknown perturbation that varies with t. Now let us assume that the solution of (2.1) can be represented by a convergent Taylor expansion such that on a set of successive instants $t_n$ (n=0,1,2,...) we have, with the remainder expressed in integral form,

$$y(t_{n+1})=y(t_n)+(t_{n+1}-t_n)\dot{y}(t_n)+\int_{t_n}^{t_{n+1}}\ddot{y}(u)(t_{n+1}-u)du \tag{2.2}$$

and by virtue of (2.1)

$$y(t_{n+1})=y(t_n)+(t_{n+1}-t_n)\dot{y}(t_n)+\int_{t_n}^{t_{n+1}}f(u,y)(t_{n+1}-u)du \\ +\int_{t_n}^{t_{n+1}}P(u)(t_{n+1}-u)du \tag{2.3}$$

Now let us consider a "reference" problem $\ddot{y}^r=f(t,y^r)$ obtained from (2.1) by dropping the unknown perturbation P and in a similar way we may write

$$y^r(t_{n+1})=y^r(t_n)+(t_{n+1}-t_n)\dot{y}^r(t_n)+\int_{t_n}^{t_{n+1}} f(u,y^r)(t_{n+1}-u)du \tag{2.4}$$

Assuming that $f(u,y)$ is a $C^1$-function of y we may write

$$f(u,y^r)=f(u,y)+\frac{\partial f}{\partial y}(u,\eta)(y^r-y) \tag{2.5}$$

with $\eta(u)\in(y^r(u),y(u))$. Let us also assume that

$$y^r(t_n)=y(t_n) \quad , \quad \dot{y}^r(t_n)=\dot{y}(t_n) \ ; \tag{2.6}$$

then substracting (2.4) from (2.3) and applying the generalised mean value theorem for integrals we obtain

$$y(t_{n+1})-y^r(t_{n+1})+\varepsilon_{n+1}(t_{n+1}-t_n)^2=\int_{t_n}^{t_{n+1}} P(u)(t_{n+1}-u)du \tag{2.7}$$

with

$$\varepsilon_{n+1}=\frac{1}{2}\frac{\partial f}{\partial y}(\xi,\eta(\xi))(y^r(\xi)-y(\xi)),\ \xi\in(t_n,t_{n+1}) \tag{2.8}$$

By taking into account (2.5) and (2.6) we find that

$$y^r(\xi)-y(\xi)\simeq\frac{1}{2}(\xi-t_n)^2P(t_n) \tag{2.9}$$

and we may finally write

$$y(t_{n+1})-y^r(t_{n+1})+\bar{\varepsilon}_{n+1}(t_{n+1}-t_n)^2=\int_{t_n}^{t_{n+1}} P(u)(t_{n+1}-u)du \tag{2.10}$$

with

$$\bar{\varepsilon}_{n+1}=\frac{1}{4}\frac{\partial f}{\partial y}(\xi,\eta(\xi))P(t_n)(\xi-t_n)^2 \tag{2.11}$$

## 3. ESTIMATIONS OF P(t)

In the sequel we shall describe some methods to obtain estimates of P(t) at the successive points $t_n$. By a linear or a quadratic representation of P(u) in (2.10) we may obtain very simple formulas as follows.

a) Linear interpolation on 2 points

For simplicity we shall put $P_n=P(t_n)$ and $R_{m,n}=y(t_m)-y^r(t_m)$ when one assumes the conditions (2.6). Similarly $R_{n,m}=y(t_n)-y^r(t_n)$ with $y^r(t_m)=y(t_m)$ and $\dot{y}^r(t_m)=\dot{y}(t_m)$. Let us indicate by $I_{n,n+1}$ the right hand member of (2.10) and replace P(u) in the interval $(t_n,t_{n+1})$ by the secant line

$$Y(u)=P_{n+1}-\frac{(P_{n+1}-P_n)(t_{n+1}-u)}{(t_{n+1}-t_n)} \tag{3.1}$$

Integrating we obtain

$$I_{n,n+1}=(t_{n+1}-t_n)^2\left(\frac{1}{3}p_n + \frac{1}{6}p_{n+1}\right) \tag{3.2}$$

and replacing in (2.10) it results

$$\bar{R}_{n+1,n}+\bar{\varepsilon}_{n+1}=\frac{1}{3}p_n + \frac{1}{6}p_{n+1} \tag{3.3}$$

where

$$\bar{R}_{n+1,n}=R_{n+1,n}/(t_{n+1}-t_n)^2 \tag{3.4}$$

and $\bar{\varepsilon}_{n+1}$ is given by (2.11).

The whole reasonment can be repeated in reverse sense from $t_{n+1}$ towards $t_n$ and one obtains, with

$$\bar{R}_{n,n+1}=R_{n,n+1}/(t_n-t_{n+1})^2 \tag{3.5}$$

and

$$\bar{\varepsilon}_n=\frac{1}{4}\frac{\partial f}{\partial y}(\xi,\eta(\xi))p(t_{n+1})(\xi-t_{n+1})^2, \tag{3.6}$$

$$\bar{R}_{n,n+1}+\bar{\varepsilon}_n=\frac{1}{6}p_n+\frac{1}{3}p_{n+1} \tag{3.7}$$

Using a matrix and vector notation we may write (3.3) and (3.7) as a system of linear equations

$$\begin{pmatrix} \frac{1}{6} & \frac{1}{3} \\ \frac{1}{3} & \frac{1}{6} \end{pmatrix} \begin{pmatrix} p_n \\ p_{n+1} \end{pmatrix} = \begin{pmatrix} \bar{R}_{n,n+1} \\ \bar{R}_{n+1,n} \end{pmatrix} \tag{3.8}$$

having neglected $\bar{\varepsilon}_n$ and $\bar{\varepsilon}_{n+1}$. Later we shall establish some upper bounds for errors caused by that neglect and by other reasons. Solving (3,8) we obtain

$$\begin{pmatrix} p_n \\ p_{n+1} \end{pmatrix} = \begin{pmatrix} -2 & 4 \\ 4 & -2 \end{pmatrix} \begin{pmatrix} \bar{R}_{n,n+1} \\ \bar{R}_{n,n+1} \end{pmatrix} \tag{3.9}$$

b) <u>Linear interpolation on 3 points</u>

Let us consider the three successive points $t_n, t_{n+1}$ and $t_{n+2}$. If we repeat the previous reasonement this time with the points $t_{n+1}$ and $t_{n+2}$ we obtain a linear system which is similar to (3.8) and putting them together we obtain the overdetermined system

$$M_L \cdot P = R \tag{3.10}$$

where

$$M_L = \frac{1}{6}\begin{pmatrix} 1 & 2 & 0 \\ 2 & 1 & 0 \\ 0 & 1 & 2 \\ 0 & 2 & 1 \end{pmatrix} \tag{3.10.a}$$

$$P^T=(P_n,P_{n+1},P_{n+2}) \tag{3.10.b}$$

$$R^T=(\bar{R}_{n,n+1},\bar{R}_{n+1,n},\bar{R}_{n+1,n+2},\bar{R}_{n+2,n+1}) \tag{3.10.c}$$

We may calculate the generalised inverse $M_L^+ =(M_L^T M_L)^{-1}M_L^T$ of $M_L$ obtaining exactly

$$M_L^+=\begin{pmatrix} -0.4 & 3.2 & 0.8 & -1.6 \\ 2.0 & -1.0 & -1.0 & 2.0 \\ -1.6 & 0.8 & 3.2 & -0.4 \end{pmatrix} \tag{3.11}$$

and then we have for P the "least squares" solution

$$P = M_L^+ R \tag{3.12}$$

c) <u>Quadratic interpolation on 3 points</u>

For simplicity we shall assume that the three points are equidistant,that is, $t_{n+1}-t_n=t_{n+2}-t_{n+1}=h$ (constant). If they were not equidistant our results would be slightly more complicated but no essential dif ficulty would arise. In the present case we replace for $P(u)$ in (2.10) a quadratic function $Y(u)=A_k+B_k(t_k-u)+C_k(t_k-u)^2$ where the coefficients depend on the reference point $t_k$.

By taking successively k=n,k=n+1 and k=n+2 and proceeding as before we may establish a system of linear equations similar to (3.10)

$$M_Q \cdot P = R \tag{3.13}$$

where

$$M_Q= \frac{1}{24}\begin{pmatrix} 3 & 10 & -1 \\ 7 & 6 & -1 \\ -1 & 6 & 7 \\ -1 & 10 & 3 \end{pmatrix} \tag{3.14}$$

and P and R are defined by (3.10.b) and (3.10.c) respectively. The generalised inverse of $M_Q$ is exactly

$$M_Q^+=\begin{pmatrix} -0.9 & 3.7 & 1.3 & -2.1 \\ 1.5 & -0.5 & -0.5 & 1.5 \\ -2.1 & 1.3 & 3.7 & -0.9 \end{pmatrix} . \tag{3.15}$$

and finally

$$P = M_Q^+ R \tag{3.16}$$

It is remarkable the relation

$$M_Q^+ - M_L^+ = \begin{pmatrix} -0.5 & 0.5 & 0.5 & -0.5 \\ -0.5 & 0.5 & 0.5 & -0.5 \\ -0.5 & 0.5 & 0.5 & -0.5 \end{pmatrix}$$

that should allow to obtain very easyly $M_Q^+$ from $M_L^+$ or viceversa.

As a final test for the estimates of P we can solve numerically the equation (2.1) replacing those estimates for the unknown function P(t) and compare the results either with the given solution y(t) in control problems or with the measurements $y_n$. More about this question will be said in the next sections.

## 4. ERROR BOUNDS

The main result of the previous section is that the unknown perturbation P can be calculated by linear formulas like (3.13) and (3.16) where the generalised inverses $M_L^+$ and $M_Q^+$ have no errors and if R is affected by an additional vector of errors $\delta R$ we have for the corresponding errors in P

$$\delta P = M^+ . \delta R \tag{4.1}$$

where $M^+$ represents either $M_L^+$ or $M_Q^+$.

In what follows we shall use for any vector $x^T = (x_1, x_2, \ldots)$ the norm $\|x\|_\infty = \max_i |x_i|$ and for any matrix the compatible norm $\|A_{ij}\|_\infty = \max_i \sum_j |Aij|$

so that we have in our case

$$\|\delta P\|_\infty \leq \|M^+\|_\infty \cdot \|\delta R\|_\infty \tag{4.2}$$

Let us put $\|M^+\|_\infty = N$ and by virtue of (3.11) and (3.15) we have N=6 for the case of linear interpolation and N=8 for the case of quadratic interpolation respectively.

Now we shall consider three main sources that contribute to the error $\delta R$.

I) We have neglected terms like $\bar{\mathcal{E}}_n$ and $\bar{\mathcal{E}}_{n+1}$ expressed by formulas (3.6) and (2.11) respectively and from this source we have

$$\|\delta R_I\|_\infty = \frac{1}{4} \max_{\xi} \left|\frac{\partial f}{\partial y}(\xi, \eta(\xi))\right| . \max_{\xi} |P(\xi)| . h^2$$

II) The integral of formula (2.10) may be written, by the generalised mean value theorem for integrals, in the form

$$I_{n,n+1}=(t_{n+1}-\xi)\int_{t_n}^{t_{n+1}} P(u)du \quad ,\xi\in(t_n,t_{n+1})$$

If we replace the integrand P(u) either by a linear or a quadratic function we introduce an error $|\delta I| = h\, h^3 P''(\xi)/12$ or $|\delta I| = h\, h^5 P^{IV}(\xi)/90$ respectively (see [1]) and by virtue of (3.5) we have either $\|\delta R_{II}\|_\infty = O(h^2)$ or $\|\delta R_{II}\|_\infty = O(h^4)$ respectively.

III) In the expression $R_{m,n}=y(t_m) - y^r(t_m)$ the solution $y^r(t)$ of the reference problem may be known analytically; if such is not the case it must be obtained by a numerical procedure where all the necessary precautions have been taken in order to make negligible the global errors. Concerning the solution y(t) it may be given exactly (control problem); in case that it is measured on a set of points it is affected by random errors and we shall assume that they have a normal distribution with zero mean and a given variance $\sigma$. Consequently it is not too difficult to prove that at practically all points we shall have

$$R_{III} = 3\sigma(1+h)/h^2$$

Finally if we put

$$\|\delta R\|_\infty = \|\delta R_I\|_\infty + \|\delta R_{II}\|_\infty + \|\delta R_{III}\|_\infty$$

we have

$$\|\delta P\|_\infty \leq N\,\|R\|_\infty$$

The errors $\delta R_I$ and $\delta R_{II}$ are <u>inherent</u> to the method itself and they are proportional either to $h^2$ or to $h^4$. In other words the inherent errors can be conveniently reduced by adopting, if possible, a sufficiently small distance between the successive points. The errors $\delta R_{III}$ depend essentially of the accuracy ($\sigma$) of the measurements and we must observe that they are magnified by a factor $1/h^2$. The effects of the different types of errors will be put in evidence in the examples that follow.

## 5. EXAMPLES

<u>Example 1</u>: Harmonic Oscillator with a Sinusoidal Perturbation

The differential equation of this problem is

$$\ddot{y} + w_0^2\, y = \mu \sin(wt) \quad , \quad y(t_0)=y_0 \ , \ \dot{y}(t_0)=\dot{y}_0 \tag{5.1}$$

being $\mu$ a small constant parameter. For $w_0 \neq w$ the analytical solution of the problem is

$$y(t)= y_0\cos(w_0 t) + \frac{\dot{y}_0}{w_0}\sin(w_0 t) + \frac{\mu}{w_0^2-w^2}\sin(wt) \tag{5.2}$$

Table 1. Example 1. Harmonic Oscillator with a Sinusoidal Perturbation

| | Measurements | | $Px10^2$ | Efficiency of Estimates | | |
|---|---|---|---|---|---|---|
| $t_n$ | $\tilde{y}_n$ | $\dot{\tilde{y}}_n$ | | EFF($\tilde{P}$) | EFF(y) | EFF($\dot{y}$) |
| 0.0 | 1.00000 | 1.00667 | 0.000 | 3.3 | - | - |
| 2.0 | 0.50440 | -1.32187 | 0.841 | 1.9 | 3.6 | 4.9 |
| 4.0 | -1.39832 | 0.10038 | 0.909 | 1.2 | 4.3 | 4.3 |
| 6.0 | 0.68266 | 1.23300 | 0.141 | 0.1 | 4.1 | 4.7 |
| 8.0 | 0.83377 | -1.13921 | -0.756 | 2.5 | 3.1 | 5.3 |
| 10.0 | -1.39589 | -0.29315 | -0.958 | 1.2 | 5.6 | 3.4 |
| 12.0 | 0.30354 | 1.38685 | -0.279 | 2.0 | 3.7 | 4.4 |
| 13.0 | 1.33049 | 0.49380 | 0.215 | 1.1 | 4.6 | 3.6 |

and

$$\dot{y}(t)= -y_o w_o \sin(w_o t) + \dot{y}_o \cos(w_o t) + \frac{\mu w}{w_o^2 - w^2} \cos(wt) \tag{5.3}$$

In this problem everything is known but we shall use it to simulate a case where by means of "measurements" of y and $\dot{y}$ made on a discrete set of points $t_n$ one obtains estimates of the true perturbation

$$P(t)=\mu \sin(wt) \tag{5.4}$$

at such points using the methods described in section 3. The simulated measurements $\tilde{y}_n$ and $\dot{\tilde{y}}_n$ were calculated from the formulas

$$\tilde{y}_n= y(t_n)+e_1 \quad , \quad \dot{\tilde{y}}_n= \dot{y}(t_n)+e_2 \tag{5.5}$$

were the true values $y(t_n)$ and $\dot{y}(t_n)$ were obtained from formulas (5.2) and (5.3) and $e_1$ and $e_2$ were "measurement errors" taken from a zero mean normal random number sequence scaled to a specific variance $\sigma$. In each interval $(t_n, t_{n+1})$ we used as a reference solution

$$y^r(t) = \tilde{y}_{n+1}\cos(w_o T)+ \frac{\dot{\tilde{y}}_{n+1}}{w_o} \sin(w_o T) \quad , \quad T= t-t_{n+1} \tag{5.6}$$

to obtain the residual $R_{n,n+1}= y_n - y^r(t_n)$ and

$$y^r(t)= \tilde{y}_n \cos(w_o T) + \frac{\dot{\tilde{y}}_n}{w_o} \sin(w_o T) \quad , \quad T= t-t_n \tag{5.7}$$

for the residual $R_{n+1,n}= y_{n+1}-y^r(t_{n+1})$. Then we proceeded to apply at the successive intervals any of the three methods described in section 3. In this example we adopted the following set of parameters involved in in the problem

$$w_o=1.0,\ w=0.5,\ \mu=0.01,\ h=0.5,\ y_o=1.0,\ \dot{y}_o=1.0,\quad \sigma= 10^{-5} \tag{5.8}$$

This and other examples that we have performed are test problems where the exact values of the perturbation are known (in this example $P(t)$ is given by formula (5.4)). To have an assessment of the efficiency of our method of finding estimates $\tilde{P}_n$ of $P(t_n)$ we have used the following empirical formula developed by ourselves for similar purposes in some earlier papers ([4],[5])

$$EFF(\tilde{P}) = -\log_{10}(|\tilde{P}_n - P(t_n)|/10^q) \qquad (5.9)$$

where q is the exponent of the floating form of $\mathrm{Min}(|\tilde{P}_n|, |P(t_n)|)$. $EFF(\tilde{P})$ is a number that when it is positive its integer part is equal to the number of figures of the q-th decimal order correctly estimated of $P(t_n)$. When it is negative it means that the estimate is poor and its integer part represents the difference between the decimal orders of magnitude of $\tilde{P}_n$ and $P(t_n)$. If the integer part of $EFF(\tilde{P}_n)$ is zero it means that at least the decimal order of magnitude of $P(t_n)$ has been correctly estimated.

As a final check we integrated numerically the differential equation (5.1) but replacing its right hand member by the estimates $\tilde{P}_n$ obtaining in this way computed values $y_n^c$ of the solution that we compared with the measured values $\tilde{y}_n$ using for this purpose the formula

$$EFF(y) = -\log_{10}(|y_n^c - \tilde{y}_n|/10^q) \qquad (5.10)$$

similar to (5.9).

To apply in this example the formulas for error bounds given in section 4 we have $f(t,y) = w_o^2 y$ and then $\frac{\partial f}{\partial y} = w_o^2$ (constant). On the other hand $P''(t) = -w^2 P(t)$ and recalling formulas from section 4 we find as an error bound for the estimates $P_n$

$$\|\delta P_n\|_\infty = 8(w_o^2/4 + w^2/12 + 3\delta/h^4) . \max_\xi |P(\xi)| . h^2$$

and with the data (5.8) it results

$$\|\delta P_n\|_\infty = 0.5 \max_\xi |P(\xi)| \qquad (5.11)$$

As this bound is on the conservative side we may expect that the efficiency values furnished by formula (5.9) may be at least of the order 0 or 1.

We solved this problem by the three methods described in section 3 obtaining in all cases practically the same results. In Table 1 we give a sample of results of the first method (that is by interpolating the unknown perturbation by a secant line at each successive interval) covering a little more than two natural periods of the free oscillator.

To obtain the computed values $y_n^c$ and $\dot{y}_n^c$ we integrated the system

$$\dot{y}_1 = y_2$$
$$\dot{y}_2 = -\omega_0^2 y_1 + \mu \sin(\omega t),$$

equivalent to (5.1), by a Runge-Kutta method of order 8th.

In the last three columns of Table 1 we give the efficiencies of the estimates for P, y and $\dot{y}$. In the case of P our estimates are better than what could be expected from the upper bound of errors (5.11) and for y and $\dot{y}$ their estimates show a good agreement with the imposed magnitude of the random measurement errors.

Example 2. Van der Pol Equation

This is the well known equation

$$\ddot{y} + \omega_0^2 y = \mu(1-y^2)\dot{y} \quad , \; y(t_0)=y_0 \; , \quad \dot{y}(t_0) = \dot{y}_0 \tag{5.12}$$

where $\mu$ is a constant parameter and it can be rewritten in the equivalent form

$$\begin{aligned} \dot{y}_1 &= y_2 \\ \dot{y}_2 &= -\omega_0^2 y_1 + \mu(1 - y_1^2)\, y_2 \end{aligned} \tag{5.13}$$

It can be considered as the equation for the free harmonic oscillator with a perturbation

$$P(t) = \mu(1 - y^2)\,\dot{y} \tag{5.14}.$$

To obtain estimates of P(t) we proceeded here in a way entirely similar to that of Example 1. The analytical solution of (5.12) might be obtained by a perturbation theory which furnishes y(t) in the form of an expansion in powers of . However to simulate the measurements of y and $\dot{y}$ we calculated a very accurate numerical solution of (5.13) and added to the results random errors as in Example 1. For this problem we adopted the following set of parameters

$$\omega_0=1.0, \; \mu=0.1, \quad h=0.25, \quad y_0=1.0, \; \dot{y}_0=0.0, \quad \sigma=10^{-5} \tag{5.15}$$

As in Example 1 the period of the free harmonic oscillator is $2\pi$ and again we extended our calculations for two periods. While in the previous example the perturbation P(t) was a smooth sinusoidal function in this case it experiences many abrupt changes in very short intervals so that its estimation becomes a much more difficult problem. However our results are again rather satisfactory as they are shown in the last three columns of Table 2. These results were obtained through our third

Table 2. Example 2. Van der Pol Equation

| $t_n$ | Measurements, $\tilde{y}_n$ | $\tilde{\dot{y}}_n$ | Px10 | Efficiency of Estimates EFF(P) | EFF(y) | EFF($\dot{y}$) |
|---|---|---|---|---|---|---|
| 0.0 | 1.00001 | 0.00003 | 0.000 | 0.8 | - | - |
| 2.0 | -0.47421 | -0.99432 | -0.770 | 1.6 | 3.4 | 4.4 |
| 4.0 | -0.72749 | 0.85618 | 0.403 | 2.1 | 4.2. | 3.6 |
| 6.0 | 1.19382 | 0.33547 | -0.143 | 2.1 | 3.9 | 3.6 |
| 8.0 | -0.21938 | -1.30286 | -0.124 | 3.7 | 4.2. | 4.4 |
| 10.0 | -1.12833 | 0.74653 | -0.020 | 2.1 | 3.9 | 3.5 |
| 12.0 | 1.25439 | 0.77385 | -0.444 | 1.8 | 4.0 | 2.8 |
| 13.0 | 1.31833 | 0.63630 | 0.470 | 2.0 | 3.8 | 3.9 |

method based on a quadratic interpolation of P on every 3 successive points.

## 6. FINAL REMARKS

We have to mention here some earlier papers by Ingram and Tapley [2] and by Tapley and Schutz [3] where they solved a similar problem concerning the motion of an artificial satellite of the Moon. They start from formulas similar to (2.2) and (2.3) and assume that the unknown perturbations may be approximated by a random vector function satisfying a Gauss-Markov process represented by a system of linear differential equations.Then the solution of the problem is carried on through a filtering standard technique. We believe that our method,which is essentially deterministic,attains the same goals with less theoretical requirements and with simpler procedures.Furthermore our formulas for error bounds show that either the given function y(t) or the measurements $\tilde{y}_n$ can be satisfied to a desired accuracy when it is possible to reduce to a convenient size the distance between successive points. Besides the examples presented here we have already made successful applications of our method to some astronomical problems [5], namely the determination of some unpredictable forces of non gravitational origin acting on the motion of comets and the perturbations from an unknown planet on a known planet.

Our last remark is that with the method presented here we do not intend to supersede the current methods based on modelling and the identification of parameters by statistical methods. However it may help to build more properly a model of any process that may contain small but systematic effects or perturbations that otherwise might be overlooked especially due to the smoothing effects of statistical methods.

REFERENCES

[1] Hildebrand,F.B.: 1956, Introduction to Numerical Analysis (McGraw-Hill),p.73.

[2] Ingram,D.S. and Tapley,B.D.: 1974,Celestial Mechanics,9,pp.191-211.

[3] Tapley,B.D. and Schutz,B.E.: 1975,Celestial Mechanics,12,pp.409-424.

[4] Zadunaisky,P.E.: 1976,Numerische Mathematik,27,pp.21-39.

[5] Zadunaisky,P.E.: 1981,Proceedings of the Conference on "Motions of Planets and Satellites",Universities of Texas,USA and Sao Paulo, Brazil (dec.1981), To appear.

# PART II

# INVERSE BOUNDARY AND EIGENVALUE PROBLEMS IN ORDINARY DIFFERENTIAL EQUATIONS

# MULTIPLE SHOOTING TECHNIQUES REVISITED

P. Deuflhard , G. Bader

## 0. Introduction

The present article is a short summary of a more extensive presentation - see [9]. Multiple shooting (MS) techniques as developed in [4,16,19,5] are one of the popular approaches for the numerical solution of (in general nonlinear) boundary value problems (BVP's) for ordinary differential equations (ODE's). For alternative approaches see e.g. [1,14]. For a recent survey on MS techniques see [6], where also further references and historical remarks can be found. The application of MS techniques to *parameter identification* has been nicely developed in [2]. In fact, the efficient and economic implementation of a parameter identification algorithm has motivated the present study. Even though, for the purpose of simplification, most of the presentation is confined to the standard BVP case (without explicit dependence on parameters), the results also apply - mutatis mutandis - to the parameter identification case.

The MS code BOUNDS(OL) due to [4,19,6] realizes a special linear system solver called *condensing algorithm* (CA). There has been a controversial discussion whether this CA is numerically stable or not - compare [16] on one side or [19] on the other side. Especially recent work [15,12] seems to indicate that the CA should be substituted by some other linear system solver - see [12,15,18,20]. On the other hand, the CA is preferable from the points of view of storage economy, theoretical transparency and ease of extension to parameter identification. In this situation the present article gives a detailed elementwise rounding error analysis (on the basis of [22]), which classifies those BVP's, for which the CA is numerically stable. In addition, a combination of the CA with a special iterative refinement technique (on the same mantissa length - in the spirit of [13]) is suggested: the limitations of the thus modified CA are theoretically shown to be the limitations of MS as a whole. Moreover, the authors took the overdue occasion of producing a more state-of-the-art MS software: the new code BVPSOL appears to compare favorably to the older code BOUNDS(OL). This behaviour is demonstrated by several illustrative examples.

## 1. Sensitivity Analysis of Boundary Value Problems

Consider the two-point BVP (with $b > a$)

(1.0.1) a) $y' = f(t,y)$ , $f \in C^2$ ,
b) $r(y(a),y(b)) = 0$ , $r \in C^2$ ,

with n first-order ODE's and n boundary conditions (BC), in general nonlinear and non-separable. This section deals with the effect of typical perturbations on the solution. Throughout the paper, let $\|\cdot\|$ denote the $\ell_\infty$-norm for vectors and any compatible norm for matrices.

### 1.1. Analytic Boundary Value Problems

For pure initial value problems (IVPs), the sensitivity can be described in terms of the *Wronskian* (n,n)-*matrices* W(t,a) of the variational equation associated with (1.0.1). For BVPs, one additionally needs the (n,n)-matrices

$$A := \frac{\partial r}{\partial y_a} \quad , \quad B := \frac{\partial r}{\partial y_b}$$

Lemma 1 [21,6]

Let the BVP (1.0.1) have at least one solution $y^*$. Let $W^*$, $A^*$, $B^*$ denote the associated matrices as defined above. Let, for some $t \in [a,b]$, the (n,n)-matrix

(1.1.1) $$E^*(t) := A^*W^*(a,t) + B^*W^*(b,t)$$

be nonsingular. Then

(I) $E^*(t)$ is nonsingular for *all* $t \in [a,b]$

(II) $y^*$ is a *locally unique* solution of (1.0.1).

If not stated differently, the above uniqueness assumption will be made throughout the paper. The matrices E(t) are called *sensitivity matrices* of the BVP.

*Sensitivity with respect to BC.* Recall from [6], p.220/221:

$$(1.1.2)\quad \text{a) } dy(t) = E(t)^{-1}dr$$

$$\text{b) } dr = Ady(a) + Bdy(b)$$

Let dy(a,b) denote the vector of all those components of dy(a) and dy(b) that actually enter into the BC. For any BVP, dy(a,b) will represent the input data, while dy(t), $t\in[a,b]$ represents the result.

*BVP condition number*:

$$(1.1.3.a)\quad \bar{\rho} = \max_{t\in[a,b]} \rho(t)$$

in terms of the pointwise condition numbers

$$(1.1.3.b)\quad \begin{aligned} &\rho(t) := \rho_a(t) + \rho_b(t) \\ &\rho_a(t) := \|E(t)^{-1}A\|,\ \rho_b(t) := \|E(t)^{-1}B\| \end{aligned}$$

With these definitions,one may write

$$(1.1.4)\quad \text{a) } \max_{t\in[a,b]} \|dy(t)\| \leq \bar{\rho}\ \|dy(a,b)\|$$

$$\text{b) } \|dy(t)\| \leq \rho(t)\ \|dy(a,b)\|\ .$$

*Remark:* Note that these affine invariant definitions naturally generalize to the case of overdetermined BVPs that are solved in a least squares sense - unlike the definition suggested in [15].

*IVP condition number* (for comparison):

$$(1.1.5)\quad \sigma(a,b) := \max_{t\in[a,b]} \|W(t,a)\| \geq 1$$

There are BVP's, which are comparatively well-conditioned even though the IVPs are ill-conditioned in both directions.

*Illustrative example* (n=2):

$$(1.1.6)\quad z'' - \lambda^2 z = 0\ ,\quad z(0) = \alpha\ ,\quad z(1) = \beta$$

One obtains

$$dy(0,1) = \begin{bmatrix} dz(0) \\ dz(1) \end{bmatrix}$$

Condition number of BVP:

$$\bar{\rho} \doteq \lambda \qquad \text{for } \lambda \to \infty \tag{1.1.7.a}$$

Condition numbers of IVP:

$$\sigma(0,1) = \sigma(1,0)$$

$$\sigma(0,1) \doteq \lambda e^{\lambda} \qquad \text{for } \lambda \to \infty \tag{1.1.7.b}$$

For IVP's, perturbations of the solution due to local perturbations df of the right-hand side f of the ODE are given by

$$dy_a(t) := \int_a^t W(t,s)\, df(s)ds \tag{1.1.8}$$

For BVP's, one obtains

$$dy(t) = \int_a^t E(t)^{-1} AW(a,s) df(s)ds - \int_t^b E(t)^{-1} BW(b,s) df(s)ds \tag{1.1.9}$$

Comparison of (1.1.8) and (1.1.9) leads to

$$||dy(t)|| \le \alpha_0(t)\, ||dy_a(t)|| + \beta_0(t)\; ||dy_b(t)|| , \tag{1.1.10}$$

where $dy_b(t)$ is the analog of $dy_a(t)$ above and

(1.1.11) a) $\alpha_0(t) := ||E(t)^{-1} AW(a,t)||$

$\beta_0(t) := ||E(t)^{-1} BW(b,t)||$

b) $\alpha_0(t) + \beta_0(t) \ge 1.$

Of course, the perturbations in (1.1.2) and (1.1.9) have to be added to cover the situation as a whole.

## 1.2. Discrete Boundary Value Problems

Any method for solving the analytical BVP (1.0.1) will provide a subdivision of the given interval, say

$$\Delta := \{a=t_1 < t_2 < \ldots < t_m = b\} \qquad (m>1).$$

The associated unknowns are

$$x_j := y(t_j) \in \mathbb{R}^n \qquad j = 1,\ldots,m.$$

It was shown in [5], that every BVP method requires the solution of a *cyclic* system of nonlinear equations

$$(1.2.1) \qquad F(x) := \begin{bmatrix} F_1(x_1,x_2) \\ \cdot \\ \cdot \\ \cdot \\ F_{m-1}(x_{m-1},x_m) \\ r(x_1,x_m) \end{bmatrix} = 0$$

for the variable $x \in \mathbb{R}^N$ with $N=n \cdot m$. The system (1.2.1) may also be regarded as a representation of a *discrete* BVP.

In the type of MS technique to be treated herein the solution is represented by sub-arcs

$$(1.2.2.a) \qquad y(t|x_j) \quad , \; t \in [t_j,t_{j+1}] \quad , \; j = 1,\ldots,m-1$$

which are obtained from "forward" numerical integration of the (m-1) IVPs

$$(1.2.2.b) \qquad y' = f(t,y) \quad , \; y(t_j) = x_j \quad , \; j = 1,\ldots,m-1 \quad .$$

Thus, in addition to the BCs, the following *continuity conditions* (CCs) must hold:

$$(1.2.3) \qquad F_j(x_j,x_{j+1}) := y(t_{j+1}|x_j) - x_{j+1} = 0 \qquad j=1,\ldots,m-1 \; .$$

Let $x_j^* = y^*(t_j)$, $j=1,\ldots,m$ denote the solution of the discrete BVP. Let $dr$, $dF_1,\ldots,dF_{m-1}$ denote perturbations of the BCs and CCs. Then, with $G_j = W(t_{j+1},t_j)$, the induced perturbations $dx_1,\ldots,dx_m$ of the solution $x^*$ will satisfy

$$(1.2.4) \qquad \text{a)} \quad G_j dx_j - dx_{j+1} = dF_j, \qquad j=1,\ldots,m-1$$

$$\text{b)} \quad A dx_1 + B dx_m = dr$$

Lemma 2 [9]
(1.2.4) can be solved to yield (j=1,...,m):

(1.2.5) a) $dx_j = d_j[dr,dF] \equiv E(t_j)^{-1} du_j$

where

b) $du_j := dr + B \sum_{l=j}^{m-1} W(b,t_{l+1}) dF_l - A \sum_{l=1}^{j-1} W(a,t_{l+1}) dF_l$

Lemma 3 [9]
Let $\hat{G}_j$ denote an approximation with

(1.2.6) $\hat{G}_j := W(t_{j+1},t_j) + \varepsilon_h \cdot dW(t_{j+1},t_j)$ ,

$|dW(t_{j+1},t_j)| \leq |W(t_{j+1},t_j)|$ elementwise.

Consider (1.2.4) with $\hat{G}_j$, $d\hat{x}_j$ replacing $G_j$, $dx_j$. Then, up to first order in $\varepsilon_h$, one obtains

(1.2.7) a) $d\hat{x}_j \doteq d_j[dr,d\hat{F}]$ ,

where

b) $d\hat{F}_l := dF_l - \varepsilon_h dW(t_{l+1},t_l) d_l[dr,dF]$ .

For the subsequent results, the following extensions of the above definitions (1.1.11) will be useful:

(1.2.8a) $\alpha_i(t) := \max_{\tau_1,\ldots,\tau_i \in [a,b]} \| |E(t)^{-1} AW(a,\tau_1)| \cdot$

$\cdot |W(\tau_1,\tau_2)| \cdot \ldots \cdot |W(\tau_i,t)| \|$

$\beta_i(t) := \max_{\tau_1,\ldots,\tau_i \in [a,b]} \| |E(t)^{-1} BW(b,\tau_1)| \cdot$

$\cdot |W(\tau_1,\tau_2)| \cdot \ldots \cdot |W(\tau_i,t)| \|$ ,

where $|M|$ denotes the matrix with entries $|m_{ij}|$, if M has the entries $m_{ij}$. The above definitions directly imply the relations

(1.2.8.b) $\alpha_0 \leq \alpha_1 \leq \alpha_2 \leq \alpha_3$ , $\beta_0 \leq \beta_1 \leq \beta_2 \leq \beta_3$ ,

$\alpha_i + \beta_i \geq 1 \quad i = 0,\ldots,3$ .

Note that all these quantities are scaling invariant.

Lemma 4 [9]

Let $d\hat{x}_j$ denote the perturbations (1.2.7) specialized to the situation $dF_1 = \ldots = dF_{m-1} = 0$, $dr \neq 0$, $\varepsilon_h \neq 0$. Then

$$\|d\hat{x}_j\| \leq \{1+\varepsilon_h[(j-1)\alpha_2(t_j)+(m-j)\beta_2(t_j)]\}\rho(t_j)\|dy(a,b)\| \tag{1.2.9}$$

*Remark.* For the sake of completeness, recall that the general discrete BVP can be characterized by replacing the identity matrices I in the Jacobian J by certain Wronskian matrices $\bar{G}_2,\ldots,\bar{G}_m$ - for details see [5]. In this situation, one has to include possible approximation errors of $\bar{G}_j^{-1}$ into the analysis. Thus, following the above lines, one essentially has to replace $\alpha_2,\beta_2$ by $\alpha_3,\beta_3$ in (1.2.9).

In what follows, any perturbation of the form d[dr,dF], where dr, dF are sufficiently small, will be accepted as natural perturbations reflecting the *condition of the discrete BVP*. This can be done without explicitly defining any condition *number* of discrete BVPs.

## 2. Modification of the Condensing Algorithm

In the MS techniques, one has to solve the linear (N,N)-system

$$\begin{aligned} &G_1\,\Delta x_1 - \Delta x_2 &&= -F_1 \\ &\quad\vdots \\ &\qquad G_{m-1}\,\Delta x_{m-1} - \Delta x_m &&= -F_{m-1} \\ &A\,\Delta x_1 \qquad\qquad + B\Delta x_m &&= -r \end{aligned} \tag{2.0.1}$$

Lemma 2 above formally yields the result

$$\text{a)}\quad \Delta x_j = -d_j[r,F] \equiv -E_j^{-1}u_j\ , \tag{2.0.2}$$

where, throughout this section,

$$\text{b)}\quad E_j := AG_1^{-1}\ \ldots\ G_{j-1}^{-1} + BG_{m-1}\ \ldots\ G_j$$

$$u_j := r + B\sum_{l=j}^{m-1} G_{m-1}\cdot\ldots\cdot G_{l+1}F_l - A\sum_{l=1}^{j-1} G_1^{-1}\cdot\ldots\cdot G_l^{-1}F_l$$

with the convention

$$G_{j-1} \cdot \ldots \cdot G_{l+1} = I \quad \text{for } l = j-1$$

*Condensing algorithm* (CA) due to [19]:

$$\text{(2.0.3)} \qquad \text{a)} \quad E\,\Delta x_1 = -u \qquad (E \equiv E_1,\ u \equiv u_1)$$

$$\text{b)} \quad \Delta x_{j+1} = G_j\,\Delta x_j + F_j \qquad j = 1,\ldots,m-1\ .$$

## 2.1. Error Analysis of the Standard Algorithm

In MS techniques, the IVP condition $\sigma(t_1,t_m)$ is reduced to the *MS condition number*

$$\text{(2.1.1)} \qquad \sigma_\Delta := \max_j \sigma(t_j,t_{j+1}) \geq 1\ ,$$

depending on the selected subdivision $\Delta$. Let $\varepsilon_y$ denote the relative *local* error tolerance of the integrator, $\varepsilon$ the relative machine precision, and $\varepsilon_x$ the user prescribed relative accuracy of the solution x. Then the following conditions have to be observed:

$$\text{(2.1.2)} \qquad \varepsilon\,\sigma_\Delta \ll \varepsilon_y\,\sigma_\Delta \leq \varepsilon_x\ .$$

Otherwise, the accuracy $\varepsilon_x$ cannot be expected to be achieved. In what follows, the results of a detailed elementwise rounding error analysis of (2.0.3)-on the basis of [22], p. 115/116 - will be given.
For proofs and derivations see [9].
Evaluation of the matrix E in (2.0.3.a):

$$\text{(2.1.3)} \qquad \begin{aligned} E_m &:= B, \\ E_j &:= E_{j+1}G_j \qquad j = m-1,\ldots,1 \\ E &:= A + E_1 \end{aligned}$$

Floating point realization:

$$\text{(2.1.3')} \qquad \begin{aligned} \tilde{E}_m &:= fl(B) = B + dB \\ \tilde{E}_j &:= fl(\tilde{E}_{j+1}G_j) = \tilde{E}_{j+1}G_j + dE_j, \qquad j = m-1,\ldots,1 \\ \tilde{E} &:= fl(a + \tilde{E}_1) = A + \tilde{E}_1 + dE_0 \end{aligned}$$

From [22], one obtains the elementwise bounds (in $O(\varepsilon)$):

(2.1.6.a) $$\|E^{-1}(\tilde{E} - E)\| \leq \varepsilon \cdot n(m-1)\beta_2(a) .$$

(2.1.6.b) $$\|dx_1\| \leq [1 + \varepsilon \cdot n(m-1)\beta_2(a)]\ \rho(a)\ \|dy(a,b)\|$$

*Interpretation:* The effect of rounding errors in the algorithm (2.1.3) is equivalent to the effect of discretization errors in the Wronskian matrices $G_j$, where $\varepsilon_h := n\varepsilon$ - compare (1.2.9). The true $\varepsilon_h$ is usually much greater. Hence, the algorithm (2.1.3) is *numerically stable.*
*Remark:* If the inner products in (2.1.3) are *accumulated* (cf. [22],p. 111), then the replacement $n\beta_2(a) \to \beta_1(a)$ is possible. In view of (1.2.9), however, this sophistication is clearly unnecessary.
For the rest of the analysis, setting $\varepsilon_h := 0$ considerably simplifies the presentation without loss of insight.
Evaluation of the vector u in (2.0.3.a):

(2.1.7)
$$v_m := r$$
$$v_j := v_{j+1} + E_{j+1}F_j \qquad j = m-1,\ldots,1$$
$$u := v_1$$

Floating point realization:

(2.1.7')
$$\tilde{v}_m := r + dr \qquad (\text{let } dr := 0)$$
$$\tilde{v}_j := fl(\tilde{v}_{j+1} + \tilde{E}_{j+1}F_j) = \tilde{v}_{j+1} + \tilde{E}_{j+1}F_j + dv_j$$

For $\varepsilon_h = 0$, some straightforward calculation yields

(2.1.8.a) $$d\tilde{x}_1 := \Delta\tilde{x}_1 - \Delta x_1 = -d_1[d\hat{r}, d\hat{F}]$$

in terms of perturbations $d\hat{r}$, $d\hat{F}$ with

(2.1.8.b)
$$|d\hat{r}| \dot{\leq} (m-1)\varepsilon |r|$$
$$|d\hat{F}_j| \dot{\leq} (n+j)\varepsilon\ |F_j| \qquad j = 1,\ldots,m-1$$

Hence, the algorithm (2.1.7) is *numerically stable.* The use of a numerically stable solver for the linear system (2.0.3.a) then assures numerical stability for (2.0.3.a) as a whole.

Floating point realization of (2.0.3.b):

$$(2.0.3'.b) \qquad \Delta\tilde{x}_{j+1} := fl(G_j\Delta\tilde{x}_j + F_j) = G_j\Delta\tilde{x}_j + F_j + d\bar{F}_j \;, \qquad j = 1,\dots,m-1$$

Theorem 5 [9]

Let $d\tilde{x}_j := \Delta\tilde{x}_j - \Delta x_j$ denote the perturbations of the exact solution due to rounding errors in the realization of the condensing algorithm (2.0.3). Then

$$(2.1.9.a) \qquad d\tilde{x}_j = -d_j\,[d\hat{r},d\hat{F}] + dy_j \qquad j = 1,\dots,m$$

where

$$(2.1.9.b) \qquad dy_j := \sum_{l=1}^{j-1} G_{j-1} \cdots G_{l+1}(d\bar{F}_l - d\hat{F}_l)$$

in terms of the small perturbations $d\hat{r}$, $d\hat{F}_j$, $d\bar{F}_j$ introduced in (2.1.8) and (2.0.3'.b)

*Interpretation:* The part $d_j[d\hat{r},d\hat{F}]$ represents the condition of the discrete BVP, whereas the part $dy_j$ depends on the condition of the IVP. Hence, in the case *when the discrete BVP is well-conditioned, but the IVP* (in the selected direction) *is ill-conditioned,* the above implementation of *the explicit recursive scheme* (2.0.3.b) *is numerically unstable.*

In order to roughly characterize the class of those BVPs, for which the standard CA is numerically stable, one may concentrate on the *critical case* $\sigma_\Delta \gg 1$. In a carefully scaled norm, one may reasonably require that

$$(2.1.10) \qquad \|dy_j\| \ll \|\Delta\tilde{x}_1\|$$

From [9], a sufficient condition to assure this requirement turns out to be

$$(2.1.10') \qquad \varepsilon g(n,m)\sigma_\Delta^{m-1} \ll 1$$
$$g(n,j) := (j-1)(2n+j-1)$$

This means that (2.1.10') is a restriction essentially independent of

the subdivision $\Delta$, i.e. just as for single shooting!
In view of *inverse problems in chemical kinetics*, this result is important : whenever the associated ODE system is *stiff*, then the quantity $\sigma_\Delta$ is of moderate size - therefore, the CA is appropriate.

## 2.2. Iterative Refinement Sweep Technique

The basic idea underlying the special iterative refinement technique to be presented here is the following: if the IVP is ill-conditioned, but the discrete BVP is well-conditioned, then the *parasitic* error part $dy_j$ from Theorem 5 is a *dominant* solution of the recursion (2.0.3.b) which can be computed from the recursion in a *numerically stable* way. This parasitic error part should show up in the residuals, which may be denoted by

(2.2.1) a) $dF_j^0 := G_j\Delta\tilde{x}_j + F_j - \Delta\tilde{x}_{j+1}$ , $j = 1,\ldots,m-1$ ,

b) $dr^0 := r + A\Delta\tilde{x}_1 + B\Delta\tilde{x}_m$ .

Notation:

$$\Delta\tilde{x}_j^0 := \Delta\tilde{x}_j \quad , \quad d\tilde{x}_j^0 := d\tilde{x}_j = \Delta\tilde{x}_j^0 - \Delta x_j$$

These definitions are extended to

$$\Delta\tilde{x}_j^\nu \ , \ d\tilde{x}_j^\nu \qquad \text{for } \nu = 0,1,\ldots$$

for the iterative refinement index $\nu$.

(2.2.2) a) $d\tilde{r}^\nu := fl(dr^\nu) = fl(r^\nu + A\Delta\tilde{x}_1^\nu + B\Delta\tilde{x}_m^\nu) =$
$= dr^\nu + d^2\tilde{r}^\nu$ ,

b) $d\tilde{F}_j^\nu := fl(dF_j^\nu) = fl(G_j\Delta\tilde{x}_j^\nu + F_j - \Delta\tilde{x}_{j+1}^\nu) =$
$= dF_j^\nu + d^2F_j^\nu$ .

From the residuals, one computes the refinements

$$\tilde{\epsilon}_j^\nu := fl(\epsilon_j^\nu)$$

by means of the condensing algorithm. Let the perturbed linear (n,n)-system be written in the form

(2.2.3.a) $\tilde{\epsilon}_1^\nu := -E^{-1}d\tilde{u}^\nu$,

$$d\tilde{u}^\nu := d\tilde{r}^\nu + d^2\tilde{r}^\nu + B\sum_{l=1}^{m-1} G_{m-1}\cdot\ldots\cdot G_{l+1}(d\tilde{F}_l^\nu + d^2\tilde{F}_l^\nu)$$

The perturbed recursion may be written in the form

$$(2.2.3.b) \qquad \tilde{\epsilon}_{j+1} = fl(G_j\tilde{\epsilon}_j + d\tilde{F}_j) = G_j\tilde{\epsilon}_j + d\tilde{F}_j + d^2\bar{F}_j \quad .$$

With these actually computed refinements one will correct the approximation for $\Delta\tilde{x}^\nu$ according to

$$(2.2.4) \qquad \Delta\tilde{x}_j^{\nu+1} := \Delta\tilde{x}_j^\nu + \tilde{\epsilon}_j^\nu \qquad \nu = 0,1,\ldots \quad .$$

<u>Theorem 6</u> [9]

With the just introduced notation, the iterates $\Delta\tilde{x}^\nu$ for $\nu = 0,1,\ldots$ contain the perturbations

$$(2.2.5.a) \qquad d\tilde{x}_j^\nu = -d_j[d\hat{r}^\nu,\ d\hat{F}^\nu] + dy_j^\nu$$

with the parasitic part

$$(2.2.5.b) \qquad dy_j^\nu := \sum_{l=1}^{j-1} G_{j-1} \cdots G_{l+1}(d\bar{F}_l^\nu - d\hat{F}_l^\nu)$$

in terms of the rounding errors

$$(2.2.5.c) \qquad \begin{aligned} d\hat{r}^o &:= d\hat{r}, \quad d\hat{F}^o = d\hat{F} \\ d\bar{F}^o &:= d\bar{F} \end{aligned}$$

$$(2.2.5.d) \qquad \begin{aligned} d\hat{r}^{\nu+1} &:= d^2r^\nu + d^2\hat{r}^\nu \\ d\hat{F}^{\nu+1} &:= d^2\hat{F}^\nu + d^2F^\nu \qquad \nu = 0,1,\ldots \\ d\bar{F}^{\nu+1} &:= d^2\bar{F}^\nu + d^2F^\nu \end{aligned}$$

with the quantities $d^2r^\nu$, $d^2\hat{r}^\nu$, $d^2F$ , $d^2\hat{F}^\nu$, $d^2\bar{F}^\nu$ as introduced in (2.2.2) and (2.2.3).

*Interpretation.* Each iterative refinement step $\nu$ substitutes the parasitic error $dy^\nu$ by $dy^{\nu+1}$.

A detailed analysis of this type of iterative refinement shows that it is successful essentially for the same class of BVPs as characterized above in (2.1.10'). To overcome this unsatisfactory behavior , a modification of the iterative refinement process is necessary. The modification to be suggested here is called *iterative refinement sweep* (IRS) *technique*. An intuitive requirement for the convergence of the iterative refinement will be that the rounding errors arising from the evaluation of the residuals should not exceed the residuals, i.e.

$$(2.2.6) \qquad \|d^2F_j^\nu\| \ll \|d\tilde{F}_j^\nu\|$$

or, equivalently,

$$(2.2.6') \qquad \|d\tilde{F}_j^\nu\| \doteq \|dF_j^\nu\| \ .$$

For some given input parameter $\bar{\varepsilon} \leq \varepsilon_x$, determine some (maximum) index $j_\nu$ - to be called *sweep index* - by

$$(2.2.7.a) \qquad \|\tilde{\epsilon}_j^\nu\| \leq \bar{\varepsilon} \qquad j = 1,\ldots,j_\nu$$

and set

$$(2.2.7.b) \qquad d\tilde{F}_j^{\nu+1} = dF_j^{\nu+1} := 0 \ , \qquad j = 1,\ldots,j_\nu-1$$

In the critical case $\sigma_\Delta \gg 1$, the refinements are *dominant* solutions of the recursion (2.0.3.b) so that

$$(2.2.8) \qquad \|\tilde{\epsilon}_{j+1}^\nu\| \doteq \sigma_\Delta \|\tilde{\epsilon}_j^\nu\| \qquad j = 1,2,\ldots \ .$$

From Theorem 6 (including further details from [9]) one may derive the rough results (with g as defined in (2.1.10')):

$$(2.2.9.a) \qquad \|\tilde{\epsilon}_j^\nu\| \doteq \|dy_j^\nu - dy_j^{\nu+1}\| \doteq \|dy_j^\nu\| \qquad j \geq 2$$

$$\text{b)} \qquad \|\tilde{\epsilon}_j^{\ +1}\| \doteq \varepsilon g(n,j) \ \|\tilde{\epsilon}_j^\nu\| \qquad j \geq 2$$

$$\text{c)} \qquad \|dy_j^{\nu+1}\| \doteq \varepsilon g(n,m) \|dy_j^\nu\| \qquad j \geq 2$$

The last result nicely demonstrates the fact that the *modified* refinement technique eventually damps out the parasitic perturbations.
Choice of $\bar{\varepsilon}$ in (2.2.7.a):

$$(2.2.10) \qquad \bar{\varepsilon} := \|\tilde{\epsilon}_1^0\| \cdot \tau \qquad \tau > 1$$

The safety factor $\tau$ is understood to compensate for possible scaling deficiencies. This $\bar{\varepsilon}$ also represents some lower bound for a reasonable relative accuracy $\varepsilon_x$. Moreover, this choice ensures that

(2.2.11.a) $$j_o \geq 1 .$$

In order to ensure

(2.2.11.b) $$j_{\nu+1} \geq j_\nu + 1 ,$$

one comes to the sufficient condition (see [9]):

(2.2.12) $$\varepsilon\ g(n,m)\sigma_\Delta < 1 ,$$

which appears to be natural compared with the basic MS condition (2.1.2). As a consequence, the number of iterations is restricted by

(2.2.11.c) $$\nu \leq m-1$$

Finally, recall from (2.2.8) that the refinement sweeps cheaply supply an estimate $\hat{\sigma}_\Delta$ for $\sigma_\Delta$ in the form

(2.2.12.a) $$\hat{\sigma}_\Delta = \max_{j,\nu} \frac{\| \tilde{\epsilon}^\nu_{j+1} \|}{\| \tilde{\epsilon}^\nu_j \|} .$$

With this estimate at hand, one may compute an approximate error tolerance

(2.2.12.b) $$\hat{\varepsilon}_y := \frac{\varepsilon_x}{\hat{\sigma}_\Delta}$$

to be prescribed in the integrator - compare (2.1.2). Thus the refinement sweeps give additional important information that can be used in a MS code.

## 3. Numerical Experiments

In this section,comparisons between the old MS code BOUNDS(OL) due to [4,19,6] and the new code BVPSOL on the basis of the results of this paper are given. As a standard integrator, the recent extrapolation code DIFEX1 due to [7] is used.

Apart from the iterative refinement sweep technique of section 2.2, the new FORTRAN code BVPSOL also contains further improvements, for example:

- workspace realization (to simulate variable dimensioning in FORTRAN, taking full advantage of storage economy),
- independent variable $x_m$ is actually used (not only formally included),
- automatic recognition and special treatment of separable linear boundary conditions at *either* boundary,
- internal scaling of boundary conditions (formally invariant under re-gauging of variables or re-scaling of BC),
- advanced treatment of rank-deficiency on the basis of [8,10],
- automatic initial rank strategy for bad initial guesses,
- automatic adaption of local error tolerance of integrator and relative deviation parameter for finite difference approximations of the Wronskian matrices $G_j$,
- improved Jacobian rank-1 updates due to [3] preserving sparsity without explicit use of [17].

Details including a listing of BVPSOL may be found in [9].

## 3.1. Artificial Test Problem (due to [19] and [12])

This test problem is of the special type (1.1.6), where the BVP is well-conditioned but the IVP is ill-conditioned:

$$(3.1) \qquad y'' = \lambda^2(y + \cos^2\pi t) + 2\pi^2\cos(2\pi t) \ , \qquad y(0) = y(1) = 0.$$

The MS approach follows the documentation in [12].
In Table 1, a comparison of the MS codes BOUNDSOL and BVPSOL with RWPM due to [12] is given. Clearly, the refinement sweep technique in BVPSOL pays off. In order to give some additional insight, a few characteristic sensivity numbers are presented in Table 2 .

Table 1. Number NFEV of evaluations of the right-hand side in Example 3.1. (F:fail) $^{1)}$ overflow , $^{2)}$ requires more nodes

| $\lambda$ | BOUNDSOL +DIFEX1 | BVPSOL +DIFEX1 | RWPM +FAW78 |
|---|---|---|---|
| 20 | 4.929 | 5.125 | 4.588 |
| 40 | 50.123 | 6.532 | 8.077 |
| 60 | F $^{1)}$ | 8.460 | 10.994 |
| 100 | | 14.266 | 16.322 |
| 180 | | 17.569 | |
| 200 | | 20.397 F $^{2)}$ | 30.439 |

Table 2. Comparison of sensitivity numbers characterizing Example 3.1: (greatest machine number 0.7D+76, compare *))

$\sigma(0,1)$ - theoretical condition number of IVP, compare the approximation (1.1.7.b)

$\|E\|$ - (scaled) sensitivity number from BVPSOL

$\sigma_\Delta$ - theoretical MS condition number, compare (2.1.1) and the approximation (1.1.7.b)

$\hat{\sigma}_\Delta$ - estimated (scaled) MS condition number from BVPSOL, compare (2.2.12.a)

| $\lambda$ | $\sigma(0,1)$ | $\|E\|$ | $\sigma_\Delta$ | $\hat{\sigma}_\Delta$ |
|---|---|---|---|---|
| 100 | 0.3 D+46 | 1. D+45 | 7. D+3 | 1. D+3 |
| 120 | 0.2 D+55 | 0.7 D+50 | 0.2 D+4 | 0.2 D+4 |
| 140 | 0.9 D+63 | 0.3 D+59 | 0.8 D+4 | 0.5 D+4 |
| 160 | 0.5 D+72 | 0.2 D+68 | 2. D+4 | 1. D+4 |
| 180 | 0.3 D+81 | 0.8 D+76*) | 0.7 D+5 | 0.4 D+5 |
| 200 | 1. D+89 | 0.4 D+85*) | 2. D+5 | 1. D+5 |

For illustration purposes, the sequence of sweep indices $\{j_\nu\}$ for $\lambda$=180 is given:

$$\{1,4,8,11,15,19,21\} \quad , \quad \nu = 6 \ .$$

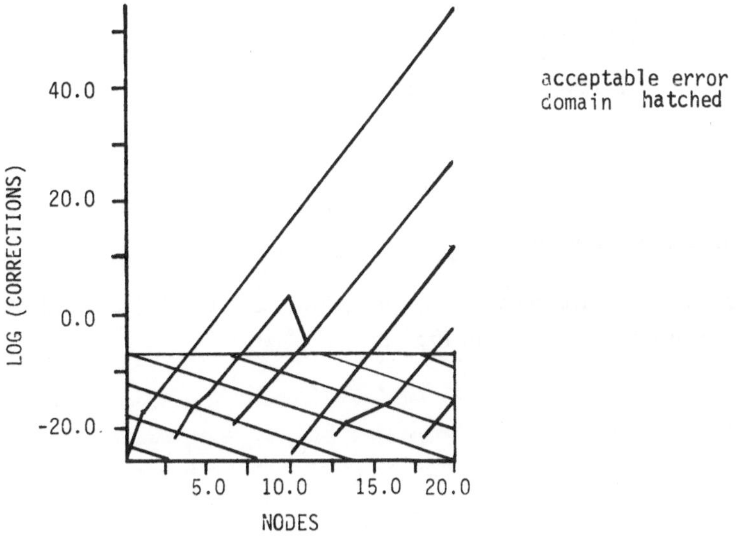


Fig. 1. Iterative Refinement sweeps

In Fig.1, a graphic representation of the sequences $\{\|\tilde{\epsilon}_1^{\nu}\|,\ldots,\|\tilde{\epsilon}_m^{\nu}\|\}$ is given for increasing index $\nu$ : the plot nicely demonstrates the theoretical relations (2.2.8) and (2.2.9.b).

## 3.2. Singular Perturbation Test Problem

Consider the following BVP:

$$\varepsilon y'' + (0.5 - t)y' = 0, \quad \varepsilon > 0, \quad y(0) = 1, \quad y(1) = 4. \tag{3.2}$$

For sufficiently small $\varepsilon$, the solution has two boundary layers at the end points $t = 0$ and $t = 1$. The MS approach with 21 equidistant nodes is documented in [9].

In Table 3, the comparative performance of BOUNDSOL and BVPSOL for decreasing values of $\varepsilon$ is presented. For $\varepsilon = 0.1$ and $\varepsilon = 0.01$ both codes yield the solution. For $\varepsilon = 0.001$, both codes fail to solve the problem with only 21 nodes. BVPSOL permits one to analyze the true reason for the failure. With 1. D+38, the sensitivity number is certainly within the range of Table 2. However, the difficulty of the BVP shows up in the MS condition number $\hat{\sigma}_\Delta = 1.$ D+9. This example seems to demonstrate the *limitations of the MS approach as a whole.*

Table 3. Comparative NFEV required to treat the singular perturbation problem (Ex. 3.2) (F:fail)

| | BOUNDSOL + DIFEX1 | BVPSOL + DIFEX1 | $\|E\|$ | $\hat{\sigma}_\Delta$ |
|---|---|---|---|---|
| 0.1 | 2.125 | 1.280 | 0.35 | < 10 |
| 0.01 | 6.670 | 5.252 | 0.75 | < 10 |
| 0.001 | 78.674 F[1] | 24.523 F[2] | 1. D+38 | 1. D+9 |

1) modified Newton method fails to converge

2) asks for new nodes

A graph of the solutions for $\varepsilon = 0.1$, 0.01, and 0.001 (obtained with more nodes) is presented in Fig. 2.

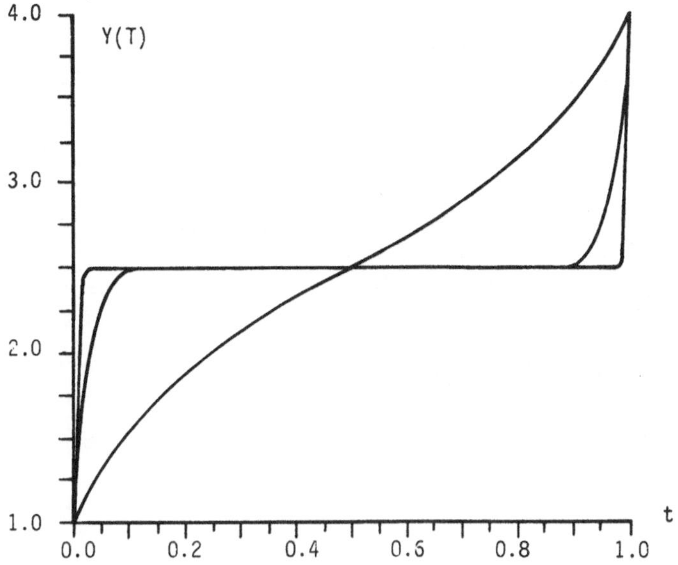


Fig. 2. Graph of solutions of Example 3.2 for $\varepsilon$=0.1,0.01,0.001.

3.3. Re-entry Problem (due to [19])

This problem has been used as a "work horse" example meanwhile. The documentation can be found in [19]. The present test uses the set of m=6 nodes including initial data as given [11], p. 460.
By varying backward or forward integration and prescribing the Hamiltonian BC at either boundary, one obtains a set of 4 test problems:

B1 : backward integration, Hamiltonian at $\tau = 1$,
(original problem from [11])
B0 : backward integration, Hamiltonian at $\tau = 0$,
F1 : forward integration, Hamiltonian at $\tau = 1$ ,
F0 : forward integration, Hamiltonian at $\tau = 0$.

Comparative results are presented in Table 4. In BVPSOL, iterative refinement turned out to have *no* effect on the computation ($\nu = 1$ throughout). Hence, the better performance in this example comes from the improvements mentioned in section 3 above.

Table 4. Comparative number of trajectories required to solve Example 3.3 (F:fail)

| | BOUNDSOL | BVPSOL |
|---|---|---|
| B 1 | 63 [1] | 63 |
| B 0 | 9 F[2] | 82 |
| F 1 | 80 F[2] | 83 |
| F 0 | 139 | 93 |

[1]) two emergency reductions of relaxation factor due to the occurrence of a singular trajectory

[2]) modified Newton method fails to converge

*Acknowledgement:* This work has been supported by the Deutsche Forschungsgemeinschaft. The authors wish to thank Mrs. I.Heitz for her quick and careful typing of the article.

## References

[1] U.Ascher, I.Christiansen, R.D.Russell:
A Collocation Solver for Mixed Order Systems of Boundary Value Problems.
Math.Comp. 33, 659-679 (1979)

[2] H.G.Bock:
Numerical Treatment of Inverse Problems in Chemical Reaction Kinetics.
Springer Series Chemical Physics 18, section 8, p.102-125 (1981).

[3] C.G. Broyden:
A class of methods for solving non-linear simultaneous equations.
Math.Comp. 19, 577-583 (1965).

[4] R. Bulirsch:
Die Mehrzielmethode zur numerischen Lösung von nichtlinearen Randwertproblemen und Aufgaben der optimalen Steuerung.
Carl-Cranz-Gesellschaft: Tech.Rep. (Oct. 1971).

[5] P. Deuflhard:
Non-linear Equation Solvers in Boundary Value Problem Codes.
Springer Lecture Notes Computer Science 76, (Childs et al., ed.), p. 40-66 (1979).

[6] P. Deuflhard:
Recent Advances in Multiple Shooting Techniques.
In: Computational Techniques for Ordinary Differential Equations (Gladwell/Sayers, ed.), Section 10, p.217-272. London, New York: Academic Press (1980).

[7] P. Deuflhard:
Order and Stepsize Control in Extrapolation Methods.
Universität Heidelberg, SFB 123: Tech.Rep. 93 (1980). To appear in Numer.Math. (1983).

[8] P. Deuflhard, V.Apostolescu:
An Underrelaxed Gauss-Newton Method for Equality Constrained Nonlinear Least Squares Problems.
Springer Lecture Notes Control Inf.Sci. 7, 22-32 (1978).

[9] P. Deuflhard, G. Bader:
Multiple Shooting Techniques Revisited.
University of Heidelberg, SFB 123: Tech.Rep. 163 (June 1982).

[10] P. Deuflhard, W. Sautter:
On Rank-Deficient Pseudo-Inverses.
J.Lin.Alg.Appl. 29, 91-111 (1980).

[11] H.J. Diekhoff, P. Lory, H.J. Oberle, H.J. Pesch, P. Rentrop, R. Seydel:
Comparing Routines for the Numerical Solution of Initial Value Problems for Ordinary Differential Equations in Multiple Shooting.
Numer.Math. 27, 449-469 (1977).

[12] M. Hermann, H. Berndt:
RWPM: A Multiple Shooting Code for Nonlinear Two-Point Boundary Value Problems.
Forschungsergebnisse der Friedrich-Schiller-Universität Jena: Nr. N/81/27 (May 1981).

[13] M. Jankowski, H. Woźniakowski:
Iterative Refinement Implies Numerical Stability.
BIT 17, 303-311 (1977).

[14] M. Lentini, V. Pereyra:
An adaptive finite difference solver for nonlinear two-point boundary value problems with mild boundary layers.
SIAM J. Numer. Anal. 14, 91-111 (1977).

[15] R.M.M. Mattheij:
The Conditioning of Linear Boundary Value Problems.
To appear in SIAM J. Numer. Anal. (1982).

[16] M.R. Osborne:
On shooting methods for boundary value problems.
J.Math.Anal.Appl. 27, 417-433 (1969).

[17] L.K. Schubert:
Modification of a quasi-Newton method for nonlinear equations with a sparse Jacobian.
Math.Comp. 24, 27-30 (1970).

[18] R.D. Skeel:
Iterative Refinement Implies Numerical Stability for Gaussian Elimination.
Math.Comp. 35, 817-832 (1980).

[19] J. Stoer, R. Bulirsch:
Einführung in die Numerische Mathematik II.
Berlin, Heidelberg, New York: Springer (1st ed., 1973).

[20] J.M. Varah:
Alternate row and column elimination for solving certain linear systems.
SIAM J. Numer. Anal. 13, 71-75 (1976).

[21] R. Weiss:
The Convergence of Shooting Methods.
BIT 13, 470-475 (1973).

[22] J.H. Wilkinson:
The Algebraic Eigenvalue Problem.
Oxford: Clarendon Press (1965).

# RECENT ADVANCES IN PARAMETERIDENTIFICATION TECHNIQUES FOR O.D.E.

H.G. Bock

Parameteridentification (PI) techniques have found a rapidly increasing interest in the past years, particularly in biology, chemistry and other sciences, where a quantitative description of a complex process can often not be derived from investigations into isolated subsystems by in vitro experiments, and observations of the complete process in vivo are the only source of information. At the same time, the development of conceptionally new numerical methods has promoted substantial improvements in the treatment of such problems.

The aim of the present paper is to review the essential lines of development of a novel class of PI algorithms and their realization in the multiple shooting code PARFIT [1] and the collocation code COLFIT [2], the common base of which is a Generalized Gauss Newton method for discretized PI boundary value problems. Several features of the implementation, that are considered important for the *efficiency*, *stability* and *applicability*, particularly of PARFIT, are presented.

## 1. Class of Problems

Both PARFIT and COLFIT are designed for the identification of $n_p$ *parameters* p in a nonlinear *model o.d.e. system* for the $n_d$ *states* x(t)

$$\begin{aligned} \dot{x} &= f(t,x,p) = \bar{f}(t,x,p;Q(t,x,p)) \\ x(t+) &= x(t-) + C(t,x(t-),p) \qquad \text{for some } Q_j(t,x(t-),p) = 0, \end{aligned} \tag{1}$$

the right hand side and solution of which may be discontinuous depending on sign changes of the *switching functions* Q.

*Observations* are given in terms of *functions* of states and parameters at data points $t_j$

$$\eta_{ij} = g_i(t_j,x(t_j),p) + \varepsilon_{ij} \tag{2}$$

subject to measurement errors $\varepsilon_{ij}$, which are (for ease of presentation) assumed to be independent, normally distributed with zero mean and known variances ($N(0,\sigma_{ij}^2)$). Thus, by minimizing a weighted least squares function

$$\begin{aligned} \min_{x,p} = \; & \| r_1(x(t_1),\ldots,x(t_k),p) \|_2^2 \\ := \; & \sum_{i,j} \sigma_{ij}^2 (\eta_{ij} - g_i(t_j,x(t_j),p))^2 \end{aligned} \tag{3}$$

a maximum likelihood estimate is obtained. Additionally, side conditions are often given to specify further model properties such as boundary conditions, initial values or parameter restrictions

$$\begin{aligned} & r_2(x(t_1),\ldots,x(t_k),p) = 0 \\ & r_3(x(t_1),\ldots,x(t_k),p) \geq 0 \;, \end{aligned} \tag{4}$$

whereas a priori information subject to error is included in (3). Note that the switching function formulation (1) allows a convenient treatment of variable data points which are only implicitly given or subject to measurement error.

Although this problem class meets most practical requirements, there are certain extensions which require special attention.

*Multiexperiment problems*

Frequently, the available data arise from several experiments under different conditions, and the PI problem consists of several independent problem sets which have possibly only the parameters p in common.

The structure of such problems is analyzed in [3]. An enhanced version PARMEX of the multiple shooting procedure PARFIT incorporating the methods developed in [3] is currently being developed, which was used for the numerical solution of two problems in section 8. An important application is e.g. the identification of geophysical structures by the tracing of seismic rays (REINERS [4], using the model of PEREYRA, KELLER and LEE [5]).

*(Semi-)infinite problems*

In a number of cases, functions space norms are more appropriate or convenient instead of the discrete norm (3)

$$\int_{t_0}^{t_f} \|r_1^c(t,x(t,p))\|_2^2 \, dt = \min \tag{5}$$

sometimes with the additional feature that functions u(t) must be identified

$$\dot{x}(t) = f(t,x(t),p,u(t)) \ . \tag{6}$$

In such cases, one may resort to optimal control methods, which in this form can be immediately treated by the algorithm described e.g. in [6], but it should be emphasized that the generalized Gauss-Newton method described below has a natural extension to the (semi-)infinite case [7]. The algorithm of NOWAK, DEUFLHARD [8] is based on the latter approach.

## 2. *The Initial Value Problem Approach*

The most intuitive and probably most widespread approach to PI in o.d.e. is the repeated solution of the initial value problem in some kind of iterative procedure to improve the fit. A well known representative of this class of IVP methods is the algorithm of MILSTEIN [9,10], which uses GEAR's [11] integrator and a refined random search technique (BREMERMANN [12]).

The IVP approach has two major *drawbacks*, which are apparent in numerical practice. Firstly, this *reinversion of the inverse problem* eliminates the states by means of (1) in favour of the parameters, and thus neglects any information on the states which is at hand in an inverse problem. There is rather impressive numerical evidence about the *deterioration of efficiency* thus caused. Secondly, the elimination of the states can cause a substantial *loss of stability* for the solution scheme. At least for bad initial guesses of the parameters, the "direct" (nonlinear) IVP may be ill-conditioned and hard to solve, or a solution may not even exist, even when the original PI problem is perfectly well conditioned. To illustrate the latter drawback, we consider

*A notorious test problem*

which is a modification of a two point boundary value problem described by BULIRSCH [13]. A model o.d.e. for two states and one unknown parameter p with fixed initial values is given by

$$\begin{aligned} \dot{x}_1 &= x_2 && , \; x_1(0) = 0 \\ \dot{x}_2 &= \mu^2 x_1 - (\mu^2 + p^2)\sin pt && , \; x_2(0) = \pi \end{aligned} \qquad t \in [0,1] \tag{7}$$

The solution of the *true parameter value* $p = \pi$ is

$$x_1(t) = \sin \pi t \quad , \quad x_2(t) = \pi \cos \pi t \; . \tag{8}$$

*Measurements* are generated by adding pseudo random noise $(N(0,\sigma^2),\sigma=0.05)$ at selected data points $t_j = j \cdot 0.1$ $(j=1,\dots,10)$. The results for single shooting as shown in fig. 1 are desastrous. Choosing $p_o = 1.$ as a starting guess, the integration routine (DIFSYS [14]) terminates at $t \doteq 0.23$ for $\mu = 60$. But even for the true value of p (correct up to 16 decimals) and highest integration accuracy the solution is properly reproduced only on the first half of the interval!

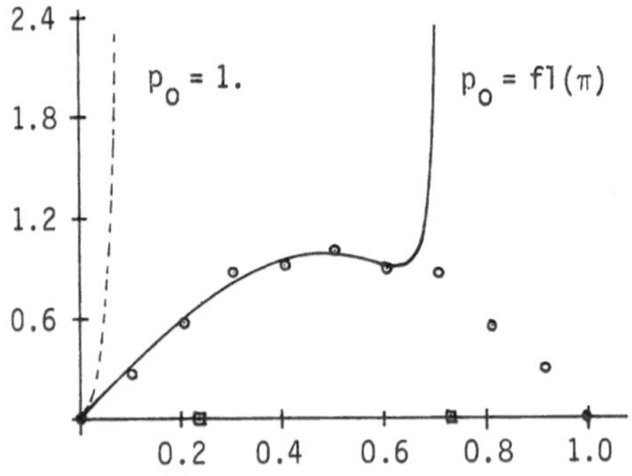


Fig. 1 Notorious test problem - single shooting ($\mu = 60$)

### *3. The Multiple Shooting Approach to PI*

An alternative to the IVP approach is derived by regarding the PI problem (1 - 4) as a constrained, overdetermined multipoint boundary value problem - independent of whether the "direct" problem is a BVP or not - as in the PI package PARFIT [1,7] which is based on an earlier multipoint BVP algorithm of the author [6] using the *multiple shooting technique* (cf. [13,15,16,17,18]).

This algorithm is briefly summarized. Using a suitable *mesh* $\pi$

$$\pi^m: \tau_1 < \tau_2 < \dots < \tau_m \quad , \quad \Delta\tau_j := \tau_{j+1} - \tau_j \quad (j=1,\dots,m-1) \tag{9}$$

e.g. a subset of the data points, one computes the solutions $x(t;s_j,p)$ of m-1 *independent* IVP

$$\dot{x} = f(t;x,p) \quad , \; x(\tau_j) = s_j \quad , \; t \in I_j = [\tau_j , \tau_{j+1}] \quad \text{a.e.} \tag{10}$$

The additional variables $(s_1,\dots,s_m)$ are estimates of the states $x(\tau_j)$. Formal insertion of this (discontinuous) parametrization of the solution into the original PI problem yields a large, constrained least squares problem for which adequate methods are outlined in the following sections.

$$\text{Problem PI1:} \quad \|R_1(s_1,\dots,s_m,p)\|_2 = \min_{(s_1,\dots,s_m,p)} \tag{11}$$

$$\text{subject to} \quad R_2(s_1,\dots,s_m,p) = 0 \, , \; R_3(s_1,\dots,s_m,p) \geq 0$$

with the additional *matching conditions*

$$h_j(s_{j+1},s_j,p) := x(\tau_{j+1};s_j,p) - s_{j+1} = 0 \tag{12}$$

to ensure continuity of the solution.

*Remark:* One obvious advantage of multiple shooting is that any a priori information about the solution trajectory can be utilized to improve the initial guess. Hence, the influence of poor estimates for the parameters is substantially reduced!

*Mesh selection*

The basic idea of multiple shooting is that by choosing a sufficiently fine mesh one can avoid drifting too far away from the solution trajectory $x(t)$. *Quantitatively accessible* bounds that do not over-estimate the growth behaviour (like global Lipschitz constants) can be derived by a first order perturbation analysis. Defining the propagation matrices by the variational differential equations (VDE)

$$\begin{aligned} d(\phi\phi^p)(t,\tau)/dt &= f_x(t,x(t),p)(\phi\phi^p) + (0 \;\, f_p(t,x(t),p)) \\ (\phi\phi^p)(\tau,\tau) &= (I \;\, 0) \end{aligned} \tag{13}$$

an *error* $\delta_p$ , $\delta x(\tau)$ at time $\tau$ is propagated according to

$$\delta x(t) \doteq \phi(t,\tau)\delta x(\tau) + \phi^p(t,\tau)\delta p \ . \tag{14}$$

If $\varepsilon_j$ is the integration tolerance used, and $\varepsilon_p$ the accuracy of the parameter, a mesh can be called *sufficiently fine* if the

Meshcondition [MC]: $\gamma_j\varepsilon_j + \gamma_j^p\varepsilon_p \leq TOL_j$ (15)

holds, where $TOL_j$ is the *required accuracy* of the solution $x(t)$ (to be selected carefully, see section 7!).
The propagation factors

$$\gamma_j := \max_{t_1 \geq t_2 \in I_j} \|\phi(t_1,t_2)\| \geq 1 \ , \ \gamma_j^p := \max_{t_1 \geq t_2 \in I_j} \|\phi^p(t_1,t_2)\| \geq 0 \tag{16}$$

are *approximated* in PARFIT by means of $\phi\phi^p(\tau_{j+1},\tau_j)$, which is computed anyway.

*Remark 1:* The meshcondition [MC] is essentially a *stability criterion* and usually allows very coarse meshsizes. Once a mesh is fixed, the integration accuracy can be adapted to satisfy [MC]. Actually, PARFIT uses a *refined strategy* which adapts the accuracy *componentwise*. If a limit accuracy is reached, the mesh must be refined (or $TOL_j$ relaxed). It may be noteworthy that [MC] is also a valuable tool for *automatic mesh selection* similar to that used in COLFIT [2], cf. sect. 4.

*Remark 2:* Throughout this paper, all norms are understood to be *scaled* due to an internal scaling of the variables in order to equilibrate them (such that $\|s_j\| \lessapprox 1$ , $\|p\| \lessapprox 1$). Consequently, all error criteria are *absolute* (but "pseudorelative") which considerably simplifies calculations.

*Remark 3:* In case of the test problem (7) one shows that for *single shooting* $\gamma_j,\gamma_j^p \approx 10^{26}$, but $\approx 10^4$ for *multiple shooting* with a uniform mesh of $\Delta\tau = 0.1$ (cf. table 1, fig. 3).

## *4. Alternative BVP Approaches*

The basic ideas of PARFIT can be applied to other BVP approaches as well. Since most methods used in practice lead to a *discretized* or *parametrized* differential equation

$$D^h x^h = f^h \ , \tag{17}$$

the insertion of the discretized $x^h$ into the PI problem yields a constrained least squares BVP analogous to problem PI1. This includes in particular *alternate multiple shooting* as in [19], *finite differences* with iterative deferred corrections [20] and *collocation* [21]. As a representative we shall sketch the collocation approach (cf. [21,22,23, 24,25]) which was realized in the code COLFIT in cooperation with SCHLÖDER and BÄR [2] (cf. [26]).

For a given mesh $\pi^m$, an additional *collocation* mesh $\rho = \{\tau_j^i\}$

$$\rho: \ \tau_j^i := \tau_j + \sigma_i \Delta\tau_j \quad ; \ 0 \le \sigma_1 < \ldots < \sigma_l \le 1 \quad (j=1,\ldots,m) \tag{18}$$

is defined, and from the class $P(l,\pi) := \{y \mid y$ is polynomial of degree $\le l$ on $[\tau_j,\tau_{j+1}]\}$ an approximation of the o.d.e. solution is determined by *collocation* on $\rho$, i.e. choosing $y \in P(l,\pi) \cap C^0[\tau_1,\tau_m]$ such that

$$\dot{y}(t) = f(t,y(t),p) \qquad \text{for all } t \in \rho \ . \tag{19}$$

Typical choices of $\{\sigma_i\}$ are the GAUSS, RADAU or LOBATTO points on [0,1]. In COLFIT, LOBATTO-collocation is used, which is especially suitable for *singular perturbation problems* - see the analysis of ASCHER, WEISS [27] using, e.g., piecewise cubic polynomials in Hermite representation - conditions (19) read

$$s_{j+1} = s_j + \frac{\Delta\tau_j}{6}\left(s'_{j+1} + s'_j + 4f\left(\tau_{j+\frac{1}{2}}, \frac{1}{2}(s_{j+1} + s_j) + \frac{\Delta\tau_j}{8}(s'_j - s'_{j+1}), p\right)\right) = 0 \tag{20a}$$

$$s'_j = f(\tau_j, s_j, p) \quad , \quad s'_{j+1} = f(\tau_{j+1}, s_{j+1}, p) \tag{20b}$$

which may be interpreted (insert (20b) into (20a)) as a symmetric A-stable *implicit Runge-Kutta* discretization

$$s_j + \Delta\tau_j \Psi_j(f; s_j, s_{j+1}, \tau_j, \tau_{j+1}) - s_{j+1} =: h_j(s_{j+1}, s_j, p) \ . \tag{21}$$

Note, that (19) can *always* be represented this way (cf. [24]). Collocation thus yields a discretized PI-BVP of the same type as in multiple shooting, with the matching conditions replaced by (19) or (21) resp.

*Mesh selection*

Firstly, in order to retain the usual superconvergence properties, and for computational simplification, all data points should belong to the basic mesh $\pi^m$. Secondly, the mesh must be sufficiently fine to guarantee the required accuracy. Since the local truncation error $e_j$ of $\Psi_j$ is of order $p = 4$ a meshcondition for collocation is given by

$$\Delta\tau_j e_j = O(|\Delta\tau_j|^{p+1}) \leq TOL_j \tag{22}$$

which is used in COLFIT for mesh selection. Since (21) may be viewed as a special multiple shooting scheme with a one step integration, the number of meshpoints required by the *accuracy criterion* (22) is typically much higher than that of multiple shooting. Note, however, that the stability condition [MC] must hold here, too.

*Remark 1:* Since global and local error are of (fixed) order 4 for cubic Lobatto collocation, the method can be effective only for moderate accuracy requirements. The flexibility and efficiency could however be greatly increased by *extrapolation* or *deferred corrections* based on the error expansions in even powers of $\Delta\tau, \Delta\tau_j$.

*Remark 2:* Discontinuities due to switching conditions in (1) are treated in [ 2 ] by a modification of the variable meshpoint technique of [28].

## 5. A Generalized Gauss-Newton Method for Discretized PI-Boundary Value Problems

The constrained approximation problem PI1 is solved in PARFIT and COLFIT by a generalized Gauss-Newton (GGN) method, which appears to be both natural and particularly effective for the considered problem.

Problem PI2: $\quad \|F_1(x)\|_2 = \min$

subject to $\quad F_2(x) = 0 \quad$ and $\quad F_3(x) \geq 0$ .

A given iterate $x_k$ is improved by

$$x_{k+1} = x_k + \lambda_k \Delta x_k \qquad (0 < \lambda_{min} \leq \lambda_k \leq 1) \tag{23}$$

where $\Delta x_k$ is the solution of the linearized system

<u>Problem PI3:</u> $\quad \|F_{1k} + J_{1k}\Delta x_k\|_2 = \min \qquad (J_{ik} = F_i'(x_k), F_{ik} = F$

subject to $\qquad F_{2k} + J_{2k}\Delta x_k = 0 \quad$ and $\quad F_{3k} + J_{3k}\Delta x_k \geq 0$

Procedures of this type are considered e.g. in [ 1 , 3 , 7 ,29,30,31] . We briefly summarize some basic results and strategies.

*Local convergence:* Since under mild regularity assumptions the active inequality constraints of PI3 remain unchanged in a solution neighbourhood [ 7 ], only the equality constrained case needs to be treated. Defining the composite function $F^T = (F_1^T F_2^T)$, $J = F'$, one easily shows that a solution of PI3 is given in terms of a *generalized inverse* $J^+$,

$$\Delta x_k = -J_k^+ F_k \; . \tag{24}$$

If in a region D where $F \in C^1(D)$

$$\|J(y)^+(J(y+t(x-y)) - J(x))(x-y)\| \leq \omega t \, \|x-y\|^2 \quad ; \; \omega < \infty \tag{25a}$$

$$\|(J(y)^+ - J(x)^+)R(x)\| \leq \kappa(x) \, \|y-x\| \quad ; \; \kappa(x) \leq \kappa < 1 \, , \tag{25b}$$

$$R(x) = (I - J(x)J(x)^+)F(x)$$

an initial guess $x_0$ is given with $\alpha_0 := \|J(x_0)^+F(x_0)\|$

$$\delta_0 := \kappa + \frac{\omega}{2}\alpha_0 < 1 \quad , \quad K(x_0, \frac{\alpha_0}{1-\delta_0}) \subset D \tag{26}$$

the full step GGN method is well-defined and converges to a stationary point $x^*$, which is a *strict local minimum* if $J_2$,J have full rank. The final convergence rate is $\kappa(x^*)$.

*Remark:* $\omega$ and $\kappa$ are two fundamental (weighted) Lipschitz constants, which characterize the *nonlinearity* of the model ($\omega$) and the *incompatibility* of the data with the model ($\kappa$). In *statistically* well-posed problems, $\kappa$ is small, and the GGN method essentially has the excellent convergence behaviour of a second order method although it requires only first order information.

*Global convergence:* In order to extend the convergence domain, the *damping factor* $\lambda_k$ is monitored by the *natural monotonicity test* (cf. [1,32,33]

$$\|\overline{\Delta x_{k+1}}(\lambda_k)\| := T_k(x_k+\lambda_k\Delta x_k) < T_k(x_k) = \|\Delta x_k\| \;\; ; \;\; T_k(x) = \|J_k^+F(x)\| \qquad (27)$$

For this (local) *level function*, a simplified line search can be derived from the estimate

$$\begin{aligned} &\|\overline{\Delta x_{k+1}}(\lambda)\| \le (1-\lambda+\tfrac{\lambda^2}{2}\,\omega_k(\lambda)\,\|\Delta x_k\|\,)\,\|\Delta x_k\| \qquad (\lambda \in [0,1]) \\ &\omega_k(\lambda) := \max_{t\in[0,\lambda]} \frac{\|J_k^+(J(x_k+t\Delta x_k)-J(x_k))\Delta x_k\|}{t\,\|\Delta x_k\|^2} \;. \end{aligned} \qquad (28)$$

An *optimal damping factor* in terms of this upper *bound* is

$$\lambda_k^* = \min\left(1\,,\frac{1}{\omega_k(\lambda_k^*)}\,\frac{1}{\|\Delta x_k\|}\right) \qquad (\omega_k \text{ mon. increasing}) \qquad (29)$$

$\omega_k(\lambda)$ can be safely estimated by the *a posteriori formula* (cf. [ 1 , 7 ])

$$\overline{\omega}_k(\lambda) := \frac{2\,\|\overline{\Delta x_{k+1}}(\lambda)-(1-\lambda)\Delta x_k\|}{\lambda^2\,\|\Delta x_k\|^2} = \omega_k(\lambda) + O(\lambda) \qquad (30)$$

This estimate is applied in two ways: $\overline{\omega}_{k-1}(\lambda_{k-1})$ of the last iteration is used as a *predictor* for $\lambda_k$ by means of (29). If the monotonicity test fails, it is used to *correct* the step length. This procedure is easily shown to be feasible.

*Remark 1:* The damping strategy (29) is also "optimal" in the sense, that the resulting (nearly constant) *steplength* $\omega_k^{-1}$ characterizes the radius of the domain in which the linearization is valid, and thus optimally exploits the computed derivative information.

*Remark 2:* Global convergence for the damped GGN method can be established by means of an $F_1$-space level function [ 1 , 7 ]. Although a similar proof for the damping strategy described here is still an open question, its use is strongly emphasized by numerical experience.

## 6. *Solution of Linear Subproblems*

The linearization of the discrete problem PI1 yields a total Jacobian of a special structure

$$J = \left[\begin{array}{ccccc|c} D_1^1 & D_1^2 & \cdots & & D_1^m & D_1^p \\ \hline D_2^1 & D_2^2 & \cdots & & D_2^m & D_2^p \\ \hline D_3^1 & D_3^1 & \cdots & & D_3^m & D_3^p \\ \hline G_1^l & -G_1^r & & & 0 & G_1^p \\ & \ddots & \ddots & & & \vdots \\ 0 & & & G_{m-1}^l & -G_{m-1}^r & G_{m-1}^p \end{array}\right] , \quad F = \left[\begin{array}{c} R_1 \\ \hline R_2 \\ \hline R_3 \\ \hline h_1 \\ \vdots \\ h_{m-1} \end{array}\right] , \quad \begin{array}{lll} \text{with} & & \\ D_i^j & := & dR_i/ds_j \\ D_i^p & := & dR_i/dp \\ G_j^l & := & dh_j/ds_j \\ G_j^r & := & -dh_j/ds_{j+1} \\ G_j^p & := & dh_j/dp \end{array} \tag{31}$$

*The Condensing Algorithm*

For *multiple shooting*, one has $G_j^r = -I$, $G_j^l := G_j = \phi(\tau_{j+1},\tau_j)$, $G_j^p = \phi^p(\tau_{j+1},\tau_j)$. Using the matrices $G_j^r$ as pivots for a block-Gaussian elimination, the *backward recursion*

$$\begin{array}{llll} u_i^m := R_i & , \; P_i^m := D_i^p & , \; E_i^m := D_i^p & (i=1,2,3) \\ u_i^{j-1} := u_i^j + E_i^j h_j & , \; P_i^{j-1} := P_i^j + E_i^j G_{j-1}^p & , \; E_i^{j-1} := D_i^{j-1} + E_i^j G_{j-1} , & \\ & & & (j=m(-1)2) \end{array} \tag{32}$$

transforms the system (31) equivalently to the *condensed problem* for initial values and parameters:

$$\begin{array}{lll} \text{CP}: & \|u_1^1 + E_1^1\Delta s_1 + P_1^1\Delta p\| = \min & \\ \text{s.t.} & u_2^1 + E_2^1\Delta s_1 + P_2^1\Delta p = 0 \quad \text{and} & u_3^1 + E_3^1\Delta s_1 + P_3^1\Delta p \geq 0 \end{array} \tag{33}$$

which is of much smaller dimension. Having solved (33), the other increments can be computed by the *forward recursion*

$$s_{j+1} = G_j\Delta s_j + G_j^p\Delta p + h_j \qquad (j=1,\ldots,m-1) . \tag{34}$$

For *collocation*, the same algorithm applies with the replacements

$$\left(G_j^r\right)^{-1}G_j^l \to G_j , \quad \left(G_j^r\right)^{-1}G_j^p \to G_j^p , \quad \left(G_j^r\right)^{-1}h_j \to h_j \tag{35}$$

Since nonsingularity of $G_j^r$ is an essential stability requirement for collocation anyway (cf. [24]), this procedure is feasible. Note, that

$$\left(G_j^r\right)^{-1} G_j^l = \phi(\tau_{j+1},\tau_j) + O(h^{p+1}) \ , \ \left(G_j^r\right)^{-1} G_j^p = \phi^p(\tau_{j+1},\tau_j) + O(h^{p+1}) \qquad (36)$$

*Solution of the condensed system*

For solution of the condensed system, several methods are available (cf. [1, 29, 34]). Efficient techniques for the multiexperiment case are developed in [3].

"Real life" PI-problems, like other inverse problems, are frequently singular or ill-conditioned, when the model is overparametrized (non-unique) with respect to the available experimental data, so that *regularizations* are needed. Since the backward recursion leaves the rank unaltered, a treatment of the condensed system is sufficient. As all algorithms basically depend on triangularisation by QR decomposition (BUSINGER, GOLUB [35]), pseudoinversion can be applied after appropriate rank selection (DEUFLHARD, SAUTTER [36]):

*Remark:* Note that in case of rank-deficiency, the condensed system allows a complete (linear) analysis of redundancies in the parameter space.

*Statistical a posteriori analysis of solution*

Parameters obtained by a PI procedure are useful only if also some estimate of their statistical reliability is given. The GGN method in combination with the *condensing algorithm* used in PARFIT and COLFIT makes this particularly easy. The covariances of the parameters (and the initial values, if they are not fixed) can be computed *cheaply* and *stable* from the decomposed condensed system, and from this, first order approximations to individual confidence intervals can be derived ([1, 7], see SCHLÖDER [3] for the multiexperiment case).

However, more statistical aids are desirable, and presently under consideration.

*Conditioning of PI-boundary value problems*

With the definitions (considering only constraints active at the solution

$$\begin{aligned} B_l &= \textstyle\sum_{j=l+1}^{m} D^j\phi(\tau_j,\tau_{l+1}) \ , \ A_l = \sum_{j=1}^{l} D^j\phi(\tau_j,\tau_{l+1}) \ , \ E_l = A_l + B_l \\ P_l &= D^p + \textstyle\sum_{j=1}^{l-1} A_j G_j^p + \sum_{j=1}^{m} B_j G_j^p \ , \ S_l = (E_l P_l)^+ \ , \end{aligned} \qquad (37)$$

an explicit representation of the generalized inverse $J^+$

$$\binom{\Delta s_1}{\Delta s_p} = S_1 (R + \sum_{j=1}^{l-1} A_j h_j + \sum_{j=l}^{m-1} B_j h_j) \tag{38}$$

is obtained by means of the condensing algorithm. With the assumptions

$$\|S_l\| \le c_l \le K_1 < \infty \ , \quad \|S_l A_j\| \le a_{lj} \le K_2 < \infty \ , \quad \|S_l B_j\| \le b_{lj} \le K_2 < \infty \tag{39}$$

one readily verifies the upper bound for $\|J^+\|$ (cf. [ 7 ,37,38]).

<u>Lemma 1:</u> $$\|J^+\|_\infty \le c_l + \sum_{j=1}^{l-1} a_{lj} + \sum_{j=l}^{m-1} b_{lj} \le K_1 + (m-1) K_2 < \infty \ \square \tag{40}$$

Asymptotic bounds for collocation are established analogously, making use of (36). Note that it is sufficient that $c_l, a_{lj}, b_{lj}$ be bounded for just one pair (l,j) to establish the bounds $K_1, K_2$ *independent* of a specific mesh.

The dependence on the *number of meshpoints* can be avoided using $1/\Delta\tau_j$ as a *weight* for $\|h_j\|$ (thus introducing an error-per-unit-interval criterion in the meshconditions (15) and (22)).

*Stability of the condensing algorithm*

There has been a lot of discussion on the stability of this type of recursive solver in the recent past. Since no pivoting is performed, it can be expected to be stable only if the o.d.e. system is *stable* - and $G_j^l$ is "smaller" than $G_j^r$ in some sense. In general, however, the errors $\delta_j := \widetilde{\Delta s}_j - \Delta s_j$ of the increment $\widetilde{\Delta s}_j$ computed in the *forward recursion*, satisfying

$$\delta_{j+1} = G_j \delta_j + G_j^p \delta p + \delta h_j \ , \tag{41}$$

may be *dominating* the exact increments $\Delta s_j$, since their growth behaviour is that of the possibly unstable IVP. The terms $\delta h_j$ incorporate the rounding errors made in the evaluation of (34), they are (elementwise) bounded by

$$\begin{aligned} |\delta h_j| &\le |G_j|\,|\widetilde{\Delta s}_j|\,n_d\varepsilon + |G_j^p|\,|\widetilde{\Delta p}|\,n_p\varepsilon + |h_j|\varepsilon \\ &\le |G_j|\,|\Delta s_j|\,n_d\varepsilon + |G_j^p|\,|\Delta p|\,n_p\varepsilon + |h_j|\varepsilon + \\ &\quad + |G_j|\,|\delta_j|\,n_d\varepsilon + |G_j^p|\,|\delta p|\,n_p\varepsilon \qquad , \ \varepsilon = \varepsilon_{mach} \ . \end{aligned} \tag{42}$$

For the sake of stability, alternative linear system solvers are often advocated, such as the algorithms of VARAH [39] or de BOOR, WEISS [40], which perform some kind of pivoting but try to utilize the structure for computational and storage economy. (A PI variant of [39] was designed by BÄR for use in COLFIT [2].)

On the other hand, the condensing algorithm has been successfully used for many sensitive real life problems including PI. Its computing and storage efficiency is outstanding, and it offers significant analytical and statistical insight by means of the condensed system. Hence, the development of modifications that maintain the efficiency of the algorithm under suitable stability assumptions is a challenging problem.

*The SHAVING algorithm makes SURE*

The SHAVING algorithm was developed in cooperation with KRÄMER-EIS after inspiring discussions with B. RUSSELL and B. MATTHEIJ. It is a special iterative refinement procedure similar to the one developed independently by DEUFLHARD and BADER [38], who also give a detailed rounding error analysis. Therefore, we may restrict the presentation of our procedure to the main lines.

The basic idea of SHAVING (combined with SUccessive REfinements) is to perform the forward recursion only as long, as the defect of the right hand side

$$\tilde{h}_j := h_j - G_j\widetilde{\Delta s}_j - G_j\widetilde{\Delta p} + \widetilde{\Delta s}_{j+1} \qquad (43)$$

can be considered to be zero. This defect *is* actually a zero, if computed numerically, but it is *known* to be exactly $\delta h_j$ of (41). From (42), one may thus conclude, that the contribution of the parasitic terms $\delta_j$ is small, if $\delta_j$ is small compared to $|\Delta s_j|$, or $|\widetilde{\Delta s}_j|$, resp.

$$|\delta_j| \le \alpha|\widetilde{\Delta s}_j| \qquad (\alpha \ll 1)\ . \qquad (44)$$

An upper bound for $|\delta_j|$ can be computed a priori by the recursion

$$d_{j+1} = |G_j|(d_j + |\widetilde{\Delta s}_j|n_d\varepsilon) + |G_j^p|(|\delta p| + |\widetilde{\Delta p}|n_p\varepsilon) + |h_j|\varepsilon \qquad (45)$$

This criterion can however be relaxed and simplified taking into account the special nature of the right hand side. In view of the mesh condition

(15), $\tilde{h}_j$ can be considered small compared to the *natural perturbation* made in the evaluation of $h_j$, if

$$\|G_j\| \, \|\widetilde{\Delta s}_j\| n_d \varepsilon + \|G_j^p\| \, \|\widetilde{\Delta p}\| n_p \varepsilon \le \beta \cdot TOL_j \qquad (\beta \ll 1) \ . \tag{46}$$

If (46) is violated, the forward recursion is stopped. Only the defect of this index, $\tilde{h}_j$, has to be computed, whereas the previous values are set to zero - they are considered to be "shaven". Repeated application of the thus modified condensing algorithm (backward recursion for the right hand sides $u_i$ only) successively refines the computed solution (and eventually removes the parasitic parts!). The performance scheme of the algorithm is visualized in Fig. 2.

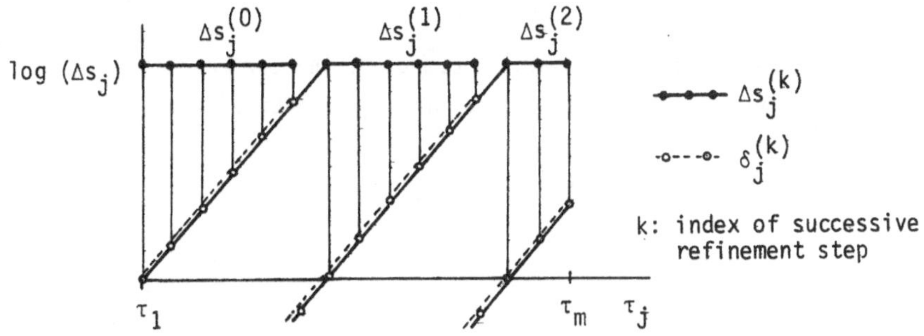


Fig. 2 Performance scheme of the "shaving" algorithm

## *7. Evaluation of Functions and Derivatives*

The main bulk of computational labour in PARFIT arises from the numerical solution of the o.d.e., and the generation of their derivatives w.r.t. parameters and initial values. Therefore some remarks seem appropriate as to how this can be done most efficiently in a multiple shooting code.

### *7.1 Solution of IVPs*

Two variable order and step *extrapolation* codes are the standard numerical integrators in PARFIT, for non-stiff systems DIFSYS (BULIRSCH, STOER [14]) based on an explicit midpoint rule, and for stiff systems METAN1(BADER, DEUFLHARD [41]), based on a semi-implicit midpoint rule.

However, other integrators can be used as well, if the necessary interfaces to PARFIT are provided.

*Adaptive accuracy selection*

A lot of computational work can be saved if the integration tolerance is adapted to the requirements of the GGN process [ 1 ]. Although this seems straightforward, a computationally effective *and* safe implementation that works for highly nonlinear as well as for ill-conditioned problems is not trivial.

Up to now, PARFIT and its companion code for optimal control problems OPCON are the only multiple shooting procedures where this concept is realized.

An *adaptive accuracy scheme* can be derived from the following considerations (details to be presented elsewhere).

First, once an iterate $x_k$, say, is computed, the error $e_k$ of the increment $\Delta x_k$ should not be smaller than $\|\Delta x_{k+1}\|$ to be efficient

$$e_k \cong \|\Delta x_{k+1}\| = [\delta_k][\delta_{k-1}] \|\Delta x_{k-1}\| \ . \tag{47}$$

Estimates of these convergence rates can be given in terms of the *local convergence theorem*.

The *global accuracy* $e_k$ can be achieved by lemma 1 choosing the *local error* tolerances $TOL_j$ according to

$$\|J_k^+\| \, TOL \leq e_k \ . \tag{48}$$

Outside the local convergence domain, condition (47) is replaced in PARFIT by the requirement, that the predictor-corrector strategy (29,30), on which the natural monotonicity test is based, be stable!

The efficiency of the overall scheme can also slightly be increased by storing additional information (about optimal initial stepsizes and order, etc.) for the integrator to be used in subsequent iterations.

## *7.2 Calculation of derivatives*

Being the most time consuming part of the calculations in PARFIT, this must be very efficiently realized. Secondly, if the algorithm shall be of use in practical applications, the chance of human error in these calculations must be strictly excluded.

*External Numerical Differentiation* (END)

A well known way to ensure the latter requirement is the use of a finite difference approximation $\Delta^\eta$ to the derivative $\Delta^0$

$$\Delta^\eta x(t;c) := \eta^{-1}(x(t;c+\eta d) - x(t;c)) = \Delta^0 x(t;c) + \varepsilon \ , \ \varepsilon = O(\eta) \qquad (49)$$

where for abbreviation, $c := (s_j, p)$ and d a specified direction. The derivatives can thus be computed by $(n_d + n_p + 1)$ full integrations with accuracy TOL varying one component of c at a time $(d = e_i)$.

Even for an optimal balance between discretization error (which is a discontinuous function of c usually) and cancellation of leading digits the best accuracy $\varepsilon$ one can expect is

$$\varepsilon = O(\sqrt{TOL}) \qquad \text{with} \quad \eta = O(\sqrt{TOL}) \qquad (50)$$

The efficiency of this scheme is low, since high accuracy integration is needed for low accuracy results. Moreover, for low tolerances as used in the adaptive accuracy scheme, it becomes totally unreliable.

*Internal Numerical Differentiation* (IND)

A remedy is provided by *internal numerical differentiation* as described in [ 1 ]. The basic idea is to compute the *derivative of the* (variable order and step) *discretization scheme* - thus getting rid of the TOL-bound - and *saving the overhead spent in the integration of* $(n_d + n_p + 1)$ *nearly identical trajectories*. We describe this procedure for the semi-implicit midpoint rule ( [41])

$$\begin{aligned} &\Delta_0 := Pf_0 &&, \ y_0 := y_0 + h\Delta_0 \ , \ h := H/N \quad &&\text{(Start)} \\ &\Delta_k = \Delta_{k-1} + 2P(f_k - \Delta_{k-1}) &&, \ y_{k+1} = y_k + h\Delta_k &&(k=1,\dots,N) \end{aligned} \qquad (51)$$

where $P = (I - hA)^{-1}$ , $A \approx f_x(t_0, y_0, p)$ , $f_j = f(t_0 + jh, y_j, p)$ .

The final smoothing step

$$x_h(c) := \frac{1}{2}(y_{N-1} + y_{N+1})$$

then yields an approximation of the initial value problem solution x(c) at $t = t_0 + H$. The discretization error has an asymptotic expansion in even powers of h, which permits the construction of higher order approximations $T_{kj}$ by polynomial extrapolation from a sequence of values $x_{h_i}(c)$.

$$\begin{aligned} x_h(c) &= x(c) + \sum_{i=1}^{M} \alpha_i(c)\, h^{2i} + O(h^{2M+2}) \\ T_{kj}(c) &= x(c) + \sum_{i=k}^{M} \gamma_{ij}^{k}(c) h^{2i} + O(h^{2M+2}) . \end{aligned} \tag{52}$$

As a consequence (under suitable assumptions) the finite difference approximation $\Delta^{\eta}$ and the derivative $\Delta^{0}$ have an asymptotic expansion, too.

Lemma 2: For $0 \le \eta < \eta_0$:

$$\begin{aligned} \Delta^{\eta} x_h(c) &= \Delta^{\eta} x(c) + \sum_{i=1}^{M} \Delta^{\eta}\alpha_i(c)\, h^{2i} + O(h^{2M+2}) \\ \Delta^{\eta} T_{kj}(c) &= \Delta^{\eta} x(c) + \sum_{i=k}^{M} \Delta^{\eta}\gamma_{ij}^{k}(c) h^{2i} + O(h^{2M+2}) \quad \square \end{aligned} \tag{53}$$

The proof involves the linearity of basic discretization and interpolation.

(53) allows a convenient *control of the discretization error* of $\Delta^{\eta} T_{kj}$ (along with that of $T_{kj}$), which may serve as an additional stability check for the discretization scheme (51). It also indicates the accuracy bound due to cancellation for $\eta > 0$.

As an interesting consequence of lemma 2, one has in comparison to END (49) (informally)

$$\varepsilon = O(TOL) + O(\sqrt{\varepsilon_{mach}}) \qquad \text{for} \quad \eta = O(\sqrt{\varepsilon_{mach}}) \tag{54}$$

i.e. the accuracy is of the same order as the accuracy TOL of the nominal trajectory.

*The analytical limit of IND*

If correct and cheap analytical derivatives of the right hand side are available, e.g. as in case of chemical reaction systems, it can be economic (and it removes the $O(\sqrt{\varepsilon_{mach}})$ bound in (54)!) to perform the limit analytically and compute $\Delta^{0} T_{kj}(c)$. It is easily shown that this is *equivalent* to the integration of the VDE (13) columnwise by the discretization scheme (51) using the same Padé-approximants, order and stepsizes as in IND - thus reflecting the identical stability behaviour. Note, that as a consequence of lemma 2, forward difference approximation and analytical limit deliver *identical* results for moderate accuracies (say, $TOL > O(\sqrt{\varepsilon_{mach}})$ ).

Summing up, *internal numerical differentiation* leads to a drastic reduction of computing time (60 - 80 p.c. especially for low tolerances [ 1 ]) compared to *external numerical differentiation* because of substantial *overhead savings* (e.g. the Padé-approximants $P(h_i)$ only have to be computed once for all trajectories), and much lower accuracy requirements for the basic discretization scheme.

Moreover, IND is stable especially for low tolerances, where most of the effort is spent in an adaptive accuracy scheme.

## *8. Numerical Results*

This last section summarizes a few numerical results to illustrate some aspects of the performance of the algorithms. For details, and more complex problems, the reader is referred to [1,7]

*Notorious test problem*

For multiple shooting with PARFIT (using the 10 data points as mesh) the problem is well conditioned. Numerical results for $10 \leq \mu \leq 120$ are summarized in table 8.1. Single shooting works for $\mu = 10,20$ only, since the propagation factor (row 4) is larger than $10^{19}$ for $\mu \geq 40$. With multiple shooting, the stability mesh condition (15) is satisfied for $TOL \leq 10^{-8}$ and all $\mu$ (row 3).

Table 1  Notorious test problem

| $\mu$ | 10 | 20 | 40 | 60 | 80 | 100 | 120 |
|---|---|---|---|---|---|---|---|
| No. of iterations | 4 | 4 | 4 | 4 | 4 | 4 | 4 |
| min. loc. error $\gamma_j \cdot \varepsilon_{mach}$ | $10^{-14}$ | $10^{-13}$ | $10^{-12}$ | $10^{-11}$ | $10^{-10}$ | $10^{-9}$ | $10^{-8}$ |
| error propagation $\|\|\phi(1,0)\|\|$ | $10^{5}$ | $10^{10}$ | $10^{19}$ | $10^{27}$ | $10^{36}$ | $10^{45}$ | $10^{54}$ |
| No. of shaving steps | 1 | 1 | 2 | 2 | 3 | 3 | 4 |

Although the initial trajectory for $\mu = 60$ looks rather weird (Fig. 3), convergence is safely obtained after 4 iterations to an accuracy of $10^{-3}$. The modified condensing algorithm needs two *shaving* steps.

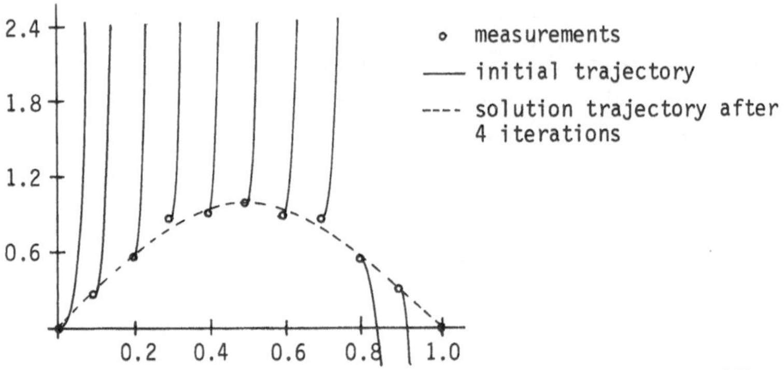


Fig. 3 Notorious test problem - multiple shooting ($\mu = 60$, $p^{(0)} = 1$)

*Denitrogenation of pyridine*

This problem (due to ZWAGA [42]) was investigated in detail in [1,7] It describes the identification of 11 unknown parameters (the rate constants) in a system of 7 differential equations

$$\begin{aligned}
\dot{A} &= -p_1A + p_9B \\
\dot{B} &= p_1A - p_2B - p_3CB + p_7D - p_9B + p_{10}DF \\
\dot{C} &= p_2B - p_3BC - 2p_4CC - p_6C + p_8E + p_{10}DF + 2p_{11}EF \\
\dot{D} &= p_3BC - p_5D - p_7D - p_{10}DF \qquad\qquad t \in [0,5.5] \qquad (55) \\
\dot{E} &= p_4CC + p_5D - p_8E - p_{11}EF \\
\dot{F} &= p_3BC + p_4CC + p_6C - p_{10}DF - p_{11}EF \\
\dot{G} &= p_6C + p_7D + p_8E
\end{aligned}$$

representing a recycling process for pyridine (A), which is transformed into ammonia (F) and pentane (G) by means of three catalysts.

The initial values are assumed to be error-free (A(0) = 1 and all others zero.) 77 Measurements of all seven state variables at times $t_j = j \cdot 0.5$ ($j = 1,\ldots,11$) are available. The pyridine problem allows some instructive numerical tests and comparisons.

*8.1 Multiple Shooting vs. Single Shooting (Efficiency)*

(A) In a first experiment the initial guess for all parameters is chosen to be $p_i^{(0)} = 1$ - a rather poor guess, since the solution values vary between $p_9 = .0201$ and $p_3 = 29.4$ . For multiple shooting, seven meshpoints are selected (0.,0.5,1.5,...,5.5). Single shooting needs 18 iterations (42 sec.), with the damping strategy active during the first 13 iterations. Multiple shooting requires only 8 iterations and a computing time of 16 sec. (IBM 370/168), and needs two damping steps only.

It is not only the drastic reduction of computing time that is noticeable: for single shooting, the initial guess is "far" from the local convergence domain, for multiple shooting, it is just on the rim!

(B) A second experiment is very instructive, here, all parameters are set to zero initially.

With multiple shooting, only 7 iterations (11 sec.) are needed. For single shooting, the initial rank of the condensed system is only 1 (excluding initial values), and it increases step by step during 22 iterations (58 sec.).

The explanation is at hand: system (55) is homogeneous, and the initial values represent a very specific steady state. Thus, only parameter $p_1$ is effective in the first iteration, followed by $p_2$ , $p_9$ in the next, etc. Multiple shooting leads to a full rank problem from the start, since the subinterval integrations do not begin in a steady state. Thus, multiple shooting has also some "regularizing" properties.

*8.2 Multiple Shooting vs. Collocation*

Table 2 Comparison PARFIT vs. COLFIT - Pyridine problem A (EPS = 1D-3)

| | iterations | comp.time(sec) | F-calls | meshsize | basic storage[c] |
|---|---|---|---|---|---|
| COLFIT | 11 (+4)[a] | 52 | 19160 | ≤ 28 | 3348 + 4900 |
| PARFIT | 8 | 16 | 33099 | 7 | 3348 + 882[b] |

[a] includes 4 mesh refinements, [b] excluding the integrator (+ 782),
[c] the second figure represents the mesh-dependent storage

Table 2 compares the performance of PARFIT and COLFIT for problem 8.1A. Obviously, PARFIT has distinct advantages over COLFIT with respect to computing time and storage requirements (and for higher accuracies the difference is even more marked.) According to the number of iterations, PARFIT also seems to have a slightly better convergence behaviour. On the other hand, COLFIT is extremely economical w.r.t. F-evaluations (recall that it is a 4th order method with two F-calls per subinterval only!). Although the results for this problem are characteristic according to our experience, it allows only for a rather preliminary conclusion.

A more detailed comparison on the basis of several examples is in preparation [2]. It seems, however, that the real advantages of collocation are in the field of singular perturbation problems.

### *8.3 Comparison of GGN and Random Search*

For quite a while, there has been a discussion in PI about the relative advantages of the (higher order) GGN method of PARFIT and low order methods such as random optimization. We therefore conclude with two examples treated previously by MILSTEIN [9 ,10] (with random optimization) and the author [ 1 , 7 ]. Both examples are multiexperiment problems and were solved by PARMEX.

*Inducible Enzyme Synthesis*

This biochemical problem originally due to HEINMETS [43] is reputed to be *extremely ill-conditioned* (SWARTZ-BREMERMANN [44]). For details, the reader is referred to [ 1 , 9 ]. The PI problem consists of a model with 12 nonlinear o.d.e. and 14 unknown parameters, which have to be estimated from artificial data generated by adding pseudo-random noise to the true solution.

Three numerical experiments are performed. (A) uses one set of initial values, with 4 data points (48 measurements), (B) is like (A) but with 12 data points (144 measurements), (C) uses a second set of initial values with 4 data points and 48 measurements *in addition* to (A).

Table 3 summarizes the results of table 8.3 in [ 1 ]. The PARMEX parameter estimates are always within the 95 % confidence intervals, whereas some of the estimates due to [9] are always outside.

Table 3 Inducible enzyme synthesis - *reliable estimates*

| | trajectories | data points [measurements] | parameters within 95 % confidence intervals [9] | PARMEX |
|---|---|---|---|---|
| A | 1 | 4 [ 48] | 8 | 14 |
| B | 1 | 12 [144] | 6 | 14 |
| C | 2 | 4 [ 96] | 12 | 14 |

*Grasseater Ecosystem*

This testproblem is based on a model of GARFINKEL [45] for an ecosystem of one herbivore feeding on two types of grass in a two trophic level ecosystem. Details can be found in [7 ,10]. The model involves three nonlinear o.d.e. and 8 parameters.
Following [10], pseudoexperimental data are generated by adding 9 % noise to three trajectories corresponding to three different sets of initial values at 20 equidistant data points each.
Three numerical problems are generated by solving the PI problem for one, two and all three data sets.
The results for the 3 critical parameters $p_1, p_4$ and $p_7$ documented in [10] are assembled in table 4 together with the corresponding results of PARMEX. Full solution details for the latter are given in [7].

Table 4 Grasseater ecosystem - comparison of results

| traj. | param. | value | estim. PARMEX | err. PARMEX | est. err.[a] | err. [10] | estim. [10] | c.time [It][b] PARMEX | c.time [It][b] [10] |
|---|---|---|---|---|---|---|---|---|---|
| I | 1 | 1.0 | .995 | .005 | .028 | .13 | .86 | 8 [6] | 277 [264] |
| | 4 | 1.2 | 1.237 | .037 | .029 | .59 | .61 | | |
| | 7 | 1.0 | 1.015 | .015 | .029 | .93 | 1.93 | | |
| I+II | 1 | 1.0 | 1.024 | .024 | .018 | .11 | .89 | 19 [5] | 210 [ 91] |
| | 4 | 1.2 | 1.215 | .015 | .016 | .42 | .78 | | |
| | 7 | 1.0 | 1.006 | .006 | .014 | .2 | 1.2 | | |
| I+II+III | 1 | 1.0 | 1.005 | .005 | .011 | .05 | 1.05 | 30 [7] | 246 [ 75] |
| | 4 | 1.2 | 1.213 | .013 | .013 | .28 | .92 | | |
| | 7 | 1.0 | .995 | .005 | .006 | .13 | 1.13 | | |

[a] to be multiplied by (approx.) 4.1 for 95 % confidence interval

[b] computing time (sec.) [Iterations]

As in the previous problem, one observes that for each problem at least two parameter estimates of the random optimizer are clearly outside the confidence region whereas the estimates of the GGN are always safely within and have an error that is overall smaller by an average factor of ten.
With respect to computing time, PARMEX seems to be more effective, too, since the two machines (IBM 370/168 (PARMEX) and IBM 360/91 [10] ) appear to be comparable.

From the results one may conclude, that the higher order GGN method has a definitely safer local convergence behaviour than random optimization, and delivers estimates which are statistically reliable. On the other hand it appears that the higher order information also pays off in the earlier iteration stages - which might be explained in terms of the "optimal stepsize" strategies discussed in sect. 5.

## *9. Conclusion*

The present article reviews recent advances in the development of constrained least squares boundary value problem algorithms for o.d.e., in particular an enhanced multiple shooting code, PARFIT, and a collocation code, COLFIT, that are apt for treatment of a broad class of inverse problems in parameter identification.

Several implementational developments are emphasized that have led to a substantial increase in efficiency, stability and applicability of these methods, such as a stabilized recursive linear system solver, adaptive accuracy selection and efficient internal numerical differentiation schemes.

The performance of the new techniques is demonstrated by several numerical test problems, which show significant advantages over previous approaches.

*Acknowledgement:* This work was supported by the Deutsche Forschungsgemeinschaft (Sonderforschungsbereich 72).

*References*


[1] H.G. Bock: Numerical Treatment of Inverse Problems in Chemical Reaction Kinetics, in K.H. Ebert, P. Deuflhard, W. Jäger (eds.), Modelling of Chemical Reaction Systems, Springer Series in Chemical Physics 18, 102, Heidelberg (1981)

[2] V. Bär: Ein Kollokationsverfahren zur numerischen Lösung von Mehrpunktrandwertaufgaben mit Schalt- und Sprungbedingungen, work done in preparation of a diploma thesis

[3] J. Schlöder, H.G. Bock: Identification of Rate Constants in Bistable Chemical Reactions, these proceedings

[4] Th. Reiners: Numerische Behandlung inverser Probleme der Geophysik, work done in preparation of a diploma thesis

[5] V. Pereyra, H.B. Keller, W.H.K. Lee: Computational Methods for Inverse Problems in Geophysics: Inversion of Travel Time Observations, Phys. Earth Planet. Inter. 21, 120 (1980)

[6] H.G. Bock: Numerical Solution of Nonlinear Multipoint Boundary Value Problems with Application to Optimal Control, ZAMM 58, 407 (1978)

[7] H.G. Bock: Randwertproblemmethoden zur Parameteridentifizierung in Systemen nichtlinearer Differentialgleichungen (in preparation)

[8] U. Nowak, P. Deuflhard: Towards Parameteridentification for Large Chemical Reaction Systems, these proceedings

[9] J. Milstein: Fitting Multiple Trajectories Simultaneously to a Model of Inducible Enzyme Synthesis, Math. Biosci. 40, 175 (1978)

[10] J. Milstein: Error Estimates for Rate Constants of Inverse Problems, SIAM J. Appl. Math. 35, 479 (1978)

[11] C.W. Gear: Numerical Initial Value Problems in Ordinary Differential Equations, Prentice Hall, Englewood Cliffs (1971)

[12] J.H. Bremermann: A Method of Unconstrained Global Optimization, Math. Biosci. 9, 1 (1970)

[13] R. Bulirsch: Die Mehrzielmethode zur numerischen Lösung von nichtlinearen Randwertproblemen und Aufgaben der optimalen Steuerung, Carl-Cranz-Gesellschaft, Tech. Rpt. (1971)

[14] R. Bulirsch, J. Stoer: Numerical Treatment of Ordinary Differential Equations by Extrapolation Methods, Numer. Math. 8, 1 (1966)

[15] H.B. Keller: Numerical Methods for Two-Point Boundary-Value Problems , London, Blaisdell (1968)

[16] M.R. Osborne: On Shooting Methods for Boundary Value Problems, J. Math. Anal. Appl. 27, 417 (1969)

[17] J. Stoer, R. Bulirsch: Einführung in die Numerische Mathematik II, Berlin, Heidelberg, New York, Heidelberg (1973)

[18] P. Deuflhard: Recent Advances in Multiple Shooting Techniques, in: Computational Techniques for Ordinary Differential Equations (Gladwell / Sayers, eds.)

[19] R. England: A Program for the Solution of Boundary Value Problems for Systems of Ordinary Differential Equations, U.K.A.E.A. Research Group, Abingdon, Tech. Rpt. CLM/PDN 3/73 (1975)

[20] M. Lentini, V. Pereyra: An Adaptive Finite Difference Solver for Nonlinear Two Point Boundary Value Problems with Mild Boundary Layers, SIAM J. Numer. Anal. 14, 91 (1977)

[21] U. Ascher, J. Christiansen, R.D. Russell: A Collocation Solver for Mixed Order Systems of Boundary Value Problems, Math. Comp. 33 (1979)

[22] C. de Boor, B. Swartz: Collocation at Gaussian Points, SIAM J. Numer. Anal. 10, 582 (1973)

[23] R.D. Russell: Collocation for Systems of Boundary Value Problems, Numer. Math. 23, 119 (1974)

[24] R. Weiss: The Application of Implicit Runge-Kutta and Collocation Methods to Boundary Value Problems, Math. Comp. 28, 449 (1974)

[25] E.D. Dickmanns, K.H. Well: Approximate Solution of Optimal Control Problems Using Third Order Hermite Polynomial Functions, Springer Lect. Notes Comp. Sci. 27, 158 (1975)

[26] H.G. Bock: Numerical Solution of Parameter Estimation Problems by Boundary Value Solvers, Workshop Numerical Boundary Value Problems, Vancouver, Collected Lecture Notes (1980)

[27] U. Ascher, R. Weiss: Collocations for Singular Perturbation Problems I : First Order Systems with Constant Coefficients, Dept. Comp. Sci., University of British Columbia, Vancouver, Tech. Rpt. 81-2

[28] H.G. Bock: Zur numerischen Behandlung zustandsbeschränkter Steuerungsprobleme mit Mehrzielmethode und Homotopieverfahren, ZAMM 57, 266 (1977)

[29] P. Deuflhard, V. Apostolescu: An Underrelaxed Gauss-Newton Method for Equality Constrained Nonlinear Least Squares Problems, Springer Lecture Notes Control Inf. Sci. 7, 22 (1978)

[30] P. Deuflhard, G. Heindl: Affine Invariant Convergence Theorems for Newton's Method and Extension to Related Methods, SIAM J. Numer. Anal. 16, 1 (1979)

[31] H.G. Bock: A Multiple Shooting Method for Parameter Identification in Systems of Nonlinear Differential Equations, GAMM Conference, Brussels, (1978)

[32] P. Deuflhard: A Modified Newton Method for the Solution of Ill-Conditioned Systems of Nonlinear Equations with Application to Multiple Shooting, Numer. Math. 22, 289 (1974)

[33] P. Deuflhard: A Relaxation Strategy for the Modified Newton Method, Springer Lecture Notes 477, 59 (1975)

[34] J. Stoer: On the Numerical Solution of Constrained Least Squares Problems, SIAM J. Numer. Anal. 8, 382 (1971)

[35] P. Businger, G.H. Golub: Linear Least Squares Solutions by Householder Transformations, Numer. Math. 7, 269 (1965)

[36] P. Deuflhard, W. Sautter: On Rank Deficient Pseudo-Inverses, J. Lin. Alg. Appl. 22, 91 (1980)

[37] R.M.M. Mattheij: The Conditioning of Linear Boundary Value Problems, to appear in SIAM J. Numer. Anal.

[38] P. Deuflhard, G. Bader: Multiple Shooting Techniques Revisited, these proceedings

[39] J.M. Varah: Alternate Row and Column Elimination for Solving Certain Linear Systems, SIAM J. Numer. Anal. 13, 71 (1976)

[40] C. de Boor, R. Weiss: SOLVEBLOK:A Package for Solving almost Block Diagonal Linear Systems, ACM Trans. Math. Softw. 6, 80 (1980)

[41] G. Bader, P. Deuflhard: A Semi-Implicit Mid-Point Rule for Stiff Systems in Ordinary Differential Equations, SFB 123, Heidelberg, Tech. Rpt. 114 (1981)

[42] P. Zwaga: Private Communication (1977)

[43] F. Heinmets: Analogue Computer Analysis of a Model System for the Induced Enzyme Synthesis, J. Theor. Biol. 6, 60 (1964)

[44] J. Swartz, J.H. Bremermann: Discussion of Parameter Estimation in Biological Modelling:Algorithms for Estimation and Evaluation of the Estimates, J. Math. Biol. 1, 241 (1975)

[45] D. Garfinkel : A Simulation Study of the Effect on Simple Ecological Systems of Making Rate of Increase of Population Density Dependent, J. Theoret. Biol. 14, 67 (1967)

# SOME EXAMPLES OF PARAMETER ESTIMATION BY MULTIPLE SHOOTING

Roland England

## 1. Introduction

This paper gives an outline of the multiple shooting code DD03, in the AERE Harwell subroutine library. Also, four parameter estimation problems are described, two from the field of fluid dynamics, and two concerned with nuclear or plasma physics, ranging from a simple eigenvalue problem on a semi-infinite interval to a complex plasma stability problem. The formulation of each is discussed, so that it may be solved by the standard program and the results obtained are outlined. For a linear Schrödinger equation on a semi-infinite interval, the efficiency of the program is quite sensitive to the boundary condition used. The Orr-Sommerfeld equation is of eighth order, with a complex eigenvalue; it consumes a great deal of computer time for large Reynold's numbers. A fifth order fluid dynamics problem suffers from singularities and multiple solutions as the Reynold's number increases, but particular solutions can be traced by a continuation method on one of the physical parameters. A class of plasma stability problems can be reduced to a singular second order differential equation, with a potential function, a complex eigenvalue, and various parameters, between which relations must be determined; with an appropriate formulation, and careful estimation of the parameters, the problem becomes quite tractable. The results show that DD03 is a versatile tool for such problems, but indicate directions in which further development should be sought, so as to improve standard software, and widen the class of problems which may be routinely solved.

## 2. The General Problem Class

The DD03 sobroutine package was originally conceived as a general purpose code for two-point boundary value problems of which a standard form might be the system of n first order differential equations:

$$dy_i/dt=g_i(t,y_1,y_2,\ldots,y_n) \quad , \quad i=1,2,\ldots,n; \tag{1}$$

with n boundary conditions:

$$h_i(y_1(a),y_2(a),\ldots,y_n(a),y_1(b),y_2(b),\ldots,y_n(b))=0, \quad i=1,2,\ldots,n; \tag{2}$$

which would in general be unseparated and non-linear. Such a problem will be written in vector form as follows:

$$dY/dt=G(t,Y(t)); \tag{3}$$

with boundary conditions:

$$H(Y(a),Y(b))=0; \tag{4}$$

where Y, G, H are n-vector functions.

However, in many two-point boundary value problems, there are constant parameters to be determined, the most classical cases being Sturm-Liouville eigenvalue problems. The generalized eigenvalue, or parameter estimation problem, has often been solved by introducing additional differential equations of the form $dy_j/dt=0(j>n)$ for the constant parameter $y_j$. However, it seems inefficient to solve differential equations for constant parameters and better to provide for an undetermined p-vector $\mu$ of parameters in the following canonical problem form, which DD03 is designed to solve:

$$dY/dt=G(t,Y(t),\mu); \tag{5}$$

with n+p boundary conditions:

$$H(Y(a),Y(b),\mu)=0; \tag{6}$$

where G,H are given vector functions of n,n+p elements respectively, while the n-vector function Y(t) ($a \leq t \leq b$) and the p-vector $\mu$ are to be determined. In standard eigenvalue problems, the normalizing conditions which are given, corresponding to each undetermined parameter, do not always have the form of boundary conditions, but can usually be replaced by conditions of that type.

## 3. The Multiple Shooting Method

Let a general solution of equation (5), for some value of $\mu$ and initial conditions Y(u)=W, be denoted by $Y(t,u,W,\mu)$; and let $M_v(t,u)=\partial Y/\partial W$ and $N(t,u)=\partial Y/\partial \mu$ be its Jacobian matrices (nxn and nxp respectively) with respect to $W,\mu$. Them $M_v,N$ satisfy the variational equations:

$$\begin{bmatrix} \frac{dM_v}{dt} & \frac{dN}{dt} \end{bmatrix} = \begin{bmatrix} \frac{\partial G}{\partial Y} & \frac{\partial G}{\partial \mu} \end{bmatrix} \begin{bmatrix} M_v(t,u) & N(t,u) \\ 0_{pn} & I_p \end{bmatrix} ; \tag{7}$$

subject to initial conditions:

$$M_v(u,u)=I_n \ , \ N(u,u)=0_{np}. \tag{8}$$

The basic idea of multiple shooting has been described by Osborne [11] and by Keller [5]. It may be slightly generalized by introducing (q+3)/2 shooting points $u_i$ ($a=u_0 < u_2 < \cdots < u_{q+1}=b$) and (q+1)/2 matching points $t_i(i=1,3,\ldots,q \ ; \ u_{i-1} \leq t_i \leq u_{i+1})$. The problem of equations (5),(6) may then be reduced to a system of algebraic equations for $W_i=Y(u_i)(i=0,2,\ldots,q+1)$ and for the vector $\mu$:

$$Y(t_i,u_{i-1},W_{i-1},\mu)-Y(t_i,u_{i+1},W_{i+1},\mu)=0, \ i=1,3,\ldots,q; \tag{9}$$

$$H(W_0,W_{q+1},\mu)=0 \ ; \tag{10}$$

where each function Y in equation (9) is the solution of an initial value problem over a single shooting interval (between adjacent shooting and matching points). The Jacobian matrix of this system with respect to $W_0, W_2, \ldots, W_{q+1}, \mu$ is the structured block matrix:

$$\begin{bmatrix} M_{10} & -M_{12} & & & & N_1 \\ & M_{32} & -M_{34} & & & N_3 \\ - & - & - & - & - & - \\ & & & M_{q,q-1} & -M_{q,q+1} & N_q \\ \partial H/\partial Y(a) & & & & \partial H/\partial Y(b) & \partial H/\partial \mu \end{bmatrix} ; \quad (11)$$

where $M_{ij}=M_v(t_i,u_j)$ $(i=1,3,\ldots,q;\ j=i\pm1)$ and $N_i=N(t_i,u_{i-1})-N(t_i,u_{i+1})$ $(i=1,3,\ldots,q)$ are the solutions of initial value problems (7),(8) over single shooting intervals. This matrix, or some approximation to it, is required by any numerical method for solving the system of equations (9),(10).

Thus there are three principal factors which constitute any multiple shooting implementation: (i) the selection of suitable shooting points $u_i$ $(i=2,4,\ldots,q-1)$ and matching points $t_i(i=1,3,\ldots,q)$; (ii) the numerical solution of initial value problems for $Y(t_i,u_j,W_j,\mu)$, $M_v(t_i,u_j)$, $N(t_i,u_j)$ $(i=1,3,\ldots,q;\ j=i\pm1)$; (iii) the numerical solution of the system of non-linear equations (9),(10). There are many choices available for each of these, and the characteristics of any implementation depend upon the choices made. No existing implementation includes an efficient choice of all three factors, for the general problem class to be tackled. Whatever the choice of these factors, the success or efficiency of each of them will be dependent on a reasonable initial estimate of $W_i$ $(i=0,2,\ldots,q+1)$ and $\mu$, at least for difficult non-linear problems.

## 4. The DD03 Implementation

The DD03 package provides the user with the option of specifying shooting and matching points, which he may do more satisfactorily than the code, or of leaving the code to choose its own shooting intervals automatically. It does this by a procedure inspired by Osborne [11], which attempts to avoid any ill-conditioning of the matrices $M_{ij}(i=1,3,\ldots,q;\ j=i\pm1)$ which might result from a wide span of the eigenvalues of the Jacobian matrix $\partial G/\partial Y$. The procedure is sensitive to the initial estimate of the solution, as it is determined by the matrices $M_v(t,u)$, $N(t,u)$ obtained along that curve. The points are chosen sequentially until a final matching point $t_s$ is found, and satisfy the following criteria along the initially estimated solution:

$$\| M_v(t_i,u_j),\ N(t_i,u_j)\|_\infty \simeq C,\ j=0,2,\ldots,q+1; \quad i=j+1(j<s)\ ,\ i=j-1(j>s), \tag{12}$$

$$u_j=t_i,\ j=2,4,\ldots,q-1;\ i=j-1(j<s),\ i=j+1(j>s). \tag{13}$$

Shooting is from the left for $t<t_s$, and from the right for $t>t_s$. Starting from $u_0=a$, equations (5),(7) are integrated forwards until the first step point at which the norm in equation (12) exceeds the constant C. (The default value for C is 10, but a larger value may often be more efficient). This point is taken as $t_1$ and $u_2$. The point $t_q=u_{q-1}$ is found in a similar way by integrating backwards from $u_{q+1}=b$. Integration then proceeds in the direction of the larger of the last two shooting intervals, the comparison being repeated after each matching point has been found, until the integrations meet in $t_s$. At each shooting point, the value of the user estimated solution is obtained by a subroutine which he must provide.

The initial value problems for $Y(t,u,W,\mu)$ are solved by a fourth order Runge Kutta method, with embedded fifth order error estimate given by England [3], the step size being chosen to maintain the error per unit step constant. The error tolerance is fixed by the user, the same for every iteration of the non-linear equation solver, although it might be advantageous to use a larger error tolerance in the earlier iterations. The variational equations (7) are solved simultaneously by the same fourth order Runge Kutta method with the same step sequence, but without the error estimate. If the user does not provide analytic Jacobian matrices

$\partial G/\partial Y$, $\partial G/\partial \mu$, each column of the right hand side of equation (7) is approximated by:

$$\{G(t,Y+M_v\delta W+N\delta\mu,\mu+\delta\mu)-G(t,Y,\mu)\}/\delta_{Y\mu} \quad ; \tag{14}$$

where the vector $\{\delta W,\delta\mu\}$ is $\delta_{Y\mu}$ times the corresponding column of $I_{n+p}$. This is equivalent to solving equation (5) a further n+p times over each shooting interval to obtain $Y(t,u,W+\delta W,\mu+\delta\mu)$, and approximating the columns of $[M_v(t,u),N(t,u)]$ by:

$$\{Y(t,u,W+\delta W,\mu+\delta\mu)-Y(t,u,W,\mu)\}/\delta_{Y\mu}. \tag{15}$$

The step size $\delta_{Y\mu}$ for numerical differentiation is specified by the user.

The non-linear equations (9),(10) must be solved by some iterative process. A straight-forward Newton-Raphson iteration frequently fails to converge on realistic problems and damped Newton iteration does not improve the situation sufficiently. To improve the global convergence properties, while maintaining rapid local convergence, DD03 uses a version of the Levenberg-Marquardt algorithm, implemented by Reid [12], to minimize the sum of squares of the residuals. This has the disadvantage that if an iteration is rejected, all the calculations corresponding to the previous iteration have to be repeated, a situation which might be avoided with some other algorithm. Reid's code takes advantage of the sparse structure of the matrix (11) in the solution of each linearized problem, without simply doing block elimination of the internal shooting points, which, without pivoting, would be equivalent to using only one shooting interval, and abandoning the conditioning criteria used in equation (12), as pointed out by Osborne [11].

A full description of DD03 is given by England [4], and one application is described by Nichols and England [10].

## 5. Advantages of Multiple Shooting

Many numerical analysts prefer a global method, such as finite differences (Lentini and Pereyra [9]) or piecewise collocation (Ascher et al.[1]) for solving two point boundary value problems. These work well when the fundamental solutions $M_v(t,u)$,$N(t,u)$ are exponentially increasing and decreasing, but may be inefficient in the presence of sharp boundary layers (singular perturbation). If the fundamental solutions are oscillatory, global methods run the danger of converting a well posed continuous problem into a singular discrete problem. Simple shooting may be considered as a global method in which the internal variables are eliminated sequentially, without pivoting, which tends to produce an ill-conditioned algebraic problem if there are exponentially increasing fundamental solutions, but works well for bounded fundamental solutions. It has the advantage that the step size may be chosen adaptively, even before the global system of equations has been set up. Multiple shooting is a compromise in which the internal variables may be eliminated selectively retaining the shooting points, but which may be identical to a global method or to a simple shooting method, depending upon the choice of shooting intervals. So far, no implementation has taken full advantage of the adaptive choice of both the step sizes and the shooting intervals, before the full system of equations has been set up.

## 6. An Eigenvalue Problem

The linear Schrödinger equation:

$$d^2\psi/dx^2+\{E-20\tanh^2(x)\}\psi=0 \; ; \qquad (16)$$

has oscillatory fundamental solutions while $|x|< \tanh^{-1}\sqrt{E/20}$ and exponential solutions elsewhere. Thus, shooting might be desirable for small values of $x$, and a global method for larger values. Multiple shooting can provide such a combination. The boundary conditions are:

$$\psi(0)=0 \; , \; \psi(x) \rightarrow 0 \text{ as } x \rightarrow \infty \; ; \qquad (17)$$

and some non-zero normalizing condition.

In standard form (5),(6) the problem becomes:

$$dy/dx=\{20\tanh^2(x)-E\}\psi,$$
$$d\psi/dx=y; \tag{18}$$

with boundary conditions:

$$y(0)=1 \text{ (for normalization)},$$
$$\psi(0)=0, \tag{19}$$

and some condition for large x. The efficiency of the code is quite sensitive to the form of this last condition. An asymptotic analysis, similar to that of Lentini and Keller [8], shows that as $x \to \infty$

$$\psi(x) \sim A\exp(x\sqrt{20-E})+B\exp(-x\sqrt{20-E}) \tag{20}$$

for some constants A,B. From the second of equations (17), $A=0$ (and $E<20$) which implies that $y(x)+\sqrt{20-E}\,\psi(x)\to 0$. This is already so close to zero in $x=10$, for the first eigenfunction $\psi(x)=\text{sech}^3(x)\tanh(x)$, that the last boundary condition may be taken as:

$$y(10)+\sqrt{20-E}\,\psi(10)=0. \tag{21}$$

With a good estimated of E slightly smaller than the first eigenvalue, $E=11$, and a reasonable estimate of $y,\psi$, DD03 shows good convergence to the frist eigensolution (four figure accuracy in around 10 iterations - see England [4]).

While it is essential that the estimates of $y,\psi$ should not both be zero in either $x=0$ or $x=10$ to avoid the initial value solver following the zero solution, it is likely that the estimate of E might be much less than 11, and even zero. From such a poor starting estimate, it is also possible that a subsequent trial value of E might be greater than 20, and the code would fail on trying to evaluate equation

(21). To avoid this, equation (21) might be replaced by $y(10)^2+(E-20)\psi(10)^2=0$. However, using the same estimates of $E,y,\psi$, the code then converges to an exponentially growing solution (E=0). Although the initial estimates of $y(10),\psi(10)$ are of opposite signs, one of them changes sign without violating the boundary condition by more than a small amount, because of the small scale of the residual. A better scaled boundary condition might be $(20-E)\psi(10)^2/y(10)^2-1=0$. However, in this case convergence is very slow, as the square of this function represents a sharp curved valley in the space of $E,y(10),\psi(10)$, along which a linear iterative method can only move in small steps, because of the steep walls. The correct sealing is such that $\partial H/\partial Y(a)$ and $\partial H/\partial Y(b)$ are of order one, which may be achieved with the boundary condition:

$$y(10)+(E-20)\psi(10)^2/y(10)=0\ . \tag{22}$$

In this case, even with an initial estimate of E=0, the code gives four figure accuracy in 7 Newton iterations. It is interesting to note that after the first 6 iterations, the signs of $y(10),\psi(10)$ are both incorrect, but change on the seventh iteration, passing through the singularity of equation (22). In all these runs, around 30 shooting intervals were chosen, while the number of integration steps ranged from 100 to 3000, according to the tolerance specified.

## 7. The Orr-Sommerfeld Equation

This a fourth order eigenvalue problem in complex variables, thus reducing to an eigth order system with two eigenvalue parameters in real variables:

$$\left\{\frac{d^2}{dx^2}-\alpha^2-i\alpha R(1-x^2-c)\right\}\left\{\frac{d^2}{dx^2}-\alpha^2\right\}\phi-2i\alpha R\phi=0\ ; \tag{23}$$

where $\alpha,R$ are real parameters and c is the complex eigenvalue. For Poiseuille flow, the boundary conditions are $d\phi/dx=d^3\phi/dx^3=0$ in $x=0$, and $\phi=d\phi/dx=0$ in $x=1$. The wave number $\alpha$ was taken to be one. It is necessary to impose two normalizing conditions, such as a non-zero value for

$\phi(0)$ (real and imaginary parts), for example $\phi(0)=1$.

The problem can be put in standard form by defining:

$$\begin{aligned} &\mu_1 + i\mu_2 = c \;, \quad \mu_3 = R \;, \quad \mu_4 = \alpha \;, \\ &y_1 + iy_5 = \phi \;, \quad y_2 + iy_6 = (1/\mu_5)\, d\phi/dx , \\ &y_3 + iy_7 = \frac{1}{\mu_6} \{\frac{d^2}{dx^2} - \alpha^2\}\phi \;, \\ &y_4 = (1/\mu_5)\, dy_3/dx \;, \quad y_8 = (1/\mu_5)\, dy_7/dx ; \end{aligned} \tag{24}$$

where $\mu_5, \mu_6$ have been introduced to permit rescaling of the dependent variables, depending on the Reynold's number R. Normally $\mu_3, \mu_4, \mu_5, \mu_6$ are fixed by four of the 14 boundary conditions, which are of the form $\mu_j - k_j = 0$ $(j=3,4,5,6)$. All the boundary conditions (elements of the vector H) were multiplied by 10 to prevent the iterative process from making substantial changes to the variables concerned, which could all be correctly estimated.

Using this formulation, it was easy to obtain a solution for R=3000; once a reasonable estimate of the eigenfunction was determined, 6 iterations were required to obtain five figure accuracy. R was then increased in 19 steps to 35000, in each case using the previous solution as an initial estimate. By this continuation process, each solution was obtained to five figures in 5 to 8 iterations, the results coinciding with those of Davey [2]. In all cases, 25 to 28 shooting intervals were chosen, almost all in the backward direction. However, the computing time was proportional to R, as the number of integration steps rose steadily, and became prohibitive on a Univac 1106 for $R > 35000$.

## 8. Another Fluid Dynamics Problem

Osborne [11] considered hydrodynamic flow between two infinite rotating discs. The formulation, transformation, and rescaling had been previously performed by Lance and Rogers [6], leaving the problem in the standard form required by DD03:

$$dx_1/dt = -2x_2,$$
$$dx_2/dt = x_3,$$
$$dx_3/dt = x_1x_3+x_2^2-x_4^2+k, \quad (25)$$
$$dx_4/dt = x_5,$$
$$dx_5/dt = 2x_2x_4+x_1x_5;$$

with boundary conditions:

$$x_1(0) = x_2(0)=0, \ x_4(0)=1,$$
$$x_1(b) = x_2(b)=0, \ x_4(b)=s; \quad (26)$$

where $b^2$ is the Reynold's number, s the relative angular velocity of the second disc, and k is an unknown parameter.

Osborne [11] used 9 equal, forward shooting intervals, and zero initial estimates. With b=9, s=0.5, his damped Newton method converged to the desired solution in 6 iterations. Using the same shooting intervals and initial estimates, DD03 failed due to overflow in the 20th. iteration. DD03 reacts badly to zero initial estimates if there is an identically zero solution to the differential equations. In this case, the first Newton iteration suggested a solution far from the desired solution. Since the system is a matrix Riccati equation, it has movable singularities, and while the Marquardt algorithm is searching for a possible solution, it falls upon a singularity.

It seems better, as the problem has a boundary condition $x_4(0)=1$, to provide that value as an initial estimate. Then, with the same shooting intervals, results were obtained for the six cases considered by Osborne [11], and to similar precision, in the number of iterations indicated:

| Case | 1 | 2 | 3 | 4 | 5 | 6 |
|---|---|---|---|---|---|---|
| b | 9 | 9 | 9 | 18 | 18 | 18 |
| s | 0.5 | 0.0 | -0.3 | 0.5 | 0.0 | -0.3 |
| Iterations | 6 | 6 | 6 | 10 | 16 | 24 |

which is at least as good as Osborne's results in every case. (In case 5, Osborne's method converged in 8 iterations, but to a solution different from that sought by Lance and Rogers [6]).

When left to choose its own shooting intervals, DD03 did not perform so well, taking 13,16,6,14,20 iterations in the first five cases repectively, and failing due to overflow in case 6. (In cases 4,5 convergence was to a different solution to that found previously). The situation was partly rectified by using the condition $x_4(b)=s$ in the initial estimate. The number of shooting intervals (all but one backwards) and of iterations, were then as follows:

| Case | 1 | 2 | 3 | 4 | 5 | 6 |
|---|---|---|---|---|---|---|
| Intervals | 6 | 5 | 6 | 12 | 11 | 11 |
| Iterations | 5 | 16 | 7 | 7 | 20 | >35 |

where in case 5, convergence was still to the wrong solution, while in case 6 convergence was not achieved. The results for b=18 were improved by using those for b=9, appropriately stretched, as initial estimates. Cases 4,5 then gave satisfactory results in 4,7 iterations respectively. Case 6 failed due to oferflow in the 6th. iteration. For this case, continuation through four intermediate values of b gave the wrong solution, and the correct solution was only obtained by continuing on the parameter s from s=0 through three intermediate values to s=-0.3.

## 9. A Plasma Stability Problem

The full statement of this problem is given by Lashmore-Davies [7] and by England [4]. It appears to be a linear eigenvalue problem on a semi-infinite interval, with jump conditions on the first derivative at two points, one of which constitutes a feedback stability control. After some simplification, and the use of an analytic solution between and outside the discontinuities, it takes the form:

$$\frac{d^2\phi}{dx^2} + \left[\frac{1}{x} + \frac{q^2}{1+q^2N(x)}\frac{dN}{dx}\right]\frac{d\phi}{dx} - \left[\frac{m^2}{x^2} + \frac{p^2}{2x(1+q^2N(x))}\frac{dN}{dx}\right]\phi=0; \quad (27)$$

with boundary conditions:

$$[\delta+|m|-p^2N(1)/2]\phi(1)+(1+q^2N(1))d\phi/dx\Big|_{x=1}=0 \ ,$$
$$|\phi(0)|<\infty \ ; \qquad (28)$$

where $p^2=-2m^2r^2/[(m+w)w]$, and $N(x)$ is a known differentiable function such that $N(x)=1+O(x^2)$ as $x\to 0$. Here, it is required to find the relation between the complex eigenvalue $w$ and the feedback parameter $\delta$, for given values of the small parameter $q^2$, the parameter $r^2$ (of order one), and the small integer $m$. This time the eigenfunction does not take complex values. Because of the singularity at $x=0$, an asymptotic analysis around that point must be performed, showing that, as $x\to 0$:

$$\phi(x)\sim Ax^{|m|}+Bx^{-|m|} \qquad (m\neq 0),$$
$$\phi(x)\sim A+B\ln(x) \qquad (m=0); \qquad (29)$$

for some constants $A,B$. Since $|\phi(0)|<\infty$, $B=0$, which implies that $\phi(0)=0$ if $m\neq 0$.

The problem can now be put in standard form by defining $y(x)=(1+q^2N(x))d\phi/dx$ giving:

$$\begin{aligned}\frac{dy}{dx} &= \left[\frac{m^2}{x^2}(1+q^2N(x))+\frac{p^2}{2x}\frac{dN}{dx}\right]\phi-\frac{1}{x}y \quad (x\neq 0),\\ &= 0 \qquad (x=0,\ m\neq 2)\ ,\\ &= 2A(1+q^2) \qquad (x=0,\ m=2)\ ,\\ d\phi/dx &= y/(1+q^2N(x));\end{aligned} \qquad (30)$$

with boundary conditions:

$$|m|\phi(0)+(|m|-1)y(0)=0,$$
$$[\delta+|m|-p^2N(1)/2]\phi(1)+y(1)=0, \qquad (31)$$

and some normalizing condition. The normalizing condition, or the constant $A$ if $|m|=2$, may be chosen to make the solution $\phi(x)$ equal to the Bessel function $J_m(px)$ for the special case of interest: $N(x)=1-x^2$, $q^2=0$. Thus $A=p^2/8$ ($|m|=2$) and:

$$(2-|m|)(\phi(1)-J_m(p))+(|m|-1)y(0)=0. \tag{32}$$

However, since w is complex, say $w=u+iv$, and p is real, the definition of $p^2$ gives rise to two additional conditions:

$$\begin{aligned} &p^2(mu+u^2-v^2)=-2m^2r^2, \\ &(m+2u)v=0; \end{aligned} \tag{33}$$

which must be added to the boundary conditions.

Using this formulation, the code DD03 has been successfully used to analyse the case $N(x)=1-x^2$, $m=1$, $q^2=0$, $r^2=1$, fixing values of p between 1 and 6, and determining $w,\delta$; and also fixing $\delta$ and determining p,w, although the choice of initial estimate for p is then important, as the solution is not unique. As before, the initial estimate of the eigenfunction should never be zero at the end points, and also, to avoid a singular Jacobian matrix (11), the initial estimates of $\phi(1), p, mu+u^2-v^2, m+2u, v$ should all be non-zero. Suitable estimates are those compatible with $\phi(x)=J_m(px)$ for some typical value of p (such as p=3) for which $J_m(p)\neq 0$, with $\delta=0$, $u=0$, $v=1$.

## 10. Conclusion

In conclusion, the results for these four parameter estimation problems show the versatility of the DD03 subroutine package, but also confirm that further development of the method is needed, along the lines mentioned, so as to make it more efficient and avoid some of the failures which can occur with the present implementation.

## 11. References

[1] U. Ascher, J. Christiansen and R. D. Russell, Math. Comp., 33 (1978), pp.659-679.

[2] A. Davey, Q.J. Mech. Appl. Math., 26 (1973), pp.401-411.

[3] R. England, Comp. J., 12 (1969), pp.166-170.

[4] R. England, A program for the solution of boundary value problems for systems of ordinary differential equations, UKAEA Culham Laboratory Report CLM-PDN3/73, Culham Laboratory, Abingdon, Oxford (1976).

[5] H. B. Keller, Numerical Methods for Two-Point Boundary Value Problems, Blaisdell (1968).

[6] G.N. Lance and M.H. Rogers, Proc. Roy. Soc., 266 (1962), pp. 109-121.

[7] C.N. Lashmore-Davies, J. Phys. A: Gen. Phys., 4 (1971), pp. 389-400.

[8] M. Lentini and H.B. Keller, SIAM J. Numer. Anal., 17 (1980), pp. 577-604.

[9] M. Lentini and V. Pereyra, SIAM J. Numer. Anal., 14 (1977), pp. 91-111.

[10] N.K. Nichols and R. England, J. Comp. Phys., 46 (1982), pp. 369-389.

[11] M.R. Osborne, J. Math. Anal. Appl., 27 (1969), pp.417-433.

[12] J.K. Reid, Fortran subroutines for the solution of sparse systems of non-linear equations, UKAEA AERE Harwell Report R7293, AERE Harwell (1972).

# UNRESTRICTED HARMONIC BALANCE III, APPLICATION TO RUNNING AND STANDING CHEMICAL WAVES

Friedrich Franz Seelig and Rainer Füllemann

Abstract. In former papers the method of "Unrestricted Harmonic Balance" (UHB) was developed for time-periodic phenomena in non-linear chemical reaction systems with mass action kinetics by Fourier expansion and solution of the non-linear algebraic system. This method is now applied to a chemical reaction-diffusion system characterized by the phenomenon of substrate inhibition that can form running and standing chemical waves depending on the choice of parameters. The system of parabolic partial differential equations with cyclic boundary conditions in time t and space s on a ring fiber of length L is transformed to a system of ODE by introduction of a constant profile wave variable $w = 2\pi(s-ct)/L$, where c is the wave velocity that is unknown in advance. The concentrations are found as a Fourier series of harmonics of w. Standing waves can be found as a limiting case of running waves or constructed directly by means of a special ansatz including only cosine-functions.

## 1. Introduction

In previous papers by one of the authors (F.F.S.) the general method of UHB was displayed [7] and an application to stiff ordinary differential equations in enzyme catalysis was given [8].

The UHB uses the representability of cyclic phenomena in time or time and space by Fourier series, e.g. for time-periodic state variables

$$x_i(t) = \bar{x}_i + \sum_{j=1}^{\infty} x_{cji}\cos(j\omega t) + \sum_{j=1}^{\infty} x_{sji}\sin(j\omega t) \qquad (1)$$

where the radian frequency $\omega$ is connected to the period T by

$$\omega = 2\pi/T. \qquad (2)$$

If all non-linearities consist of products of state variables $x_i$, these products are again of the form (1), but now the amplitudes of the cosine- and sine-functions are complicated expressions in terms of the factor variables. In practical cases the infinite series have to be truncated at some highest harmonic, say N, and the amplitudes $\{x_{cji}\}$, $\{x_{sji}\}$ together with the $\{\bar{x}_i\}$ and frequency $\omega$ have to be determined from a system of non-linear algebraic equations that can be solved iteratively, e.g. by the method of Powell [4]. In order to achieve low computation times the process is run in steps of increasing N until a sufficient accuracy is reached. The procedure is particularly competitive in cases of mass production, e.g. if the same problem is to be investigated ever and again for slightly different parameters, because then the result of the last problem is a good starting point for the next problem and only few iterations are needed. This method is competing with simulation techniques on one side and the shooting method [2] on the other.

## 2. Formulation of PDE for the Chemical System and Their Transformation of ODE for Running Waves with Constant Profile

At present a number of systems with the possibility of chemical waves is known ranging from pure mathematical models that are simple, but transferable to chemical reality with great difficulties, if at all, like Prigogine's [5] to very particular experimental arrangements that show huge kinetic complexities like the Belousov-Zhabotinsky-system as unraveled by Field, Körös and Noyes [1].

We chose one model system that is simple enough to allow a complete discussion of the domain of oscillations [6], but is reasonable from an experimental point of view and shows great flexibility in realization. For this model system Mimura and Murray [3] found already standing waves. The system with substrate inhibition is depicted in Fig. 1 and is described by the following set of mass-action equations:

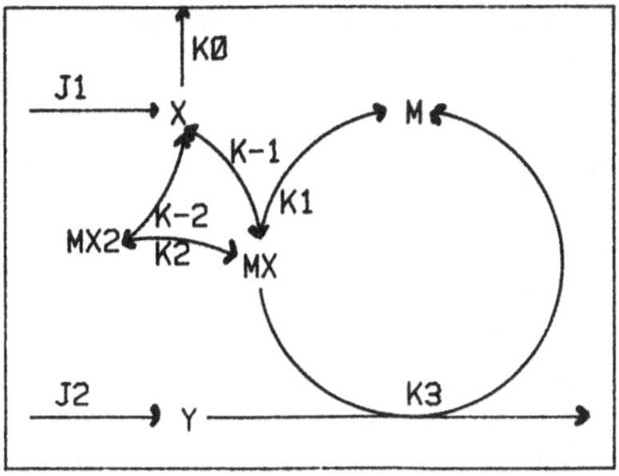


FIGURE 1:
Chemical Reaction Scheme

$$\frac{\partial X}{\partial t} = j_1 - k_0X - k_1X\cdot M + k_{-1}MX - k_2X\cdot MX + k_{-2}MX_2 + D_X\frac{\partial^2 X}{\partial s^2} \quad (3)$$

$$\frac{\partial Y}{\partial t} = j_2 - k_3Y\cdot MX + D_Y\frac{\partial^2 Y}{\partial s^2} \quad (4)$$

$$\frac{\partial M}{\partial t} = -k_1X\cdot M + k_{-1}MX + k_3Y\cdot MX + D_M\frac{\partial^2 M}{\partial s^2} \quad (5)$$

$$\frac{\partial MX}{\partial t} = k_1X\cdot M - k_{-1}MX - k_2X\cdot MX + k_{-2}MX_2 - k_3Y\cdot MX + D_{MX}\frac{\partial^2 MX}{\partial s^2} \quad (6)$$

$$\frac{\partial MX_2}{\partial t} = k_2X\cdot MX - k_{-2}MX_2 + D_{MX_2}\frac{\partial^2 MX_2}{\partial s^2} \quad (7)$$

Symbols X,Y... of the chemical species designate their concentrations, $j_1$ and $j_2$ are constant influxes to X and Y, respectively, $k_0, k_1, k_{-1}, k_2, k_{-2}, k_3$ are rate constants, $D_X \ldots D_{MX_2}$ are diffusion constants.

If the concentration of a catalyst or a compound of the catalyst with one of the substrates is much smaller than that of all substrates, the catalytic species are in a partial steady state relation to the substrates which expresses itself in the source terms $-k_1X\cdot M + k_3Y\cdot MX$ etc. being equal to zero. The sum of all source terms of (5) through (7) equals to zero, too, articulating the fact that there is another parameter, namely the total concentration of catalyst

$$M_{tot} = M + MX + MX_2, \quad (8)$$

which is constant. By means of the partial steady state approximation

M, MX and $MX_2$ are functions of X and Y and so does the net source term entering (3) and (4). So we get

$$\frac{\partial X}{\partial t} = j_1 - k_0 X - f(X,Y) + D_x \frac{\partial^2 X}{\partial s^2} \tag{9}$$

$$\frac{\partial Y}{\partial t} = j_2 - f(X,Y) + D_y \frac{\partial^2 Y}{\partial s^2} \tag{10}$$

$$f(X,Y) = k_3 Y \cdot MX = k_{-1} M_{tot} \frac{\dfrac{k_1 X}{k_{-1}} \cdot \dfrac{k_3 Y}{k_{-1}}}{1 + \dfrac{k_3 Y}{k_{-1}} + \dfrac{k_1 X}{k_{-1}} \left(1 + \dfrac{k_2 X}{k_{-2}}\right)} \tag{11}$$

The boundary conditions of running and standing waves on a ring fiber of length L are

$$\begin{aligned} X(t,s=0) &= X(t,s=L) \\ Y(t,0) &= Y(t,L) \end{aligned} \tag{12}$$

$$\begin{aligned} \frac{\partial X}{\partial s}(t,0) &= \frac{\partial X}{\partial s}(t,L) \\ \frac{\partial Y}{\partial s}(t,0) &= \frac{\partial Y}{\partial s}(t,L) \end{aligned} \tag{13}$$

There are now 2 state variables (X and Y), 2 independent variables (t and s) and 12 parameters, 4 of which ($j_1, j_2, M_{tot}$ and L) are adjustable in certain limits, whereas the rest ($k_0, k_1, k_{-1}, k_2, k_{-2}, k_3, D_x, D_y$) is fairly fixed under isothermal conditions. In case of running waves a 13th parameter, velocity c, emerges. The formulation can be simplified without loss of rigor by the introduction of dimensionless variables and parameters:

$$\begin{aligned} \tau &= k_0 t & \phi_1 &= k_1 j_1/(k_0 k_{-1}) \\ \sigma &= s/L & \phi_2 &= k_1 j_2/(k_0 k_{-1}) \\ \xi &= k_1 X/k_{-1} & \alpha &= k_3/k_1 \\ \eta &= k_3 Y/k_{-1} & \kappa &= k_{-1} k_2/(k_1 k_{-2}) \\ \delta_x &= D_x/(k_0 L^2) & \mu &= k_1 M_{tot}/k_0 \\ \delta_y &= D_y/(k_0 L^2) & \gamma &= c/(k_0 L) \end{aligned} \tag{14}$$

yielding (9), (10), (11) in the form

$$\frac{\partial\xi}{\partial\tau} = \phi_1 - \xi - g(\xi,\eta) + \delta_x \frac{\partial^2\xi}{\partial\sigma^2} \tag{15}$$

$$\frac{\partial\eta}{\partial\tau} = \alpha(\phi_2 - g(\xi,\eta)) + \delta_y \frac{\partial^2\eta}{\partial\sigma^2} \tag{16}$$

$$g(\xi,\eta) = \mu \frac{\xi\eta}{1+\eta+\xi(1+\kappa\xi)} \tag{17}$$

and reducing the set of parameters from 12 to 8.
If there are running waves with constant profile, $\xi(\tau,\sigma) = \xi(w)$ and $\eta(\tau,\sigma) = \eta(w)$ with

$$w = 2\pi(s-ct)/L = 2\pi(\sigma-\gamma\tau) \tag{18}$$

so that (15) and (16) transform to

$$-2\pi\gamma \frac{d\xi}{dw} = \phi_1 - \xi - g(\xi,\eta) + 4\pi^2\delta_x \frac{d^2\xi}{dw^2} \tag{19}$$

$$-2\pi\gamma \frac{d\eta}{dw} = \alpha(\phi_2-g(\xi,\eta)) + 4\pi^2\delta_y \frac{d^2\eta}{dw^2} \tag{20}$$

$$0 \le w \le 2\pi \tag{21}$$

with the boundary conditions

$$\xi(0) = \xi(2\pi), \quad \eta(0) = \eta(2\pi) \tag{22}$$

$$\frac{d\xi}{dw}(0) = \frac{d\xi}{dw}(2\pi), \quad \frac{d\eta}{dw}(0) = \frac{d\eta}{dw}(2\pi) \tag{23}$$

The velocity c or $\gamma$, respectively, is a characteristic parameter of the solution.

## 3. UHB-Treatment

We want to find solutions of (19) through (23) in the framework of the UHB-method. Therefore we start from

$$\xi(w) = \bar{\xi} + \sum_{j=1}^{N} \xi_{cj} \cos(jw) + \sum_{j=1}^{N} \xi_{sj} \sin(jw) \tag{24}$$

$$\eta(w) = \bar{\eta} + \sum_{j=1}^{N} \eta_{cj} \cos(jw) + \sum_{j=1}^{N} \eta_{sj} \sin(jw) \tag{25}$$

(19) and (20) can be formulated as a purely linear ODE

$$4\pi^2(\delta_x \alpha \frac{d^2\xi}{dw^2} - \delta_y \frac{d^2\eta}{dw^2}) + 2\pi\gamma(\alpha \frac{d\xi}{dw} - \frac{d\eta}{dw}) + \alpha(\phi_1 - \phi_2 - \xi) = 0 \tag{26}$$

and a non-linear one

$$(4\pi^2\delta_x \frac{d^2\xi}{dw^2} + 2\pi\gamma \frac{d\xi}{dw} + \phi_1 - \xi)(1 + \eta + \xi + \kappa\xi^2) - \mu\xi\eta = 0 \tag{27}$$

Insertion of (24) and (25) into (26) and equating all expressions containing cos(jw), sin(jw) and all constant terms separately yields

$$\bar{\xi} = \phi_1 - \phi_2 \tag{28}$$

which is exactly $\xi_{cr}$, the value of $\xi$ at the single critical point, and

$$\eta_{cj} = \frac{\alpha}{2\pi j} \cdot \frac{2\pi j\delta_y a_j + \gamma b_j}{4\pi^2 j^2 \delta_y^2 + \gamma^2} \tag{29}$$

$$\eta_{sj} = \frac{\alpha}{2\pi j} \cdot \frac{2\pi j\delta_y b_j - \gamma a_j}{4\pi^2 j^2 \delta_y + \gamma^2} \tag{30}$$

with

$$a_j = (1+4\pi^2 j^2 \delta_x)\xi_{cj} - 2\pi j\gamma\xi_{sj} \tag{31}$$

$$b_j = (1+4\pi^2 j^2 \delta_x)\xi_{sj} + 2\pi j\gamma\xi_{cj} \tag{32}$$

Thus $\{\eta_{cj}\}$ and $\{\eta_{sj}\}$ can be expressed by $\{\xi_{cj}\}$ and $\{\xi_{sj}\}$ for each $j=1...N$, and $\bar{\xi}$ by two parameters.

Utilizing the self-reproducing property of cosines and sines under multiplication investigated in detail in [7] which allows for the formulation of pseudo-linear functions for products of state variables in the form (1), we introduce intermediate variables in this order

$$u = 4\pi^2\delta_x \frac{d^2\xi}{dw^2} + 2\pi\gamma \frac{d\xi}{dw} + \phi_1 - \xi \tag{33}$$

with

$$\bar{u} = \phi_2, \; u_{cj} = -a_j, \; u_{sj} = -b_j \tag{34}$$

$$v = \kappa\xi^2 \tag{35}$$

$$r = 1 + \eta + \xi + v \tag{36}$$

with

$$\bar{r} = 1 + \bar{\eta} + \bar{\xi} + \bar{v}, \; r_{cj} = \eta_{cj} + \xi_{cj} + v_{vj}, \; r_{sj} = \eta_{sj} + \xi_{sj} + v_{sj}$$

(37

$$s = ur$$

and

$$p = \mu\xi\eta \tag{38}$$

where the constant terms and cosine- and sine-amplitudes of the harmonics of the pseudo-linear functions v,s,p are complicated non-linear functions of the corresponding quantities of their factors.

So the implicit non-linear equations to be solved iteratively by the Powell algorithm are

$$\bar{s} - \bar{p} = 0 \tag{39}$$

$$\left.\begin{array}{l} s_{cj} - p_{cj} = 0 \\ s_{sj} - p_{sj} = 0 \end{array}\right\} \; j=1...N \tag{40}$$

This is a system of 2N+1 highly nested non-linear equations of 2N+1 unknowns $\bar{\eta}, \gamma, \{\xi_{cj}\} \setminus \xi_{c1}, \{\xi_{sj}\}$ (j=1...N). $\xi_{c1}$ is set = 0 to fix the zero of the independent wave variable w.

## 4. Results

In order to demonstrate the strength of the UHB method, the existing FORTRAN programs for bigger computers were rewritten in BASIC to match the limited resources of a personal computer. The particular formulas of the treated problem needed a particular version of the subroutine CALFUN. The total working storage of 32 K bytes of the used Hewlett-Packard HP 85 allowed a highest N of 12 (achieved in steps over n=6) for the treatment of this case of a running wave. But a refinement by an especially written, less storage consuming pure Newton-Raphson routine made it possible to double N to 24. The results for the running wave are shown in Fig. 2. Since the problem is extremely nonlinear, the jumps are very hard so that even for N=24 the drop of amplitudes from j=1 to j=24 is not big enough which results in a slightly wavy $\xi(\tau)$ at low values of $\xi$.

In the case of the standing wave, as shown in Fig. 3, a pure cosine expansion is possible which reduces storage needs and computation time roughly by a factor of 4, but needs a special routine for the formation of the product of two Fourier series. Here N=24 could be realized (in steps over N=3,6 and 12) and the result is completely smooth. Since in case of a pure cosine expansion not only $\frac{d\xi}{dw}(0) = \frac{d\xi}{dw}(2\pi)$, but also both are = 0 (the same holds for $\frac{d\eta}{d\tau}$), we have a zero flux boundary condition which characterizes a reaction-diffusion system in a rod, too, so that the results are much more general than for the case of a ring fiber.

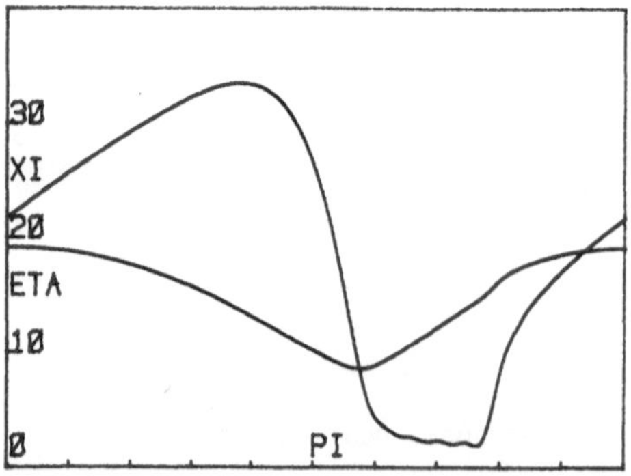


Figure 2: $\xi$ and $\eta$ vs. $\tau$ for a running wave. Parameters: $\phi_1 = 46.5$, $\phi_2 = 26.5$, $\mu = 33.125$, $\alpha = 0.1$, $\kappa = 1$, $\delta_x = \delta_y = 0.0001$; resulting $\gamma = 0.048653$.

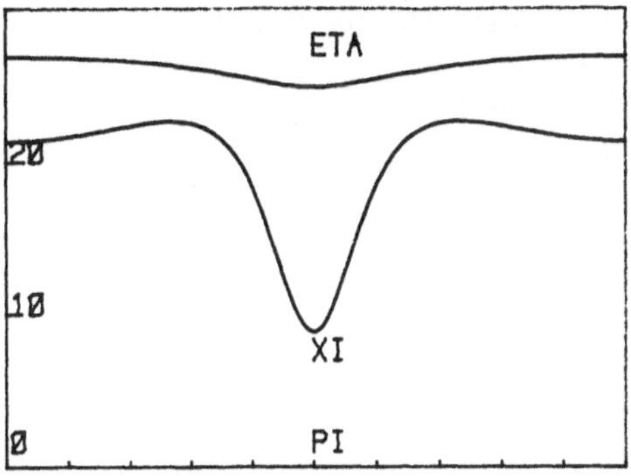


Figure 3: $\xi$ and $\eta$ vs. $\tau$ for a standing wave. Parameters: $\phi_1 = 44$, $\phi_2 = 24$, $\mu = 20$, $\alpha = 0.1$, $\eta = 1$, $\delta_x = 0.001$, $\delta_y = 0.01$

## 5. Acknowledgement

This work was generously supported by the German "Verband der Chemischen Industrie", which is gratefully acknowledged.

## 6. References

[1] Field, R.J.,E. Körös, and R.M. Noyes, J.Amer.Chem.Soc. 94, 8649 (1972)

[2] Gladwell, I., In: Modern numerical methods for ordinary differential equations (Hall, P., Watt, J.M., eds.) p. 216, Oxford: Clarendon Press 1976

[3] Mimura, M. and J.D. Murray, J.theor. Biol. 75, 249 (1978); Z.Naturforsch. 33c, 580 (1978)

[4] Powell, M.J.D., In: Numerical Methods for Nonlinear Algebraic Equations, Ed. P. Rabinowitz, Gordon & Breach Science Publ., London 1970

[5] Prigogine, I. and G. Nicolis, J.Chem.Phys. 46, 3542 (1967)

[6] Seelig, F.F., Z.Naturforsch. 31a, 731 (1976)

[7] Seelig, F.F., Z.Naturforsch. 35a, 1054 (1980)

[8] Seelig, F.F., J.Math.Biol. 12, 187 (1981)

# INVERSE EIGENVALUE PROBLEMS FOR THE MANTLE

Ole H. Hald

We represent the earth as a sphere with radius $R$ and assume that the material is perfectly elastic and isotropic. Thus we ignore ellipticity, rotation, damping, lateral inhomogeneities and anisotropy. The eigenvalue problem for the torsional modes is

$$-(r^4\beta^2\rho u')' + (\ell + 2)(\ell - 1)r^2\beta^2\rho u = \omega^2 r^4 \rho u \ ,$$

$$u'(R_{core}) = u'(R) = 0 \ .$$

Here $r$ is the distance to the center, $\beta$ is the velocity of the shear waves, $\rho$ is the density, $\omega/2\pi$ the eigenfrequencies and $u/r$ the displacement of a particle. The constant $\ell = 1, 2, 3, \ldots$ . For each $\ell$ the eigenfrequencies satisfy

$$\omega_0^{(\ell)} < \omega_1^{(\ell)} < \omega_2^{(\ell)} < \ldots$$

The inverse problem amounts to determining $\rho$ from the overtones $\omega_j^{(\ell)}$. This problem can have more than one solution (multiply $\rho$ by 7). Moreover, if $\beta = constant \cdot r$ then two different density distributions can have exactly the same torsional overtones, see [2]. Assume the velocity is known in the mantle and in the crust. If the density is given in the lower mantle then the density is uniquely determined in the upper mantle and in the crust by the eigenfrequencies corresponding to a fixed value of $\ell$, see [1]. Here the lower mantle is defined such that the travel time of the vertical wave through the lower mantle is equal to the travel time through the upper mantle and the crust.

*Proof* for $\ell = 1$. Set $y(x) = r^2\sqrt{\rho\beta}u(r)$, $x = K^{-1}\int_r^R \beta^{-1}\,dr$ and $K = \pi^{-1}\int_{R_{core}}^R \beta^{-1}\,dr$. Then

$$\begin{aligned} -y'' + q(x)y &= \lambda y \ , \\ y'(0) - hy(0) &= y'(\pi) + Hy(\pi) = 0 \ , \end{aligned} \tag{1}$$

where $q = f''/f$, $f = r^2\sqrt{\rho\beta}$ and $\lambda = K^2\omega^2$. Hochstadt and Lieberman [3] have shown that if $q$ is known in $(\pi/2, \pi)$ and $H$ is given then $q$ and $h$ are uniquely determined by the eigenvalues $\lambda_0, \lambda_1, \ldots$ . Thus $f$ and hence $\rho$ are uniquely determined by the eigenfrequencies $\omega_j^{(\ell)}$.

To solve the direct problem we consider the Rayleigh quotient

$$\lambda = \frac{\int_0^\pi (y'^2 + qy^2)dx + hy^2(0) + Hy^2(\pi)}{\int_0^\pi y^2\,dx} = R[y] \ .$$

Let $y = \sum_0^n w_i u_i$ where $u_i(x)$ satisfy equation (1) with $q = 0$. We denote the corresponding eigenvalues by $\mu_i$ and set $u_i(0) = 1$. Then $R[y] = w^T A w / w^T w$ where

$$A = \begin{pmatrix} \mu_0 & & \\ & \mu_1 & \\ & & \cdot \\ & & & \cdot \end{pmatrix} + \left( \int_0^\pi q(x) \frac{u_i}{\|u_i\|} \cdot \frac{u_j}{\|u_j\|}\,dx \right)$$

and $w^T = (w_0, \ldots, w_n)$. The eigenvalues of $A$ approximate the eigenvalues of the differential equation. If $q$ is small then

$$\lambda_j \sim \lambda_j(A) \sim \mu_j + q_0 + \frac{1}{\|u_j\|^2}\int_0^{\pi/2} (q - q_0)(u_j^2 - 1/2)\,dx$$

$$+ \frac{1}{\|u_j\|^2}\int_{\pi/2}^{\pi} (q - q_0)(u_j^2 - 1/2)\,dx \ .$$

Observe that $\mu_j = j^2 + (2/\pi)(h + H) + O(j^{-2})$ and that $(2/\pi)(h + H) + q_0 = C$ is the leading term in the asumptotic expansion of the eigenvalues of equation (1). We can now solve the inverse problem.

Let $h$ and $H$ be fixed and set $q_0 = C - (2/\pi)(h + H)$. Then

$$\frac{1}{\|u_j\|^2} \int_0^{\pi/2} (q - q_0)(u_j^2 - 1/2)\, dx$$

$$\sim \lambda_j - \mu_j - q_0 - \frac{1}{\|u_j\|^2} \int_{\pi/2}^{\pi} (q - q_0)(u_j^2 - 1/2)\, dx \qquad (= \gamma_j) \quad .$$

The right hand side can be computed. Hence

$$q - q_0 \sim \sum \gamma_j \cdot \textit{biorthogonal functions to } u_j^2 - 1/2$$

in $(0, \pi/2)$ and we let *new* $q_0 = \pi^{-1} \int_0^{\pi} q\, dx$ and *new* $h = (\pi/2) \times (C -$ *new* $q_0) - H$. If the off-diagonal terms in $A$ are included then $\gamma_0, \gamma_1, \ldots$ satisfy a set of nonlinear equations which we solve by iteration. Finally we solve $f'' = qf$ and reconstruct $\rho$.

*Example.* If the density and the velocity in Bullen's model $A$ are smoothed over $100$ km and the asymptotic constant $C$ is given then we can reconstruct the density in the upper mantle with an accuracy of $.05\ g\ cm^{-3}$ from four eigenvalues and $.01\ g\ cm^{-3}$ from eight eigenvalues. The density changes by $.03\ g\ cm^{-3}$ if we perturb one of the eigenvalues by $.1$ per cent.

*Remarks.* (1) The density in the lower mantle can be determined by Adams-Williamson's equation. Thus the density at the core-mantle boundary and the mass of the core must be given. However, these values can be replaced by the density at the Mohorovičić discontinuity and the total mass below this discontinuity. (2) The velocity of the shear wave in the upper mantle and in the crust can be uniquely determined by an additional spectrum (Catherine Willis). (3) The algorithm works because $\mu_i$ and $u_i$ are the leading terms in the asymptotic expansion of the eigenvalues and the eigenfunctions of equation (1). (4) The obvious finite element or finite difference approach leads to tridiagonal matrices whose eigenvalues do not have the correct asymptotic behavior. One can modify the eigenvalues (John Paine) and solve the inverse problem (C. de Boor and G.H. Golub or F.W. Biegler-König), but it is difficult to prove that the method converges.

*Open problems.* (1) Make a better algorithm (easy) and prove that it converges (difficult). (2) Interpret your results by the Backus-Gilbert technique. (3) Give a global existence proof for the inverse problem for equation (1). (4) Include discontinuities (partially done). (5) Include ellipticity, rotation, damping, lateral inhomogeneities and anisotropy. (6) The constant $C$ cannot be measured. What should be done? (7) Prove that missing data can be replaced by data from nearby spectra. (8) Reconstruct $\rho$ from the fundamentals alone -- if it is possible (*cf.* inverse scattering at fixed energy). (9) Consider the inverse problem as a least squares problem and give a statistical discussion of the solution.

*Warnings.* (1) No geophysicist has used the approach presented here. (2) There is only one Earth.

*Acknowledgment.* The author is an Alfred P. Sloan Research Fellow.

*References*

[1] Hald, O.H., Inverse eigenvalue problems for the mantle, Geophys. J. R. astr. Soc., 62 (1980), 41-48.

[2] Hald, O.H., Inverse eigenvalue problems for the mantle, II., Geophys. J. R. astr. Soc., to appear.

[3] Hochstadt, H. and Lieberman, B., An inverse Sturm-Liouville problem with mixed given data, SIAM J. appl. Math., 34 (1978), 676-680.

# INVERSE PROBLEM OF QUANTAL POTENTIAL SCATTERING AT FIXED ENERGY

H. Fiedeldey, R. Lipperheide, and S. Sofianos

## 1. Introduction

The inverse problem of quantal scattering is a problem of long standing. However, despite an extensive literature (cf. the reviews [1, 5]) it seems that only recently has interest been directed toward practical applications, at least as far as the fixed-energy case is concerned [6, 8, 10, 12, 14]. The present work is a contribution to these efforts.

The scattering problem we are concerned with here is defined by the partial wave Schrödinger equation

$$\frac{d^2}{dr^2}\phi_\ell(r) - \left(\frac{\ell(\ell+1)}{r^2} + V(r) - E\right)\phi_\ell(r) = 0 \qquad (1)$$

for the regular partial wave function of angular momentum $\ell$ with boundary condition $\phi_\ell(r)\to r^\ell$ for $r\to 0$. The (in general complex) potential $V(r)$ determines the scattering function $S_\ell(E)$ as a coefficient in the asymptotic wave function

$$\phi_\ell(r) \underset{r\to\infty}{\sim} e^{-i(kr - \ell\pi/2)} - S_\ell(E)\, e^{i(kr - \ell\pi/2)}. \qquad (2)$$

The inverse problem is to determine the potential $V(r)$ from the scattering function $S_\ell(E)$. We are here looking at the case where $E$ is fixed and the scattering function is regarded as a function of $\ell$.

The general solution of this inversion problem at fixed energy may be based on the Gel'fand-Levitan-type integral equations [1, 4] or matrix versions of it [15, 16, 18]. In the present work we use a more special approach: the given scattering function is interpolated by a simple parametrized expression for which the corresponding potential is known analytically. In the fixed-$\ell$ case a simple class of scattering functions, which are rational in $k = \sqrt{E}$, has long been known to be associated with a class of potentials of simple analytic form: the Bargmann potentials [3]. For the case of fixed energy, a similar but somewhat more complicated class of scattering functions as functions of $\lambda = \ell+1/2$ has been established recently [11] which again corresponds to simple potentials. The inversion method is then simply to fit a scattering function of this class to a given scattering function to be "inverted", by fixing a number of parameters. These parameters then determine the corresponding potential as well. In the next section we describe the main features of the inversion method, while in sect. 3 some numerical applications are reported. It is felt on the basis of these results that, although in using an "interpolation scattering function" one foregoes a general solution of the scattering problem, the interpolation is flexible enough to produce and hence, to invert, any reasonable scattering function.

## 2. The Inversion Method

### 2.1 Inversion for Rational Scattering Functions

In the fixed-$\ell$ case, the Bargmann potentials correspond to a rational scattering function, for which the integral kernel of the general Gel'fand-Levitan theory becomes degenerate, making an algebraic solution possible. In this case the form of the potential, the wave function, and the scattering function are readily established directly [20]. Translating this procedure (with modifications) to the fixed-E case (for a different approach, cf. refs. [12]), we arrive at the following recipe (setting $E = k^2 = 1$):

Consider the Jost solutions $\psi_\lambda^{(0)\pm}(r)$ to a "reference potential" $V_0(r)$ with

$$\psi_\lambda^{(0)\pm}(r) \to e^{\mp ir} \qquad \text{for} \qquad r \to \infty, \tag{3}$$

and define the potential

$$V^{(1)}(r) = \frac{2i(\alpha_1^2-\beta_1^2)}{r}\frac{d}{dr}\left[\frac{1}{r}\,\frac{1}{L_{\alpha_1}^{(0)+}(r) - L_{\beta_1}^{(0)-}(r)}\right], \tag{4}$$

where

$$L_\lambda^{(0)\pm}(r) = \pm i \,\frac{d\psi_\lambda^{(0)\pm}(r)}{dr}\Big/\psi_\lambda^{(0)\pm}(r) \tag{5}$$

is the logarithmic derivative of the corresponding Jost solution, and $\alpha_1$, $\beta_1$ are arbitrary complex values of the angular momentum $\lambda$. Then it turns out that $V_1(r) = V_0(r) + V^{(1)}(r)$ is a potential which yields the scattering function

$$S^{(1)}(\lambda) = S^{(0)}(\lambda)\,\frac{\lambda^2-\beta_1^2}{\lambda^2-\alpha_1^2}\ . \tag{6}$$

The potential $V^{(1)}(r)$ is thus a "Bargmann" potential for the fixed-E case, leading to a rational scattering function in angular momentum space. If $V_0 = 0$, the $\psi_\lambda^{(0)\pm}(r)$ are simply Hankel functions of complex order. In general, the quantities $L_\lambda^{(0)\pm}(r)$ are most convenie :ly computed numerically by solving the Riccati equation

$$\mp i\frac{d}{dr}L_\lambda^{(0)\pm} - \left(L_\lambda^{(0)\pm}\right)^2 + 1 - V_0 - \frac{\lambda^2-1/4}{r^2} = 0 \tag{7}$$

with boundary condition $L_\lambda^{(0)\pm} \to 1$ for $r \to \infty$.

Introducing a second pair $\alpha_2$, $\beta_2$ and regarding the previous $V_1$ a..d $S^{(1)}$ as reference potential and scattering function in a second step, one finds new functions $V_2$ and $S^{(2)}$ of analogous form; repeating this process N times one arrives at a potential

$$V_N(r) = V_0(r) + V^{(1)}(r) + \ldots + V^{(N)}(r)\ , \tag{8}$$

which is associated with the scattering function

$$S_\ell = S^{(N)}(\lambda) = S^{(0)}(\lambda) \prod_{n=1}^{N} \frac{\lambda^2 - \beta_n^2}{\lambda^2 - \alpha_n^2} . \tag{9}$$

The inversion procedure therefore consists in the following: determine the parameters $\{\alpha_n, \beta_n\}$ by fitting the rational scattering function (9) (including a known background, e.g. Coulomb scattering function $S^{(0)}(\lambda)$) to a given scattering function for real (in practice: half-integer) values of $\lambda$, and then calculate the potential $V_N(r)$ by applying eqs. (7), (4), and (8) in N iterative steps. Since the determination of the parameters $\{\alpha_n, \beta_n\}$ merely involves the simple task of interpolating a given function by the ratio of two N-th order polynomials, the method is simple and straightforward. However, there are pitfalls. The function (9) represents an analytic continuation of the scattering function from the real axis into the complex $\lambda$-plane. Now general theorems [2] require e.g. that Im $\alpha^2>0$ for a unitary scattering function with $\beta=\alpha^*$, in order that the corresponding potential be regular. A fit of a given unitary scattering function by expression (9) will usually yield some parameters $\alpha_n$ with Im $\alpha_n^2<0$, in order to account for "repulsive" features in the scattering. Nonetheless, the corresponding potential (8) can be computed - but it will be singular. An acceptable potential must be regular, however. Therefore we abandon the "Bargmann-type" potentials associated with the rational scattering function (9) and look for another class of potentials yielding a different type of scattering function which have no such "wrong Regge poles" $\alpha_n$.

## 2.2 Inversion for a Class of Nonrational Scattering Functions

The preceding procedure is readily modified by considering, instead of the Jost solutions (3), the regular solutions $\phi_\lambda^{(0)}(r)$ with

$$\phi_\lambda^{(0)}(r) \to r^{\lambda+1/2} \qquad \text{for} \quad r \to 0 \tag{10}$$

and their logarithmic derivatives

$$L_\lambda^{(0)}(r) = \frac{d}{dr}\phi_\lambda^{(0)}(r) / \phi_\lambda^{(0)}(r) \tag{11}$$

satisfying

$$\frac{d}{dr} L_\lambda^{(0)} + \left(L_\lambda^{(0)}\right)^2 + 1 - V_0 - \frac{\lambda^2 - 1/4}{r^2} = 0 \tag{12}$$

with the boundary condition

$$\begin{aligned} L_\lambda^{(0)}(r) &\to \frac{\lambda + 1/2}{r} \text{ for } r \to 0;\ \text{Re}\,\lambda > 0; \\ L_\lambda^{(0)}(r) &\to -\cot\left[r - (\lambda - 1/2)\pi/2 + \ln S^{(0)}(\lambda)\right] \\ &\qquad \text{for } r \to \infty;\ \text{Re}\,\lambda < 0. \end{aligned} \tag{13}$$

It can be shown that with the potential

$$V^{(1)}(r) = \frac{2(\alpha_1^2 - \beta_1^2)}{r} \frac{d}{dr}\left[\frac{1}{r} \frac{1}{L_{\alpha_1}^{(0)}(r) - L_{\beta_1}^{(0)}(r)}\right], \tag{14}$$

the potential $V_1(r) = V_0(r) + V^{(1)}(r)$ produces the scattering function

$$S^{(1)}(\lambda) = S^{(0)}(\lambda) \frac{\dfrac{(\sigma_{\beta_1}^{(0)} - \sigma_{\alpha_1}^{(0)})}{(\beta_1^2 - \alpha_1^2)} - \dfrac{(\sigma_\lambda^{(0)} - \sigma_{\alpha_1}^{(0)})}{(\lambda^2 - \alpha_1^2)} \cdot \dfrac{\sigma_{\beta_1}^{(0)}}{\sigma_\lambda^{(0)}}}{\dfrac{(\sigma_{\beta_1}^{(0)} - \sigma_{\alpha_1}^{(0)})}{(\beta_1^2 - \alpha_1^2)} - \dfrac{(\sigma_\lambda^{(0)} - \sigma_{\alpha_1}^{(0)})}{(\lambda^2 - \alpha_1^2)}} \tag{15}$$

where

$$\sigma_\lambda^{(0)} = e^{-i\pi(\lambda - 1/2)} S^{(0)}(\lambda). \tag{16}$$

Iteration again leads to a potential $V_N(r)$ as in eq. (8), with the $V_n(r)$ calculated by an iterative application of eqs. (12) to (16) for successive pairs $\{\alpha_n, \beta_n\}$, n=1...N. The associated scattering function $S^{(N)}(\lambda)$, obtained by iteration of eqs. (15) and (16), can also be written in the concise form

$$S^{(N)}(\lambda) = S^{(0)}(\lambda)\,\frac{\det\left\{\frac{(\sigma^{(0)}_{\beta_n}-\sigma^{(0)}_{\alpha_m})}{(\beta_n^2-\alpha_m^2)} - \frac{(\sigma^{(0)}_{\lambda}-\sigma^{(0)}_{\alpha_m})}{(\lambda^2-\alpha_m^2)}\cdot\frac{\sigma^{(0)}_{\beta_n}}{\sigma^{(0)}_{\lambda}}\right\}}{\det\left\{\frac{(\sigma^{(0)}_{\beta_n}-\sigma^{(0)}_{\alpha_m})}{(\beta_n^2-\alpha_m^2)} - \frac{(\sigma^{(0)}_{\lambda}-\sigma^{(0)}_{\alpha_m})}{(\lambda^2-\alpha_m^2)}\right\}} \cdot \tag{17}$$

The corresponding iterated potential $V_N(r)$ can be represented in "closed" form as well:

$$V_N(r) = V_0(r) + \frac{2}{r}\frac{d}{dr}\left[\frac{1}{r}\sum_{m,n=1}^{N} y^{-1}_{mn}(r)\right], \tag{18}$$

where $y^{-1}$ is the inverse of the NxN matrix

$$y_{mn}(r) = \frac{L^{(0)}_{\alpha_m}(r) - L^{(0)}_{\beta_n}(r)}{\alpha_m^2 - \beta_n^2}. \tag{19}$$

The scattering function (17) is no longer rational in $\lambda^2$; on the other hand, it has been proven [11] that for $V_0=0$ and N=1, $\alpha_1=\beta_1^*$, $S^{(1)}(\lambda)$ has no "wrong Regge poles" for <u>any</u> choice of $\alpha_1$, and the corresponding potential is always regular. Although a similar proof for more general cases is still outstanding, the parametrization (17) of the scattering function is clearly preferable to that of eq. (9) for the purpose of inversion. Numerical studies support this.

Again the inversion procedure consists in finding a set of parameters $\{\alpha_n,\beta_n\}$ by fitting the expression (17) to a given scattering function, and then calculating the potential $V_N(r)$ from a repeated application of eqs. (12) to (16), replacing the superfixes (0),(1) by (1),(2) as the next pair $\alpha_2,\beta_2$ is included, until in the N-th step $V_N(r)$ and $S^{(N)}(\lambda)$ are reached. This iteration procedure is to be used for numerical reasons, since the direct application of the "closed" formulas (17) to (19) would involve the calculation of numerically unstable determinants.

Fitting the nonrational expression (17) to a given function is, of course, not so easy as for the rational function (9). However, it turns out that in all but pathological cases one can arrange the fitting parameters such that all $|\sigma\alpha_m| \ll 1$ and all $|\sigma\beta_n| \gg 1$; then, for real $\lambda$, expression (17) approximately reduces again to the rational form

$$S^{(N)}(\lambda) \approx S^{(0)}(\lambda) \prod_{n=1}^{N} \frac{\lambda^2 - \beta_n^2}{\lambda^2 - \alpha_n^2} \,. \tag{20}$$

Thus we fit expression (20) to a given scattering function $S^{(N)}(\lambda)/S^{(0)}(\lambda)$ and verify that for the parameter values $\{\alpha_n, \beta_n\}$ found, the values of expression (17) do not differ appreciably from those of expression (20) (and hence of the given scattering function to be interpolated) for real $\lambda$; if needed, a least-square procedure can be added on to improve the fit with eq. (17). Then we use these parameters $\{\alpha_n, \beta_n\}$ in eq. (17) to continue $S^{(N)}(\lambda)$ into the complex $\lambda$-plane, and to compute the corresponding potential $V_N(r)$ by the indicated iterative process.

While the original Bargmann potentials with their associated scattering functions correspond to a degenerate kernel in the general Gel'fand-Levitan inversion theory, it is not clear how the present class of potentials and nonrational scattering functions is related to the Gel'fand-Levitan procedure (or more precisely, to its fixed-E version [4]). There is, however, a connection with the matrix method of Newton and Sabatier [15, 16, 18] (for a finite number of fitted phase shifts). In fact the form of expression (17) is the same as that of the scattering function used by Newton and Sabatier, but it can be shown (cf. ref. [7]) that no choice of parameters $\{\alpha_n, \beta_n\}$ exists which makes the two scattering functions identical. Correspondingly, the classes of potentials in the two inversion schemes are not the same. From a practical point of view, the present scheme differs from that of refs. [15, 16, 18] in that the number of parameters $\{\alpha_n, \beta_n\}$ needed to describe a given scattering function is determined by the number of salient features (oscillations, inflection points, etc.) of the latter, which may be very few if the function is smooth. In refs. [15, 16, 18], the number of parameters, and hence the dimensionality of the matrices involved in solving the inversion problem, is determined by the number of nonvanishing pase shifts with integer $\ell = \lambda - 1/2$.

## 3. Applications

The inversion scheme of sect. 2.2 has been applied to several cases. Before discussing these, we consider some schematic examples from the class of scattering functions and potentials introduced in sect. 2.2, in order to see to what extent it can represent realistic cases. In Figure 1 the phase shift $\delta(\lambda) = (2i)^{-1} \ln S(\lambda)$ and the deflection function $\theta(\lambda) = 2d\delta(\lambda)/d\lambda$ are shown for the unitary scattering function (15) with $S^{(0)}(\lambda) = 1$ and $\alpha_1 = \beta_1{}^* = -6-i2$. The corresponding potential (14) (Fig. 2) is real and attractive. In Figure 3 the phase shift and the potential for the analogous repulsive case with $\alpha_1 = \beta_1{}^* = 35+i30$ is presented. A last, nonunitary case is shown in Figure 4, where the pair of parameters $\alpha_1 = -35-i30$, $\beta_1 = -30+i20$ describes attractive, absorptive scattering. These examples indicate that even with a single pair of parameters, the discussed class of scattering functions and potentials have a quite realistic shape.

Examples for the true reconstruction of a potential from its scattering function are given in the next figures. In Figure 5 the complex potential for the scattering of neutrons on $^{16}O$ at 52.5 MeV is shown (solid curves). Its scattering function has been fitted by the expression (17) at the (physical) half-integer values $\lambda = \ell+1/2$ ($\ell=0,1,2,\ldots$) using three pairs $\{\alpha_n,\beta_n\}$ with goodness of fit $\chi^2(N=3) = 0.47\times10^{-3}$ and six pairs with $\chi^2(N=6) = 0.16\times10^{-3}$. The corresponding potentials $V_N$, represented by the dotted and dashed curves, respectively, are also shown in Figure 5. It is seen that the reconstruction of the input potential is almost perfect in this case.

Another example (Fig. 6) is the reconstruction of the Gaussian potential

$$V(r) = -14\, e^{-(r/3.5)^2} , \qquad (21)$$

for which a comparison with an inversion by the Newton-Sabatier method [6] is possible. The solid curve is the input potential, the dash-dotted curve represents the reconstructed potential of ref. [6] at the energy E = 10 MeV, and the dashed curve gives our reconstructed potential using three pairs of parameters for the fit of the scattering function of the potential (21). Presumably our reconstructed potential could be improved further by making a better fit to the scattering function using more parameters.

Not all inversions attempted so far with the present method have yielded satisfactory results. Difficulties arise in particular when the fitting procedure yields parameter values $\{\alpha, \beta\}$ with relatively small imaginary parts. This question is under investigation. Another aspect is the inclusion of a reference potential $V_0(r)$. Since the scattering function (17) involves the reference scattering function $S^{(0)}(\lambda)$ at real and complex values of $\lambda$, it is easiest to consider only functions $S^{(0)}(\lambda)$ which are given in analytic form (not as a numerical solution to a given reference potential $V_0(r)$). This is the case for the point-Coulomb scattering function, where of course the corresponding potential is also known analytically. Other smooth forms of the reference scattering function $S^{(0)}(\lambda)$ could also be used, and the corresponding potential $V_0(r)$ be obtained by the simple classical inversion method [13]. This aspect is also being studied with the aim of solving the inverse problem for charged-particle scattering.

In closing we mention that two inversions of practical interest in nuclear heavy-ion scattering have actually been carried out earlier using the rational (Bargmann) class of scattering functions. Here it is easy to include an arbitrary reference potential. In one instance [10] the reference potential was complex and the rational scattering function contained only one pole $\alpha_1$, which was a "correct Regge pole" with Im $\alpha_1^2 > 0$. In the other example [8], with the point Coulomb field as reference potential, the rational scattering function included several "wrong Regge poles"; these had to be converted into correct poles by an approximate classical procedure in order to obtain a non-singular potential.

References

[1] Z.S. Agranovich and V.A. Marchenko, The Inverse Problem of Scattering Theory (Gordon and Beach, New York, 1963).

[2] V. de Alfaro and T. Regge, Potential Scattering (North-Holland, Amsterdam, 1965).

[3] V. Bargmann, Revs. Modern Phys. 21, 488 (1949).

[4] G. Burdet, M. Giffon, and E. Predazzi, Nuovo cimento 26, 1337 (1965).

[5] K. Chadan and P.C. Sabatier, Inverse Problems in Quantum Scattering Theory (Springer, New York, Berlin, Heidelberg, 1977).

[6] C. Coudray, Lett. Nuovo cimento 19, 319 (1977)

[7] J.R. Cox and K.W. Thompson, J. Math. Phys. 11, 805 (1970).

[8] P. Fröbrich, R. Lipperheide, and H. Fiedeldey, Phys. Rev. Lett. 43, 1147 (1979.

[9] R. Lipperheide and H. Fiedeldey, Z. Physik A 286, 45 (1978).

[10] R. Lipperheide, H. Fiedeldey, H. Haberzettl, and K. Naidoo, Phys. Lett. 82B, 39 (1969).

[11] R. Lipperheide and H. Fiedeldey, Z. Physik A 301, 81 (1981).

[12] V.V. Malyarov, I.V. Poplavskii, and M.N. Popushoi, Sov. J. Nucl. Phys. 22 , 445 (1976) and 25, 38 (1977); Sov. Phys. JETP 41, 210 (1975).

[13] W.H. Miller, J. Chem. Phys. 51, 3631 (1969).

[14] M. Münchow and W. Scheid, Phys. Rev. Lett. 44, 1299 (1980).

[15] R.G. Newton, J. Math. Phys. 3, 75 (1962).

[16] R.G. Newton, J. Math. Phys. 8, 1566 (1967).

[17] T. Regge, Nuovo cimento 9, 491 (1958) and 14, 951 (1959).

[18] P.C. Sabatier, J. Math. Phys. 7, 1515 and 2079 (1966); 8, 905 (1927) and 9, 1241 (1968).

[19] P.C. Sabatier and F. Quyen Van Phu, Phys. Rev. D 4, 127 (1971).

[20] W.R. Theis, Z. Naturforschung 11a, 889 (1956).

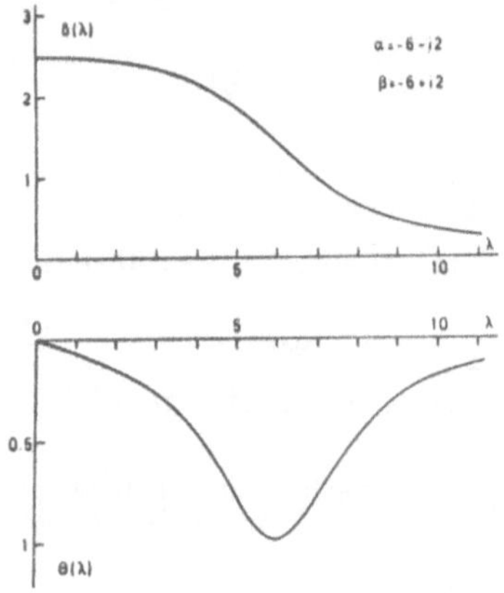


Figure 1

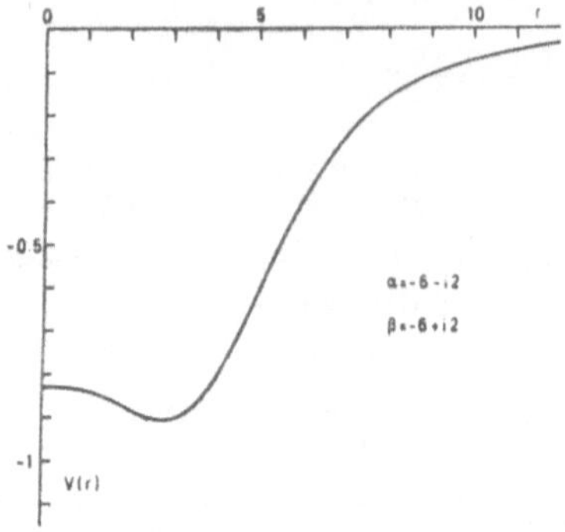


Figure 2

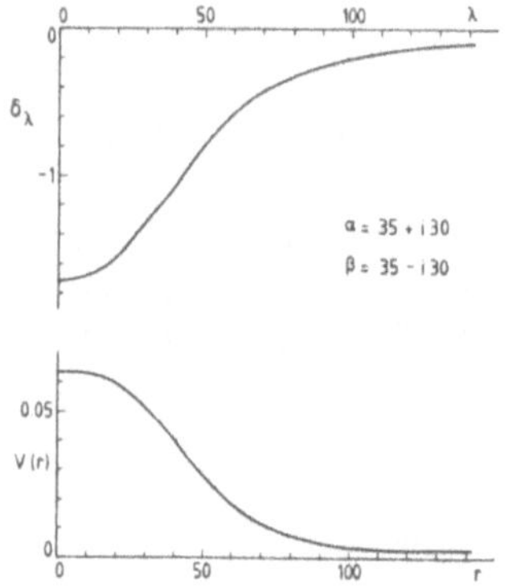


Figure 3

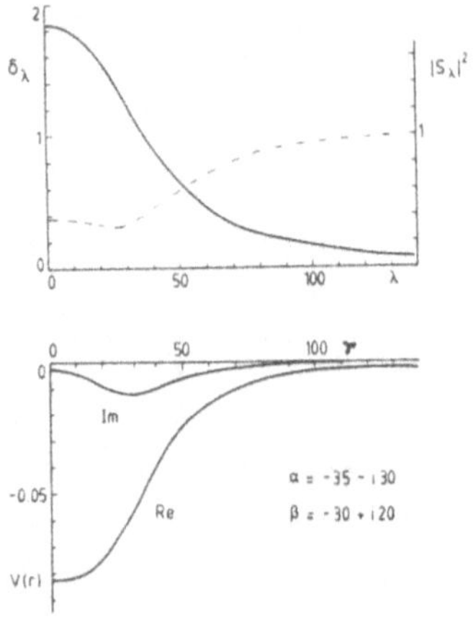


Figure 4

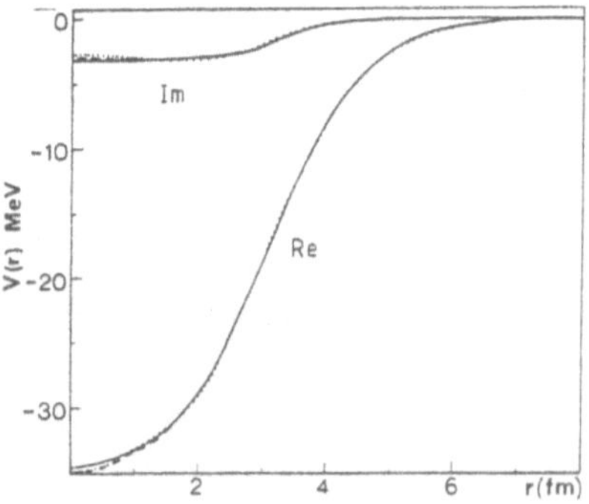


Figure 5

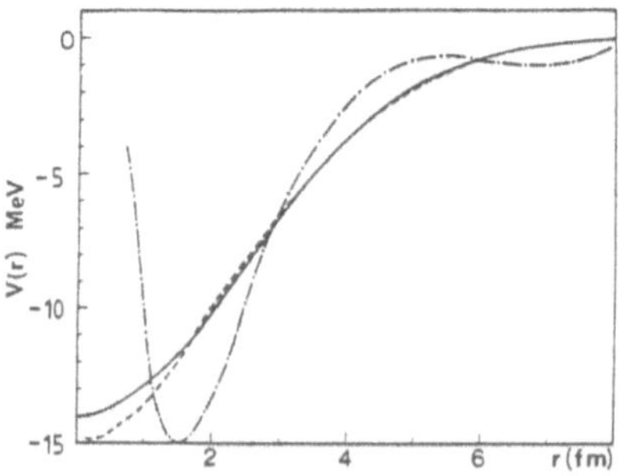


Figure 6

# AN INVERSE EIGENVALUE PROBLEM FROM CONTROL THEORY

L.R. Fletcher

Abstract

The basic pole assignment problem for time-invariant linear multivariable control systems is described together with a number of variants. The crucial concept of controllability is defined and the seminal theory of Wonham stated. Three formally constructive proofs of this result are outlined, though with no attempt to adjudicate between them on the grounds of numerical suitability. Finally some further open questions of potential interest to numerical analysts are mentioned.

## 1. Introduction

The basic system of equations studied in linear multivariable control theory is

$$\dot{x}(t) = Ax(t) + Bu(t) \qquad x(0) = x_0 \tag{1}$$

Here $x(t)$ is an n-vector of differentiable functions of time called the state vector and $u(t)$ is an m-vector of differentiable functions of time called the input vector. A and B are real constant matrices of the appropriate sizes. The aim of the part of multivariable control theory which I shall be concerned with is to determine if and how the input vector $u(t)$ may be chosen so that the solution $x(t)$ of (1) behaves in some predetermined "acceptable" manner.

An important special case of this general problem arises when the input vector is restricted to be a linear function of the state vector

$$u(t) = -Fx(t) \tag{2}$$

(the negative sign is included for sentimental reasons). In this case the $m \times n$ real matrix $F$ is to be chosen so that the solution $x(t)$ of (1), (2) behaves appropriately - this is called control by state feedback. It is of course easy to eliminate $u(t)$ from (1) and (2) to obtain an autonomous system of equations for $x(t)$ namely

$$\dot{x}(t) = (A-BF)x(t) \qquad x(0) = x_0 \tag{3}$$

There is a variety of types of "appropriate" behaviour which might be asked of (3). I shall speak about "modal control"; namely, to what extent is it possible to control the modes of the system (3) by suitable choice of $F$? This leads directly to the inverse eigenvalue problem of my title:

Given a set $\Lambda$ of $n$ complex numbers closed under complex conjugation when and how may a real $m \times n$ matrix $F$ be found so that

$$\sigma(A-BF) = \Lambda \ ?$$

Control theorists call this problem pole assignment by state feedback and a survey of engineering approaches to it is contained in a paper by Munro[4]. We shall restrict ourselves to one approach to the pole assignment problem, namely that which attempts to answer the questions by use of linear algebra -- what control theorists call the time domain or state space or geometric approach (in contrast to the frequency domain approach in which the basic mathematical tool is the transfer function matrix, whose $(i,j)$-th element is the Laplace Transform of $x_j(t)$ when $u_i(t) = \delta(t)$ and $u_k(t) = 0 \quad (k \neq i)$).

In Section 2 below we mention some generalisations of the basic pole assignment problem. In Section 3 we describe the basic concept of controllability and state Wonham's theorem which settles the theoretical state of affairs regarding pole assignment by state feedback. In Section 4 we outline three different ways of proving Wonham's Theorem. In Section 5 we mention some unsolved problems regarding pole assignment which might be of interest to numerical analysts -- in addition, that is, to the fundamental unsolved problem

Suppose there exists a matrix $F$ such that

$$\sigma(A-BF) = \Lambda \ .$$

What is a numerically sound method of computing $F$?

## 2. Some More General Problems

The most widely studied generalisation of pole assignment by state feedback is pole assignment by output feedback in which the basic system of equations is

$$\dot{x}(t) = Ax(t) + Bu(t) \qquad x(0) = x_0$$

$$y(t) = Cx(t)$$

The $p$ vector $y(t)$ (where $p < n$) is called the output vector and now the input $u(t)$ is taken to be a linear function of the output:

$$u(t) = -Ky(t).$$

Now the pole assignment problem is to assign the eigenvalues of the matrix

$$A - BKC$$

by suitable choice of the matrix $K$. Some theoretical results concerning this problem are known but much remains to be discovered.

A more difficult problem is pole assignment by decentralised output feedback. In this problem the control system has a number, $q$ say, of inputs:

$$\dot{x}(t) = Ax(t) + \sum_{i=1}^{q} B_i u_i(t)$$

and the same number of outputs:

$$y_i(t) = C_i x(t) \quad i = 1, \ldots, q.$$

The feedback is now restricted by the requirement that input $i$ be a linear function of output $i$:

$$u_i(t) = -K_i y_i(t) \quad i = 1, \ldots, q.$$

and we are thus faced with the question:

Given a set $\Lambda$ of $n$ complex numbers closed under complex conjugation when and how may real matrices $K_1, \ldots, K_q$ be found such that

$$\sigma(A - \sum_{i=1}^{q} B_i K_i C_i) = \Lambda?$$

Very few worthwhile theoretical results on this problem have been published. This problem is especially interesting as practical applications usually involve large sparse $A$ ($50 \times 50$ or $100 \times 100$ say) with each $B_i$ and $C_i$ of much smaller rank (2 or 3 say).

Another direction in which generalisation may be made is to make (1) a (possibly) degenerate system

$$J\dot{x}(t) = Ax(t) + Bu(t). \tag{4}$$

This leads, as in Section 1, to an inverse generalised eigenvalue problem - at least, for square systems (it is not clear how the modes of non-square systems should be defined and to what extent they are related to the control of the system.) Recently I have proved a generalisation of Wonham's theorem, giving necessary and sufficient condition for the existence of a real matrix so that the roots of the polynomial

$$\det(A - BF - \lambda J) = 0$$

are precisely a given arbitrary set of $q$ complex numbers.

There are, of course, generalised eigenvalue versions of the output feedback and decentralised output feedback pole assignment problems.

## 3. Controllability and Wonham's Theorem

In this section we describe the key theoretical concept which leads to a complete resolution of the pole assignment problem for state feedback. Suppose $\mu \in \mathbb{C}$ and $x$ is a non-zero (constant) vector such that

$$x'A = \mu x', \; x'B = 0 \qquad (5)$$

Then, whatever the matrix $F$

$$x'(A-BF) = \mu x'$$

so $\mu$ is an eigenvalue of $A - BF$ for all $F$. In these circumstances we must have $\mu \in \Lambda$ if there is to exist $F$ such that

$$\sigma(A-BF) = \Lambda.$$

Thus the mode $\mu$ cannot be altered by state feedback. It is therefore reasonable to study systems without such fixed modes; this is the point of the following:

DEFINITION The pair (A,B) is said to be controllable if for every $\mu \in \mathbb{C}$ the only vector $x$ satisfying (5) is the zero vector.

There are numerous equivalent definitions of this key concept - Chris Paige identifies three different ones in his paper (1981) [5] on the numerical determination of controllability and there are several others of a more directly system theoretic nature.

A controllable system has no modes which are unalterable by state feedback -- the following remarkable result shows that much more is true.

THEOREM (Wonham 1967) A pair (A,B) is controllable if and only if given any set $\Lambda$ of $n$ complex numbers closed under complex conjugation there exists a real matrix $F$ such that

$$\sigma(A-BF) = \Lambda.$$

We have proved the "if" part of this theorem above and we will outline three ways of proving the "only if" part in Section 4 below. Let us first mention that this theorem also settles the uncontrollable case as well. More precisely let $\mathcal{X}$ be the largest $A'$-invariant subspace in ker $B'$. Then by the same argument as above the characteristic polynomial of $A'$ on $\mathcal{X}$ is a divisor of the characteristic polynomial of $A-BF$ for every $F$. On the other hand $\mathcal{X}^{\perp}$ is $A$-invariant and contains the range of $B$ so $(A,B)$ may be considered as defining a control system with state vector an $\mathcal{X}^{\perp}$-valued function of time. It is easy to see that this system is controllable so, by

Wonham's Theorem the characteristic polynomial of A - BF can, by appropriate choice of F, be made any monic polynomial of degree $n$ which is divisible by the fixed polynomial we noted above.

## 4. Some Proofs of Wonham's Theorem

In this section we outline three proofs of Wonham's theorem. Let us reiterate that we are now dealing with questions in linear algebra - differential equations, dependence on time and other dynamical features have disappeared. We will assume throughout that (A,B) is controllable.

The first proof of Wonham's theorem is based on the following lemma

LEMMA (Heyman 1968) If $b = Bu$ for some $u \in \mathbb{R}^m$ then there exists a real matrix $F_0$ such that the pair $(A+BF_0, b)$ is controllable.

The proof of this result if straight forward but messy. Using it we may restrict ourselves to controllable pairs (A,b) with $b$ a vector. We have already noted that controllability is equivalent to no A'-invariant subspace being orthogonal to $b$ and this is equivalent in turn to there being no proper A-invariant subspaces of $\mathbb{R}^n$ containing $b$. Thus

$$A^{n-1}b,\ A^{n-2}b,\ \ldots,\ Ab,\ b \tag{6}$$

is a basis of $\mathbb{R}^n$. Let the characteristic polynomial of $A$ be

$$\alpha(\lambda) = \lambda^n - (a_1+a_2\lambda+\ldots+a_n\lambda^{n-1}).$$

Define auxiliary polynomials $\alpha^{(i)}(\lambda)$ $i = 0, \ldots, n$ by

$$\alpha^{(0)}(\lambda) = \alpha(\lambda)$$

$$\alpha^{(1)}(\lambda) = \lambda^{n-1} - (a_2+a_3\lambda+\ldots+a_n\lambda^{n-2})$$

$$\vdots$$

$$\alpha^{(n-1)}(\lambda) = \lambda-a_n$$

$$\alpha^{(n)}(\lambda) = 1$$

and vectors $e_i$ by

$$e_i = \alpha^{(i)}(A)b \qquad (i=1, \ldots, n) \tag{7}$$

Then it is easy to see that (7) is a basis of $\mathbb{R}^n$ because (6) is and that relative to the basis (7)

$$A = \begin{pmatrix} 0 & 1 & 0 & \dots & 0 & 0 \\ 0 & 0 & 1 & \dots & 0 & 0 \\ \vdots \\ 0 & 0 & 0 & \dots & 0 & 1 \\ a_1 & a_2 & & \dots & a_{n-1} & a_n \end{pmatrix} \qquad b = \begin{pmatrix} 0 \\ \vdots \\ 0 \\ 1 \end{pmatrix}$$

Thus $A - bf^T = \begin{pmatrix} 0 & 1 & 0 & \dots\dots & 0 & 0 \\ 0 & 0 & 1 & \dots\dots & 0 & 0 \\ \vdots \\ 0 & 0 & 0 & \dots\dots & 0 & 1 \\ a_1-f_1 & a_2-f_2 & & \dots\dots & a_{n-1}-f_{n-1} & a_n-f_n \end{pmatrix}$

so the characteristic polynomial of $A - bf^T$ can be arbitrarily assigned by a suitable choice of $f$.

The other two methods to be described involve the eigenvectors as well as the eigenvalues of the matrix $A - BF$. These methods are most easily explained in the case in which $\Lambda$ consists of $n$ distinct complex numbers so we will assume that this is the case.

If $s$ is a (right) eigenvector of $A - BF$ with eigenvalue $\Lambda$ then

$$(A-BF)s = \lambda s$$

so

$$(A-\lambda I)s = BFs$$

Thus, whatever the matrix T, such an eigenvector must be in the subspace

$$\mathcal{S}(\lambda) = \{s \mid (A-\lambda I)s \in \mathcal{B}\}$$

where $\mathcal{B}$ denotes the space spanned by the columns of $B$. The crucial properties of $\mathcal{S}(\lambda)$ are summed up in the following:

LEMMA 2 The following are equivalent

(i) $(A,B)$ is controllable

(ii) $\dim \mathcal{S}(\lambda) = m$ for all $\lambda \in \mathbb{C}$

(iii) $\dim \mathcal{S}(\lambda_1) + \dots + \mathcal{S}(\lambda_r) \geq \min(n,r)$ for all sets $\{\lambda_1, \lambda_2, \dots \lambda_r\}$ of distinct complex numbers.

Rado's Theorem ([1], Theorem 4.6) gives (i) of the next lemma directly and (ii) follows with a little more effort.

LEMMA 3 (i) If $\lambda_1, \dots \lambda_n$ are distinct complex numbers then there exist $s_1, \dots, s_n$ which are linearly independent and satisfy $s_i \in \mathcal{S}(\lambda_i)$ $i = 1, \dots, n$.

(ii) The vectors $s_1, \dots, s_n$ in (i) can be chosen to satisfy in addition

$$s_k = \overline{s}_\ell \text{ if } \lambda_k = \overline{\lambda}_\ell \qquad 1 \leq k, \ell \leq n.$$

We can easily construct $F$ to satisfy Wonham's Theorem from vectors $s_1, \ldots, s_n$ satisfying Lemma 3 as follows. Define vectors $v_i$ $(i=1,\ldots,n)$ by

$$(A-\lambda_i I)s_i = Bv_i \tag{8}$$

write $S = [s_1,\ldots,s_n]$, $V = [v_1,\ldots,v_n]$ for the matrices with the columns indicated and let $\Lambda$ be the diagonal matrix with $\lambda_1, \ldots, \lambda_n$ along the diagonal. Then the equations in (8) may be consolidated in the single matrix equation

$$AS - S\Lambda = BV$$

which can be re-arranged

$$(A-BVS^{-1})S = S\Lambda$$

so

$$F = VS^{-1}$$

is the required $F$.

Finally we describe an iterative method for constructing $F$ related to the method just described. Suppose $\{\lambda_1,\ldots,\lambda_r\}$ is a non-empty subset of $\Lambda$ closed under complex conjugation and assume that there exist $s_i \varepsilon \mathcal{S}(\lambda_i)$ $i = 1, \ldots, r$ such that $s_1, \ldots, s_r$ are linearly independent and satisfy $s_k = s_\ell$ if $\lambda_k = \lambda_\ell$ for $1 \leqslant k, \ell \leqslant \sigma$. If $\lambda_1$ is real then $\{\lambda_1\}$ is a suitable non-empty subset and we may take $s_1$ to be any non-zero vector in $\mathcal{S}(\lambda_1)$. If $\lambda_1$ is not real then we may assume that $\lambda_1 = \overline{\lambda}_2$ so that $\{\lambda_1, \lambda_2\}$ is a suitable non-empty subset. It is not difficult to show, without resort to Lemma 3, that $\mathcal{S}(\lambda_1) = \mathcal{S}(\lambda_2)$ if and only if $B$ is a non-singular square matrix and that this is a trivial situation. Thus there exists $s_1 \varepsilon \mathcal{S}(\lambda_1)$ such that $\overline{s}_1 = s_2 \notin \mathcal{S}(\lambda_1)$ and so the conditions mentioned above can readily be satisfied in this case also.

Define $v_1, \ldots, v_r$ by

$$(A-\lambda_i I)s_i = Bv_i \qquad i = 1, \ldots, r$$

and let $F_1$ be a real $m \times n$ matrix satisfying

$$F_1 s_i = v_i \qquad i = 1, \ldots, r \tag{9}$$

Then

$$(A-BF_1)s_i = \lambda_i s_i$$

Now let $Q$ be a real $n \times (n-r)$ matrix satisfying

$$Q'Q = I_{n-r}, \quad Q's_i = 0 \qquad i = 1, \ldots, r$$

Then it is easy to see that the pair $(Q'(A-BF_1)Q, Q'B)$ is controllable so arguing by induction we may assume that there exists an $m \times (n-r)$

matrix $F_2$ such that

$$\sigma(Q'(A-BF_1)Q-Q'BF_2) = \{\lambda_{r+1}, \ldots, \lambda_n\}$$

Then

$$F = F_1 + F_2Q'$$

is the required F.

We make two comments about the implementation of this algorithm. First we note that if $r = 1$, which we may assume to be the case if $\lambda_1$ is real, then the rank 1 matrix

$$F_1 = v_1z_1'$$

where $z_1$ is any vector such that $z_1's_1 = 1$, satisfies (9). If $r = 2$, the only other value of $r$ which it is necessary to consider, it is easy to obtain a similar expression for a suitable $F_1$ as the sum of two rank 1 matrices. Secondly these observations indicate that the present algorithm may be re-formulated to avoid the matrix $Q$; there may be some numerical advantage in this as the reduction in system order implied by the use of this matrix is related to the familiar process of "matrix deflation". There is certainly a system theoretic advantage in that it enables the algorithm to proceed on the basis of choosing eigenvectors for the original system rather than of the somewhat mysterious reduced order systems.

The modified algorithm is as follows:
suppose $s_1, \ldots, s_q$ have been chosen and $F_q$ found such that

$$(A-BF_q)s_i = \lambda_i s_i \qquad i = 1, \ldots, q.$$

Suppose for simplicity that $\lambda_{q+1}$ is real. Then find $s_{q+1} \in \mathcal{S}(\lambda_{q+1})$ such that

$$s_{q+1} \notin \text{span}\{s_1,\ldots,s_q\}$$

(Such an $s_{q+1}$ exists by Lemma 3(iii)) and let $z_{q+1}$ be a vector such that

$$z_{q+1}' s_{q+1} = 1$$

$$z_{q+1} s_i = 0 \qquad i = 1, \ldots, q$$

Then put

$$F_{q+1} = F_q + s_{q+1} z'_{q+1} .$$

With this iteration $F_n$ is the required F.

## 5. Some Unanswered Questions

In this section we mention some numerical questions, so far unanswered, to which answers would be of great practical value. The basic numerical question is to provide a numerically sound method of computing the feedback matrix F. A preliminary question to this may be the determining of criteria for the numerical well-posedness of the pole assignment problem for a given controllable pair (A,B). Some work on numerical measures of controllability has been done by C Paige [5] and this may be relevant here.

The matrix F assigning a given spectrum $\Lambda$ is not unique (unless m = 1) and so the question arises as to how this non-uniqueness is to be parameterised and how it is to be exploited in practice. Two questions of particular interest in this connection are first the obtaining an F of minimum norm assigning a given spectrum $\Lambda$ (see for example [3] for some relevant work on this) and second the existence and determination of a suitable F in some given linear subspace of $\mathbb{R}^{m\times n}$.

Finally there is the question of how best to exploit any freedom of choice in regard to the eigenvalue spectrum $\Lambda$, which in practice is usually required to be contained in some region of the complex plane rather than to be precisely some given set of complex numbers. Numerical experimentation indicates that some eigenvalue spectra are much more sensitive than others to perturbations in A, B and F so that choosing $\Lambda$ to minimise this sensitivity is an important practical issue about which virtually nothing is known.

## 6. Conclusion

We have reviewed the problem of pole assignment by state feedback, mentioning the theory due to Wonham, and a number of open numerical questions. All of these numerical questions may be applied to more subtle pole assignment questions such as those mentioned in our Section 2. For most of these questions the theory is not very well developed at present so it would be appropriate if theoretical and numerical investigations proceeded side by side.

References

[1] V Bryant and H Perfect (1980) "Independence Theory in Combinatorics" Chapman and Hall, London.

[2] M Heyman (1968) IEEE Trans. Aut. Control, AC-13, 748.

[3] B Kovaritakis and R Cameron (1980) IEE Proc. 127, Part D, 32.

[4] N Munro (1979) IEE Proc. 126, 549.

[5] C Paige (1981) IEEE Trans Aut. Control AC-26, 130.

[6] W M Wonham (1967) IEEE Trans Aut. Control AC-12, 660.

# NUMERICAL METHODS FOR ROBUST EIGENSTRUCTURE ASSIGNMENT IN CONTROL SYSTEM DESIGN

J. Kautsky, N.K. Nichols, P. van Dooren and L. Fletcher

## 1. Introduction

The state feedback pole assignment problem in control system design is essentially an inverse eigenvalue problem, which requires the determination of a matrix having given eigenvalues (cf. Fletcher, in these proceedings). A number of formally constructive methods for eigenvalue assignment by feedback are described in the literature [13] [11], [1], but these procedures are not in general *stable* for numerical computation, and do not necessarily lead to *robust*, or *well-conditioned*, solutions of the problem, that is, to solutions which are insensitive to perturbations in the system. Stable numerical methods for inverse eigenvalue problems have been developed in other contexts (compare for instance, references [2], [5], [6]), but these procedures are designed to handle only very specific classes of matrices and are not directly applicable to the forms arising in control theory.

The basic difficulty in developing an algorithm for the inverse eigenvalue problem is that the solution is not uniquely determined. In the special case of a single-input control system, only one solution to the eigenvalue assignment problem may exist, and a numerically stable technique for computing the feedback is available [9]. For the multi-input problem additional criteria must be imposed to restrict the degrees of freedom in the problem. In this paper we describe algorithms for computing solutions to the pole assignment problem which satisfy certain *robustness* criteria. These criteria guarantee that the assigned eigenvalues are as insensitive to perturbations as is feasible, and also that the resulting feedback matrix and corresponding transient response are as reasonably bounded as may be expected, given the original system.

In the next section the pole assignment problem is defined in detail, and theoretical considerations are discussed. In Section 3 we describe the numerical algorithm. Applications and results are presented in Section 4, and concluding remarks follow in Section 5.

## 2. The State Feedback Pole Assignment Problem

We consider the completely controllable time-invariant, linear multivariable system with dynamic state equation

$$\dot{\underline{x}} = A\underline{x} + B\underline{u} \ , \tag{1}$$

where $\underline{x}$, $\underline{u}$ are n- and m-dimensional vectors, respectively, and A, B are real constant matrices of compatible orders. Matrix B is assumed to be of full rank. To modify the *poles* of the system, that is, the eigenvalues of matrix A, a state feedback control $\underline{u} = F\underline{x} + \underline{v}$ may be used, where the *gain* matrix F is chosen such that the modified dynamic system

$$\dot{\underline{x}} = (A + BF)\underline{x} + B\underline{v} \ , \tag{2}$$

now with input $\underline{v}$, has the desired poles. The state feedback pole assignment problem for system (1) is stated precisely by

Problem 1 Given real matrices A, B of orders $n \times n$ and $n \times m$, respectively, and a set of $n$ complex numbers $\Delta = \{\lambda_1, \lambda_2, \ldots, \lambda_n\}$, closed under complex conjugation, find a real $m \times n$ matrix F such that the eigenvalues of $A + BF$ are $\lambda_j$, $j = 1, 2, \ldots, n$.

Conditions for the existence of solutions to Problem 1 are well-established [13]. If we restrict the choice of feedback matrices such that the resulting system matrix $A + BF$ is non-defective (diagonalizable), then Problem 1 is equivalent to

Problem 1' Given A, B and $\Delta$, as in Problem 1, find real matrix F such that

$$(A + BF)X = X\Lambda \tag{3}$$

for some *non-singular* matrix X, where $\Lambda = \text{diag}\{\lambda_1, \lambda_2, \ldots, \lambda_n\}$.

We note that system matrices which are defective are necessarily less robust than those which are non-defective, and that this restriction of the problem simply implies that eigenvalues of multiplicity greater than m cannot be assigned.

From (3) it is clear that the columns $\underline{x}_j$, $j = 1, 2, \ldots, n$ of matrix X are the right eigenvectors of $A + BF$ corresponding to the assigned eigenvalues $\lambda_j$, and that the rows $\underline{y}_j^T$, $j = 1, 2, \ldots, n$ of matrix $Y^T \equiv X^{-1}$ are the corresponding left eigenvectors. The sensitivity of the eigenvalue $\lambda_j$ to perturbations in the components of A, B, F is known [12] to depend upon the magnitude of $1/c_j$ where $c_j \equiv |\underline{y}_j^T \underline{x}_j| / \|\underline{y}_j\|_2 \|\underline{x}_j\|_2$. A general measure and upper bound for the sensitivities is given by the *condition number* $K_2(X) = \|X\|_2 \|X^{-1}\|_2$ of the matrix X. Thus the sensitivities of the assigned poles can be controlled to a restricted extent by an appropriate choice of the eigenvectors comprising X, and we are interested in robust assignment of the entire eigen*structure*. The robust pole assignment problem may therefore be formulated as

Problem 2 Given A, B and $\Delta$, as in Problem 1, find real matrix F and real, non-singular matrix X satisfying (3) such that $K_2(X)$ is minimized.

We prove elsewhere [4] that necessary and sufficient conditions for specific eigenstructure assignment are given by

Theorem 1 Given $\Lambda = \mathrm{diag}\{\lambda_j\}$ and X non-singular, then $\exists$ F, a solution to (3) *if and only if*

$$U_1^T(A - X\Lambda X^{-1}) = 0 \tag{4}$$

where $B = [U_0, U_1]\begin{bmatrix}\Sigma\\0\end{bmatrix}V^T.$ (5)

Then F is given explicitly by

$$F = V\Sigma^{-1}U_0^T(X\Lambda X^{-1} - A). \tag{6}$$

Two immediate consequences follow from the theorem.

Corollary 1 The eigenvector $\underline{x}_j$ of A + BF corresponding to the assigned eigenvalue $\lambda_j$ must belong to the null space $\mathcal{S}_j \equiv \mathcal{N}[U_1^T(A - \lambda_j I)]$.

Corollary 2 The gain matrix F and the transient response $\underline{x}(t)$ of the modified system (2), where $\underline{x}(0) = \underline{x}_0$ and $\underline{v}(t) \equiv 0$, satisfy the inequalities

$$\|F\|_2 \le (\|A\|_2 + \max_j\{\lambda_j\} \cdot K_2(X))/\sigma_m\ , \tag{7}$$

where $\sigma_m$ is the smallest, non-zero singular value of B, and

$$\|\underline{x}(t)\|_2 \le K_2(X) \cdot \|\underline{x}_0\|_2 \cdot \max_j\{|e^{\lambda_j t}|\}\ . \tag{8}$$

The second corollary shows that $F$ and $\underline{x}(t)$ can be bounded in terms of the condition number $K_2(X)$, the assigned eigenvalues and the given matrices $A, B$. Hence for given $A, B$ and $\Delta$, optimizing the conditioning of $X$ also minimizes the feedback gains and the magnitude of the transient response.

For any given set $\Delta$ of eigenvalues, the minimal conditioning that can be achieved is limited. If we let the columns of matrix $S_j$ be an orthonormal basis for the space $\mathcal{S}_j$, defined in Corollary 1, and define $S = [S_1, S_2, \ldots, S_n]$, then a lower bound on the achievable value of $K_2(X)$ is given by $K_2(S)/\sqrt{n}$, where $K_2(S)$ is the ratio of the largest singular value of $S$ to the smallest [14]. Although this lower bound cannot necessarily be realized, the conditioning of $S$ does give a measure of the suitability of the set $\Delta$ of poles for assignment.

## 3. Numerical Algorithms for Robust Eigenstructure Assignment

We now consider the practical implementation of the theory of Section 2. We describe two methods for obtaining approximate solutions to Problem 2. The procedures consist of three basic steps:

Step A: Compute the SVD of matrix B, given by (5), to obtain $U_0$, $U_1$, $\Sigma$ and $V$; construct orthonormal bases, comprised of the columns of matrices $S_j$ and $\tilde{S}_j$, for the space $\mathcal{S}_j = \mathcal{N}[U_1^T(A - \lambda_j I)]$ and its complement $\tilde{\mathcal{S}}_j$, $\lambda_j \in \Delta$, $j = 1, 2, \ldots, n$.

Step X: Select $\underline{x}_j \in \mathcal{S}_j$ with $\|\underline{x}_j\|_2 = 1$, $j = 1, 2, \ldots, n$, such that $X = [\underline{x}_1, \underline{x}_2, \ldots, \underline{x}_n]$ is *well-conditioned.*

Step F: Find matrix $M \equiv A + BF$ by solving $MX = X\Lambda$ and compute $F = V\Sigma^{-1}U_0^T(M - A)$.

The first and third steps are identical for both methods and are achieved using standard library software for computing the QR and SVD (singular value) decomposition of matrices and for solving systems of linear equations [3].

The key step is Step X. We discuss here two methods for accomplishing this step, each based on a different principle.

### 3.1 Method I

The objective of this method is to determine an *orthonormal* set of vectors $\tilde{\underline{x}}_j$, $j = 1, 2, \ldots, n$, such that some measure of the distance between the vectors $\tilde{\underline{x}}_j$ and the subspaces $\mathcal{S}_j$ is *minimized*, or such that a measure of the distance between the vectors and the complementary spaces $\tilde{\mathcal{S}}_j$ is *maximized*; then the required eigenvectors $\underline{x}_j$, $j = 1, 2, \ldots, n$ are taken as the normalized projections of $\tilde{\underline{x}}_j$ into $\mathcal{S}_j$. The resulting $\underline{x}_j$ are then approximately orthogonal to each other and the conditioning of $X$ is expected to be reasonably close to unity.

The measure of distance to be minimized here is a weighted sum of the squares of the sines of the angles between the vectors and the subspaces, given by $\tilde{\nu} = \Sigma w_j \| \tilde{S}_j^T \tilde{\underline{x}}_j \|_2^2$ with specified weights $w_j > 0$; the corresponding complementary measure, to be maximized, is $\nu = \Sigma w_j \| S_j^T \tilde{\underline{x}}_j \|_2^2$. The complementary measure is the more efficient when $m < n - m$.

The vectors $\tilde{\underline{x}}_j$ are determined by applying a sequence of rotations to some initial set of orthonormal vectors in such a way that each rotation reduces (or increases) the given measure $\tilde{\nu}$ (or $\nu$) by an optimal quantity. The procedure continues until the improvement in the measure is less than a specified tolerance. The projections of the resulting vectors $\tilde{\underline{x}}_j$ into the subspaces $\mathcal{S}_j$, for $j = 1, 2, \ldots, n$, are then determined explicitly by $\underline{x}_j = S_j S_j^T \tilde{\underline{x}}_j / \| S_j^T \tilde{\underline{x}}_j \|_2$.

## 3.2 Method II

The objective here is to choose vectors $\underline{x}_j \in \mathcal{S}_j$, $j = 1, 2, \ldots, n$ such that each vector is as orthogonal as possible to the space spanned by the remaining vectors; that is, such that the angle between vector $\underline{x}_j \in \mathcal{S}_j$ and the space $X_j = \langle \underline{x}_i, i \neq j \rangle$ is maximised, for all $j$. Equivalently, for $j = 1, 2, \ldots, n$, we choose $\underline{x}_j \in \mathcal{S}_j$ to minimize the angle between $\underline{x}_j$ and the normalized vector $\underline{y}_j$ orthogonal to the space $X_j$.

The solution is found by an iteration in which each vector $\underline{x}_j$ is replaced by a new vector with maximum angle to the current space $X_j$ for $j = 1, 2, \ldots, n$ in turn. The process is then repeated as long as $K_2(X)$ is reduced by more than a given tolerance. The new vectors may be obtained by using either an SVD or a QR procedure. In effect, the method chooses, at each step, the vector in $\mathcal{S}_j$ which minimizes the sensitivity of the eigenvalue $\lambda_j$. The conditioning of the remaining eigenvalues is disturbed, however, when the new vector replaces the old vector $\underline{x}_j$. Thus the process does not necessarily converge to a fixed point, and the condition number $K_2(X)$ is not necessarily improved by the iteration. This method is generally more expensive than Method I, but it can lead to more robust solutions.

### 3.3 Implementation

The numerical methods described here have been implemented using the system MATLAB [10], and a small package of executive files has been developed to carry out the three basic steps of the algorithms, with various options. Details of the methods and a description of the package, with listings, are given in [7] and [8].

## 4. Application

The procedures described in Section 3 have been applied to a number of realistic examples collected from the literature. To illustrate the behaviour of the methods, we give here the results obtained for a simple test problem. Other results are given in [7].

Test Example $n = 3 \quad m = 2$

$$A = \begin{bmatrix} 0 & 1 & 0 \\ 0 & 0 & 1 \\ 6 & -11 & 6 \end{bmatrix} \qquad B = \begin{bmatrix} 1 & 0 \\ 0 & 1 \\ 1 & 1 \end{bmatrix} .$$

The eigenvalues of $A$ are $\{1.0, 2.0, 3.0\}$. We first assign the eigenvalue set $\Delta = \{1, 1, 3\}$, as in [1]. The condition $K_2(S) = 8.32$ and therefore we expect to be able to obtain a reasonably well-conditioned solution. After two sweeps with Method I, (taking approximately 736 flops), we obtain the solution

$$F = \begin{bmatrix} -1.6053 & 3.0941 & -1.4887 \\ -2.0941 & 4.2907 & -2.1966 \end{bmatrix} .$$

Here $K_2(X) = 7.8098$, and the sensitivities of the assigned eigenvalues are $\{3.92, 1.40, 3.95\}$. Using two sweeps of Method II, (approximately 1321 flops), we obtain almost the identical solution

$$F = \begin{bmatrix} -1.6073 & 3.0972 & -1.4899 \\ -2.0972 & 4.2955 & -2.1984 \end{bmatrix},$$

with $K_2(X) = 7.7772$. The eigenvalue sensitivities are here {2.72, 3.03, 3.95}.

These solutions compare favourably with the solution

$$F = \begin{bmatrix} -3 & 4 & -1 \\ -3 & 4 & -1 \end{bmatrix},$$

derived in [1], which has very poor conditioning. If errors $\pm 0.001$ are introduced into the resulting feedback system matrix $A + BF$, perturbations of up to 14% occur in the assigned poles. Introducing the same errors into the system matrices obtained by Methods I and II leads to errors of at most 0.2% in the assigned eigenvalues.

We have also assigned the eigenvalue set $\Delta = \{-0.2, -0.2, -10.0\}$, which produces a *stable* system. The conditioning of $S$, $K_2(S) = 3.65$, is again satisfactory. With two sweeps of each method very similar results are again obtained. For Method I, $K_2(X) = 3.2827$, $\|F\|_2 = 16.54$, and the sensitivities of the eigenvalues are {1.59, 1.41, 1.79}. For Method II, $K_2(X) = 3.2732$, $\|F\|_2 = 16.46$, and the sensitivities are {1.43, 1.47, 1.79}.

## 5. Conclusions

The importance of the concept of *robustness* for the pole assignment problem in control system design is demonstrated here, and it is shown that the criterion of well-conditioning may be applied to the inverse eigenvalue problem to restrict the degrees of freedom in the solution. Two efficient numerical methods are described for constructing robust, well-conditioned solutions to the state feedback pole assignment problem. Generalizations of these methods for degenerate systems and for the output feedback problem are expected to be easy to develop. The necessary theoretical results, corresponding to Theorem 1, have already been derived [4], and numerical experiments are presently being conducted.

## 6. References

[1] Barnett, S. Introduction to mathematical control theory. Oxford University Press, Oxford (1975).

[2] Boley, D. and Golub, G.H. Inverse eigenvalue problems for band matrices, Proc. of Biennial Conf. on Numerical Analysis, Dundee 1977, Springer-Verlag Lecture Notes in Mathematics, 630, 23-31 (1978).

[3] Dongarra, J.J., Moler, C.B., Bunch, J.R. and Stewart, G.W. LINPACK User's Guide. SIAM, Philadelphia (1979).

[4] Fletcher, L.R., Kautsky, J., Kolka, G.K.G. and Nichols, N.K. Some necessary and sufficient conditions for eigenstructure assignment. University of Salford Department of Mathematics Rpt. (to appear).

[5] Golub, G.H. and Welsch, S.W. Calculation of Gauss quadrature rules, Math. Comp. 23, 221-230 (1969).

[6] Golub, G.H. and Kautsky, J. Calculation of Gauss quadratures with multiple free and fixed knots. University of Flinders School of Mathematical Sciences Rpt. (1982).

[7] Kautsky, J., Nichols, N.K. and Van Dooren, P. Robust eigenstructure assignment in state feedback control. University of Reading Dept. of Mathematics Numerical Analysis Rpt. NA/82 (1982).

[8] Kautsky, J. and Nichols, N.K. MEAP-1: MATLAB Eigenstructure Assignment Package - Mark 1. Flinders University School of Mathematical Sciences Rpt. (1982).

[9] Minimis, G.S. and Paige, C.C. An algorithm for pole assignment of time invariant linear systems. McGill University School of Computer Science Rpt. (1982).

[10] Moler, C.B. MATLAB User's Guide. University of New Mexico Dept. of Computer Science (1981).

[11] Munro, N. Pole assignment, Proc. IEE 126, 549-555 (1979).

[12] Wilkinson, J.H. The algebraic eigenvalue problem. Oxford University Press, Oxford (1965).

[13] Wonham, W.M. On pole assignment in multi-input controllable systems, IEEE Trans. Auto. Control AC-12, 660-665 (1967).

[14] Van Dooren, P.M. and De Wilke, P. Minimal cascade factorization of real and complex rational transfer matrices, IEEE Trans. Circ. and Syst. CAS-28, 390-400 (1981).

# PART III

# INVERSE PROBLEMS IN PARTIAL DIFFERENTIAL EQUATIONS

# SOME INVERSE PROBLEMS IN ELECTROCARDIOLOGY

Piero Colli Franzone

## 1. Introduction

We shall first say something about the clinical framework of the problems. In many countries automated instruments are employed to record potential Body Surface Maps (BSM); for instance one of these instruments [13] records 240 potential values of the electric cardiac field on the human body surface for about 400 time instants of the cardiac beat.

The spreading of this techniques of potential recording is motivated by the fact that BSM provide diagnostic informations which could not be obtained by conventional electrocardiograms in many heart diseases. In fact the features of the potential surface field, like the number, location and time course of potential maxima and minima on the chest surface are correlated to the shape of the depolarization wavefronts and to the spatial distribution of the repolarization processes in the heart which are the phenomena generating bioelectrical currents.

In recent years in Italy in the frame of a Special Program on Biomedical Engineering of C.N.R. (National Research Council) a reduced mobile instrument was built, specially designed to record BSM "on line" for monitoring the infarct size in the intensive care units.
BSM exhibit a highly changeable surface pattern as characterized by equipotential lines; moreover there is a high variability of the surface signal magnitude over the entire heart beat. The diffusion of BSM technique has raised the problem of the interpretation of potential distribution on the thorax. The problem of the best use of the information content of the BSM may be attempted by solving the so-called Direct and Inverse problems of Electrocardiology. Two approaches are possible:

(S) In terms of cardiac sources

(P) In terms of potential alone

The first approach (S) consists in modelling cardiac sources by means of so called equivalent cardiac generators and the Direct Problem consists:

in simulating the body surface maps by means of an "adequate" Equivalent Generator Model

and the Inverse problem consists:

in identifying an "adequate" Equivalent Generator of the heart from body surface maps.

In the second approach (P) we have the following:

Direct problem: from the "epicardial" potential distribution simulate the BSM.

Inverse problem: estimate the "epicardial" potential distribution from the BSM.

The utility of BSM or of the traditional electrocardiogram in diagnosing and quantifying cardiac abnormalities is at present limited by an incomplete understanding of the relationships between BSM and intracardiac electric events. In fact the direct problem in terms of sources, also called the electrocardiology forward problem, i.e. the prediction of BSM from a knowledge of intracardiac electric events and of the geometry and conductivity of the thorax and its contents, is at present unsolved.

Many attempts to solve this problem have been made for the excitation phase of the myocardium and were based on the following classical model. During this phase a layer of cardiac cells, changing with time undergoes the so called "depolarization process" i.e. a change occurs in the intracellular electric potential $u_i(x,t)$, in a short time interval, with an approximately monotone variation from a resting value $u_r$ to a plateau value $u_a$. This temporal function is called intracellular action potential. The following assumption is usually made:

The depolarization of a cell is instantaneous and is the same for all cells ($u_r < u_a$ constants).

Then at any time instant t we denote by:

$H^+=\{x\in H : u_i(x,t)=u_a\}$, (activated heart tissue)

$H^-=\{x\in H : u_i(x,t)=u_r\}$, (resting heart tissue)

$S = \partial H^+ \cap \partial H^-$, (the excitation wavefront surface)

Supposing S a regular surface then the classical model consists in representing the cardiac sources as:

a dipole layer on S, normal to S, oriented toward the resting heart tissue with density (moment) $u_a$-$u_r$.

This uniform dipole layer model used for predicting BSM has been questioned on the basis of experimental evidence collected on the animal by Corbin-Scher [12] and Baruffi et al. [2]. These experiments revealed a strong influence of the cardiac fiber direction on the extracellular potential distribution.

In the first part of the paper we deal with the direct and inverse problem related to a new model of cardiac sources; in the second part we discuss the direct and inverse problem in terms of potential alone. We remark that this last approach is justified by experimental studies on animals both "in vitro" and "in vivo" which show that epicardial maps contain a great amount of information more directly readable in terms of the underlying cardiac events than BSM; moreover in this approach it is possible to compare quantitatively computed epicardial maps with the measured ones collected in experiments on animals.

## 2. Mathematical Model of the Cardiac Electric Sources

Let $\Omega$ be the body volume and H the heart muscle. We suppose that $\Omega$ and H are open connected sets of class $C^2$ and $\bar{H}\subset\Omega$. The excitable cardiac tissue H is conceived as the superposition of two continuous media: the extra (e) and intra (i) cellular medium. We denote by $D_e$ and $D_i$ the extra and intra conductivity tensor of these media. The anisotropic conduction properties characterized by the tensors $D_i$, $D_e$ are related to the fiber structure of the heart muscle. We consider the spreading of the excitation wavefront in a part of the myocardium, which we will still denote by H, for which the following assumptions are made:

(i) the fibers are mathematically described by regular curves;

(ii) through any point of H passes a unique fiber;

(iii) on those parts of $\partial H$, which lie on the heart surface (epi or endocardium), the fibers are tangent to $\partial H$. In the following only these parts of $\partial H$ are considered.

For $x \in H$ let $\sigma_\ell^i(x)$, $\sigma_\ell^e(x)$ be the (i)-(e) conductivity coefficients along the fiber direction at point x. Assuming axial symmetry around the fiber direction let $\sigma_t^i(x)$, $\sigma_t^e(x)$ be the (i)-(e) conductivity coefficient in any direction perpendicular to the fiber direction at point x.

If $A=(\underline{a}_1,\underline{a}_2,\underline{a}_3)$, where $\{\underline{a}_1,\underline{a}_2,\underline{a}_3\}$ is a local basis with $\underline{a}_3$ parallel to the fiber direction, then the tensors are defined by:

$$D_i=A\ \mathrm{diag}(\sigma_t^i,\sigma_t^i,\sigma_\ell^i)A^T=\sigma_t^i(\underline{a}_1\ \underline{a}_1^T+\underline{a}_2\underline{a}_2^T)+\sigma_\ell^i a_3 a_3^T$$

$$D_e=A\ \mathrm{diag}(\sigma_t^e,\sigma_t^e,\sigma_\ell^e)A^T$$

The axis $\underline{a}_1\underline{a}_2$ are defined up to a rotation around $\underline{a}_3$. Moreover $D_i,D_e$, do not depend on the orientation of $\underline{a}_3$ parallel to the longitudinal fiber direction. If $\underline{\nu}$ is a unit vector at point x then:

$$D_i\underline{\nu}=\sigma_t^i(\underline{\nu}\cdot\underline{a}_t)\underline{a}_t+\sigma_\ell^i(\underline{\nu}\cdot\underline{a}_\ell)\underline{a}_\ell$$

where $\underline{a}_\ell=\underline{a}_3$ and $\underline{a}_t$ is a unit vector coplanar with $\underline{\nu},\underline{a}_\ell$ and perpendicular to $\underline{a}_\ell$; moreover $D_i\underline{\nu}$ is independent of the orientation of $\underline{a}_\ell$ and $\underline{a}_t$. In the following we shall make the simplifying assumption that:

the extra-cellular cardiac medium and the body volume are isotropic homogeneous with constant conductivity $\sigma_0(D_e=\sigma_0 I)$

Assuming an instantaneous depolarization process at any time instant t we denote by:

$H^+=\{x\in H:u_i(x,t)=u_a\}$, $H^-=\{x\in H:u_i(x,t)=u_r\}$, with $\bar{H}^+\cup\bar{H}^-=\bar{H}$ $H^+\cap H^-=\emptyset$ and $S=\partial H^+\cap\partial H^-$, the excitation wavefront surface. We also assume that $u_r$, $u_a$ are constant and S is an orientable regular surface. Denoting by u(x) the potential in $\Omega$ and by $\underline{I}=-\sigma_0\nabla u$, $\underline{I}_i=-D_i\nabla u_i$ the current density in $\Omega$ and in the intracellular medium H, the current conservation implies:

$$\begin{cases} \mathrm{div}(\underline{I}+\underline{I}_i)=0 & \text{, in } H\\ \mathrm{div}\ \underline{I}=0 & \text{, in } \Omega-\bar{H}\text{(no sources)}\\ \underline{n}^T\underline{I}=0 & \text{, on } \Gamma=\partial\Omega\text{(insulating boundary)}\end{cases}$$

The boundary condition on $\Gamma=\partial\Omega$ describes the fact that the thorax is surrounded by air which is an insulating medium. In terms of potential we have:

$$\begin{cases} \sigma_0\Delta u=\begin{cases}-\mathrm{div}\ D_i\nabla u_i, & \text{in } H\\ 0 & \text{, in } \Omega-\bar{H}\end{cases}\\ \dfrac{\partial u}{\partial n}=0 & \text{, on } \Gamma\end{cases}$$

Denoting by $\delta_S$ the Dirac measure on S:

$$\langle\delta_S,\phi\rangle=\int_S\phi dS \quad , \quad \phi\in\mathcal{D}(\Omega)$$

and observing that $-\nabla u_i=(u_a-u_r)\underline{\delta}_S\underline{n}_S$ where $\underline{n}_S$ is the normal to S directed toward $H^-$ we obtain the following boundary value problem:

$$\begin{cases} \Delta u = \dfrac{u_a - u_r}{\sigma_o} \operatorname{div} D_i \delta_S \underline{n}_S, \text{ in } \Omega \\ \dfrac{\partial u}{\partial n} = 0 \text{ , on } \Gamma \end{cases} \tag{2.1}$$

Setting $M = \dfrac{u_a - u_r}{\sigma_o} D_i$ and using the results of [17](for the Sobolev space $H^S$ see e.g. [17]) the following existence result holds:

Theorem 2.1

Assuming that:

- the elements of M can be extended in the whole $\Omega$ as functions of $C^1(\Omega)$.
- S is a surface of class $C^2$ and if S is open there exists an open subset $\Omega'$ of $\Omega$ of class $C^2$ such that $\bar{\Omega}' \subset \Omega$.

Then there exists a solution $u \in H^{\frac{1}{2}-\varepsilon}(\Omega), \forall \varepsilon > 0$ of problem (2.1) defined up to an additive constant.

Using Green's formula it is also possible to prove the following ([7]):

Theorem 2.2

The solution of (2.1) in $H^{\frac{1}{2}-\varepsilon}(\Omega)$, $\forall \varepsilon > 0$, satisfies the following boundary value problem:

$$\begin{cases} \Delta u = 0 \quad \text{in } \Omega - \bar{S} \\ \dfrac{\partial u}{\partial n} = 0 \quad \text{on } \Gamma & \text{, (in the sense of } H^{-1-\varepsilon}(\Gamma)) \\ [u]_S = u_{|S^+} - u_{|S^-} = \underline{n}^T M \underline{n} & \text{, (in the sense of } H^{-\varepsilon}(S)) \\ [\dfrac{\partial u}{\partial n}]_S = \dfrac{\partial u}{\partial n}\Big|_{S^+} - \dfrac{\partial u}{\partial n}\Big|_{S^-} = \operatorname{div}_S \underline{\beta} & \text{, (in the sense of } H^{-1-\varepsilon}(S)) \end{cases} \tag{2.2}$$

where $\underline{\beta} = M\underline{n} - (\underline{n}^T M \underline{n})\underline{n}$ is a tangential field on S and $\operatorname{div}_S$ is the divergence operator on S.

In order to investigate the structure of the source model on the wavefront S generating the potential u and also for the numerical approximation of u we state the following useful theorem ([6],[7]).

Theorem 2.3

The solution of (2.1) admits the following boundary integral representation:

$$u(x) = -\int_\Gamma v(\xi) \frac{\partial s(x,\xi)}{\partial n_\xi} d\Gamma_\xi + \int_S \underline{n}_\xi^T M \nabla_\xi s(x,\xi) dS_\xi, \ \forall x \in \Omega - \bar{S} \tag{2.3}$$

where v is the unique solution in $L^2(\Gamma)/R$ of the following Integral Equation on $\Gamma$:

$$\frac{1}{2}v(x)+\int_\Gamma v(\xi)\frac{\partial s(x,\xi)}{\partial n_\xi}\, d\Gamma_\xi = \int_S \underline{n}_\xi^T M\nabla_\xi s(x,\xi)dS_\xi, \quad \text{a.e. on } \Gamma \qquad (2.4)$$

and $s(x,\xi) = \frac{1}{4\pi|x-\xi|}$ is the fundamental solution of the Laplace operator $\Delta$ in $\mathbb{R}^3$. Moreover $u_{|\Gamma} = v$. In the integral representation given by (2.3) there appear two terms of which one takes into account the presence of the insulating boundary $\Gamma$, the other is related to the source model on the wavefront S. Since we have :

$$\int_S \underline{n}^T M\nabla s dS = \int_S \|M\underline{n}\| \frac{\partial s}{\partial \nu}\, dS, \quad \underline{\nu} = \frac{M\underline{n}}{\|M\underline{n}\|}$$

we derive that the mathematical model of the excitation wavefront S consists of:

an oblique dipole (double) layer on S having the same orientation of the vector $M\underline{n}$ and moment density on S given by $\|M\underline{n}\|$

Hence M can be viewed as a dipolar moment tensor. Introducing the unit vectors $\underline{a}_\ell, \underline{a}_t$ such that $\underline{a}_\ell$ is parallel to the longitudinal fiber direction, $\underline{a}_t$ is perpendicular to $\underline{a}_\ell$ and coplanar with $\underline{n}, \underline{a}_\ell$ and orienting $\underline{a}_t, \underline{a}_\ell$ toward the resting tissue so that $\underline{a}_t \cdot \underline{n} \geq 0$ $\underline{a}_\ell \cdot \underline{n} \geq 0$ it is easy to verify:

$$\int_S \underline{n}^T M\nabla s dS = \int_S m_t(\underline{n}\cdot\underline{a}_t)\frac{\partial s}{\partial a_t}\, dS + \int_S m_\ell(\underline{n}\cdot\underline{a}_\ell)\frac{\partial s}{\partial a_\ell}\, dS$$

The source can be viewed as the superposition of a transverse dipole layer with density $m_t(\underline{n}\cdot\underline{a}_t)$ and an axial dipole layer with density $m_\ell(\underline{n}\cdot\underline{a}_\ell)$ where, in this context, transverse and axial means respectively perpendicular and parallel to the local fiber direction.

Moreover the following decomposition holds:

$$\int_S \underline{n}^T M\nabla s dS = \int_S m_t \frac{\partial s}{\partial n}\, dS + \int_S (m_\ell - m_t)(\underline{n}\cdot\underline{a}_\ell)\frac{\partial s}{\partial a_\ell}\, dS$$

which shows that the oblique dipole layer potential can be thought of as the superposition of a normal dipole layer with density $m_t$ and of an axial dipole layer with density $(m_\ell - m_t)(\underline{n}\cdot\underline{a}_\ell)$. If $m_t = m_\ell$ constant we recover the classical model i.e. the source is represented by a normal dipole layer. We remark that $m_t = m_\ell$ implies $\sigma_t^i = \sigma_\ell^i$ i.e. the intracellular medium is isotropic; however experimental measurements have shown that $\sigma_t^i \neq \sigma_\ell^i$ and the experimental data of Corbin and Scher and of Baruffi et al. [12,2] did not agree with the prediction of the normal dipole layer model thus questioning the general validity of the classical model. Our model generalizes the "classical model" and also the "axial model"

introduced by Corbin and Scher [12] for explaining the discrepancy with the prediction of the classical model in dog heart experiments.
Our source model is an anisotropic dipole model where the anisotropy derives from the properties of the intracellular medium since we have:

$$m_t = \frac{u_a - u_r}{\sigma_o} \sigma_t^i \quad , \quad m_\ell = \frac{u_a - u_r}{\sigma_o} \sigma_\ell^i$$

## 2.1 Boundedness of the Potential

If the wavefront S is a closed surface the solution u of problem (2.1) belongs to $L^\infty(\Omega)$. We now discuss the condition for boundedness of the potential field u in the case of an open surface S.
From the integral representation we have:

$$u(x) = -\int_\Gamma v \frac{\partial s}{\partial n} d\Gamma + P(x), \text{ with } P(x) = \int_S \underline{n}^T M \nabla s dS$$

If S is an open surface the following decomposition holds:

$$P(x) = \int_S \alpha \frac{\partial s}{\partial n} dS - \int_S \operatorname{div}_S \underline{\beta}\, s\, dS + \int_{\partial S} \underline{\beta} \cdot \underline{n}_b\, s\, d\partial S$$

where $\alpha = \underline{n}^T M \underline{n}$, $\underline{\beta} = M\underline{n} - \alpha \underline{n}$ and $\underline{n}_b$ is the normal to $\partial S$ contained in the tangent plane to S. This decomposition shows that the presence of the simple layer on $\partial S$ introduces a logarithmic singularity in the potential. Since the physical potential is bounded in $\Omega$ we must impose that the line integral must be zero $\forall x \in \Omega - \bar{S}$ hence

$$\underline{\beta} \cdot \underline{n}_b = 0 \quad \text{on } \partial S \tag{2.5}$$

On electrophysiological grounds we assume that:

a) the fiber direction on $\partial H$ (the epi-endo cardial heart surface) is tangent to $\partial H$ i.e. $\underline{a}_\ell \cdot \underline{n}_H = 0$ on $\partial H$ (where $\underline{n}_H$ is the unit normal to $\partial H$) and if S is open $\partial S \subset \partial H$.

It is not difficult to verify that ([6]):

Proposition 2.1
Under the assumption a) and if $m_t(x) \neq m_\ell(x)$, $\forall x \in \partial S$ then:

$$\begin{cases} \underline{\beta} \cdot \underline{n}_b = 0 \text{ on } \partial S \text{ iff } \underline{n}_S \text{ is parallel to } \underline{n}_H \text{ or} \\ \underline{n}_S \text{ is perpendicular to } \underline{n}_H \text{ or } \underline{a}_\ell \text{ is tangent to } \partial S. \end{cases} \tag{2.6}$$

That is the wavefront surface S and the heart surface $\partial H$ are, respectively, tangent or perpendicular or else intersect at an arbitrary angle in the point of $\partial S$ in which $\underline{a}_\ell$ is tangent to $\partial S$.

## 2.2 Numerical Approximation

The numerical approximation of the potential u in $\Omega-\bar{S}$ was obtained considering the boundary integral representation of u given by Theorem 2.3 i.e.

$$u(x) = -\int_\Gamma v\frac{\partial s}{\partial n}\, d\Gamma + \int_S \underline{n}^T M\nabla s dS \ , \ \forall x \in \Omega-\bar{S}$$

where v is solution of:

$$\tfrac{1}{2}v(x) + \int_\Gamma v\ \frac{\partial s}{\partial n} d\Gamma = \int_S \underline{n}^T M\nabla s\ dS \ , \text{ on } \Gamma$$

The numerical approximation of the integral equation on $\Gamma$ can be performed using:

i) the Galerkin finite element method

ii) the collocation finite element method

We remark that for i) it is possible to prove an optimal error estimate in the case of linear finite element approximation of the surface and function: $\|v-v_h\|_{L^2(\Gamma)/R} = O(h^2)$ (see [11]).

For ii) we do not know error estimate results. However the numerical results that we present were obtained using ii) due to the high cost in computer time of the i) procedure.

The surfaces $\Gamma$, S are approximated by means of two polyhedral surfaces $\Gamma^h$, $S^h$ with triangular elements.

The potential v is approximated by a piecewise linear continuous function $v_h(x)$ on $\Gamma^h$ i.e.

$$v_h(x) = \sum_{j=1}^{N} v_j p_j(x), \quad \text{with} \quad p_j(x_k) = \delta_{jk},\ v_k = v_h(x_k)$$

where $x_k$ denote the nodes on $\Gamma^h$. Applying the collocation method at the nodes of $\Gamma^h$ the integral equation is approximated by:

$$\omega(x_k)v_k + \int_{\Gamma^h} v_h(\xi)\ \frac{\partial s(x_k,\xi)}{\partial n_{h,\xi}}\ d\Gamma^h_\xi = \int_{S^h} \underline{n}_h^T M\nabla_\xi s(x_k,\xi)dS_\xi \quad \text{for } k=1,\dots,N \qquad (2.7)$$

where

$$\omega(x_k) = \frac{\alpha(x_k)}{4\pi} \quad \text{with} \quad \alpha(x) = -\int_{\Gamma_h} \frac{\partial s(x,\xi)}{\partial n_{h,\xi}} d\Gamma^h_\xi$$

i.e. $\alpha(x)$ is the solid angle under which the surface $\Gamma^h$ is seen from the point x, $\underline{n}_h$ is the normal to the triangular elements which make up

$\Gamma^h$ or $S^h$. Setting

$$A = \{a_{kj} = \omega(x_k)\delta_{kj} + \int_{\Gamma^h} p_j(\xi)\,\frac{\partial s(x_k,\xi)}{\partial n_h}\,d\Gamma^h_\xi\}$$

$$\underline{b} = \{b_k = \int_{S^h} \underline{n}_h^T M \nabla_\xi s(x_k,\xi)\,dS^h_\xi\} \quad , \quad \underline{v} = (v_1,\ldots,v_N)^T$$

the solution of (2.7) is equivalent to the solution of the linear system

$$A\underline{v} = \underline{b} \tag{2.8}$$

With $\omega(x)$ defined as above, matrix A admits the zero eigenvalue associated to the right eigenvector

$$\underline{e} = (1,\ldots,1)^T$$

We remark that the integral operator associated to the first side of equation (2.4) admits zero as a simple eigenvalüe; in the following we assume that zero is a simple eigenvalue also for the approximate operator A. Hence the solution of the linear system is defined up to an additive constant; moreover system (2.8) admits a solution iff $\underline{\ell}^T\underline{b} = 0$, where $\underline{\ell}^T A = 0$ i.e. $\underline{\ell}$ is the left eigenvector of A, associated to the zero eigenvalue. While in the continuous case the corresponding compatibility condition is exactly satisfied by the term $\int_S \underline{n}^T M\nabla s\,dS$, on the other hand due to approximation errors, the term $\underline{b}$ does not satisfy exactly the condition $\underline{\ell}^T\underline{b} = 0$. For solving the singular linear system we used the following deflation method. Let $\underline{p}$ be a vector such that $\underline{p}^T e=1$ it is easy to verify that:

if $\underline{\ell}^T\underline{b}=0$ then the solution $\underline{w}$ of the system $B\underline{w} = (A+ep^T)\underline{w}=\underline{b}$

(where B is non singular)satisfies the equations $A\underline{w}=\underline{b}$, $\underline{p}^T\underline{w}=0$

For $\underline{\ell}^T\underline{b}\simeq 0$ we solve the linear system

$$(A+\underline{ep}^T)\underline{v} = \underline{b}$$

and we consider $\underline{v}$ as an approximation of $v_h(x)$.

In practice we choose $p_i=\delta_{ij}$. We remark that A is a full matrix. The determination of the coefficients requires the computation of integrals of the form

$$\int_T p(\xi)\,\frac{\partial s(x,\xi)}{\partial n_\xi}\,dT_\xi \;, \quad \text{where T is a triangle of } \Gamma^h$$

These integrals are computed by means of Gaussian quadrature if x is "not very near" to T otherwise the computation is performed analytically.

## 2.3 Validation of the Oblique Dipole Layer Model

The representation of the depolarization wavefront by means of the oblique dipole layer model was used to simulate the potential field elicited by paced dog hearts in a volume conductor. Here we report some results about the comparison between experimental data and simulated results.

In these experiments on isolated paced dog hearts the potential distribution was measured in a homogeneous cylindrical volume conductor surrounding the heart in particular we had the potential maps on a surface at about .5 cm from the epicardium.

In the simulation a mesh of 504 elements and 254 nodes was used for the cylindrical tank surface $\Sigma$ and a mesh of 528 elements and 266 nodes was chosen for the wavefront surface S. The fields considered were those elicited by an excitation wavefront related to epicardial, mid-wall and endocardial stimulation in the left ventricular few msec. after the stimulus. The wavefront S was suitably shaped as a semi-ellipsoidal or ellipsoidal surface according as the stimulus was endo/epicardial or mid-wall and S was imbedded in a curved fiber layer, taking into account the different orientation of cardiac fibers at different depths in the myocardium (see [3,4,5] for details). In the mid-wall stimulus we have a closed excitation wavefront and a non uniform potential field was observed together with a reentrant current flow ahead of the wavefront in a direction transverse to the fibers. According to the classical model one would expect constant potential field.

In Fig. 1(A) we report the comparison between measured and computed potential, on a surface lying at 0.5 cm from the heart surface, 8 msec. after a midwall stimulus; the simulated potential field shows a pattern which is in good agreement with the measured one with a correlation coefficient of 0.84. We remark that, since we assume $m_t$, $m_\ell$ constant, this simulation of the oblique layer model corresponds in effect to the axial layer model i.e. the dipole are parallel to the local fiber direction. Analogous simulations were performed for the epi/endocardial stimulations (see Fig. 1(B), 1(C)) setting $m_\ell/m_t = 20$; the simulated potential field yielded a correlation coefficient of 0.82 and 0.87 respectively.

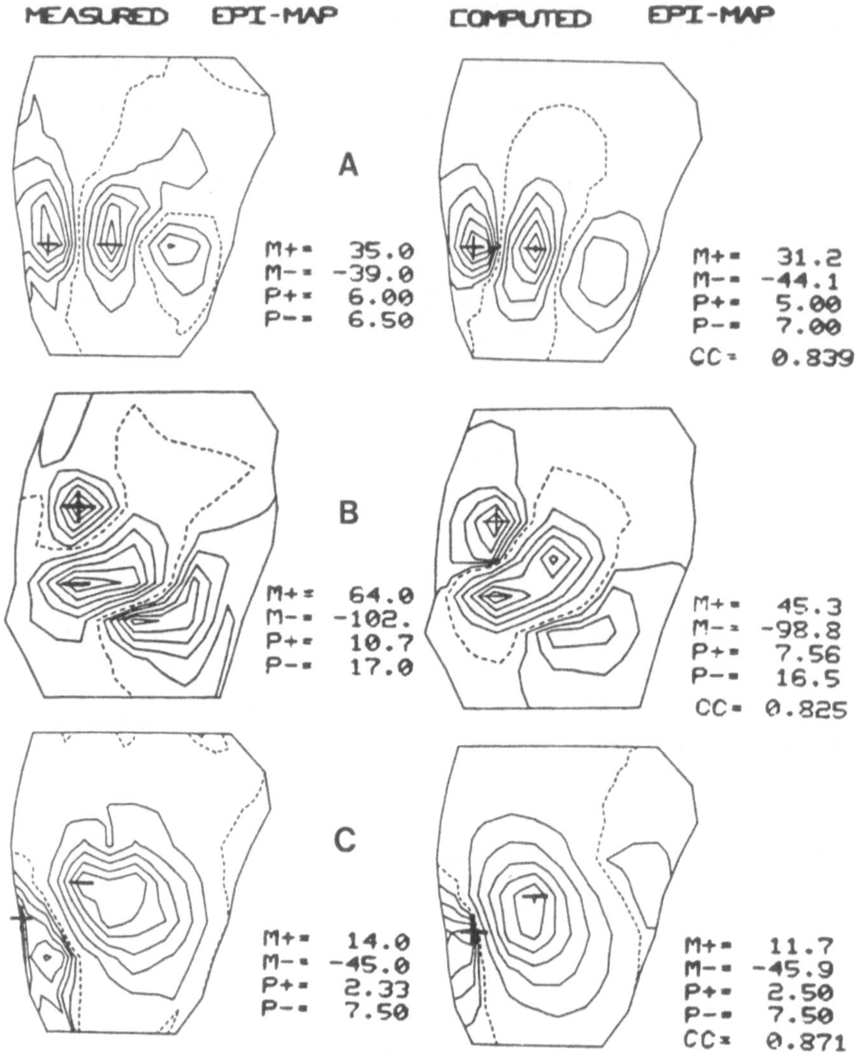


Fig. 1

## 2.4 Inverse Problem in Terms of Wavefront

Now we shall deal with the inverse problem in terms of the wavefront using the oblique dipole layer model as a representation of the depolarization wavefront.

The problem can be roughly formulated as follows: is it possible to determine the wavefront S from the knowledge of the potential on the

<u>thorax surface</u> $\Gamma$? We recall that the knowledge of:

the myocardial fibers direction ($\underline{a}_\ell$), the jump of the intracellular action potential ($u_a-u_r$), the coefficients of the longitudinal and transverse intracellular conductivity ($\sigma_\ell^i, \sigma_t^i$) and the extracellular and extracardiac conductivity ($\sigma_0$)

characterizes the matrix M in H given by:

$$M=A\ \mathrm{diag}(m_t,m_t,m_\ell)A^T,\quad m_t = \frac{u_a-u_r}{\sigma_0}\sigma_t^i,\quad m_\ell=\frac{u_a-u_r}{\sigma_0}\sigma_\ell^i$$

where $A=(\underline{a}_1,\underline{a}_2,\underline{a}_3)$, $\{\underline{a}_1,\underline{a}_2,\underline{a}_3\}$ is a local basis with $\underline{a}_3=a_\ell$ and $\underline{a}_1,\underline{a}_2$ defined up to a rotation around $\underline{a}_3$.

In the following we assume that $\sigma_\ell^i,\sigma_t^i$ are positive constants in H and $\sigma_t^i\neq\sigma_\ell^i$, i.e. we have homogeneous intracellular anisotropy, or equivalently we assume:

$$\begin{array}{l}\text{the eigenvalues of } M \text{ are independent of } x \text{ in } H\\ \text{and } m_t\neq m_\ell \text{ with } m_t,m_\ell>0,\ \ (m_t \text{ double eigenvalue})\end{array}\tag{2.9}$$

We introduce the following <u>family</u> W <u>of admissible wavefronts</u> S:

W is the set of the open orientable connected surfaces S of class $C^2$ contained in H with boundary $\partial S\subset\partial H$ and S can be considered as a part of a $C^2$ closed surface contained in $\Omega$ moreover $\partial S$ satisfies condition (2.6).

Then for $S\in W$ we consider the function (defined up to an additive constant):

$$S\to U_S = u|_\Gamma$$

where u is the solution of problem (2.1) of Theorem 2.1 or equivalently

$$S\to U_S = v$$

where v is the solution of the integral equation (2.4) of Theorem 2.3. That is, given S we consider the potential $U_S$ on the insulating boundary $\Gamma$ generated by the oblique dipole layer on S. We remark that in the special case in which $\underline{a}_\ell$ is always perpendicular or tangent to a given surface S i.e. $\underline{a}_\ell=\underline{n}_s$ or $\underline{a}_\ell\cdot\underline{n}_s=0$ then the oblique dipole layer on S is a normal double layer; if $S_1$ and $S_2$ are two surfaces of W satisfying the preceding constraint then it is well known that $U_{S_1}=U_{S_2}$ implies $\partial S_1=\partial S_2$ i.e. the potential characterizes only the boundary $\partial S$ of S. Assuming that the wavefront is generally oblique w.r.t. the fiber direction (i.e. $\underline{a}_\ell$ is not always perpendicular or tangent to the wavefront) the following uniqueness result holds ([7]):

Theorem 2.4

If $S_1$, $S_2 \in W$ and on at least one of the surfaces $S_i$, i=1,2, e.g. $S_1$,$\underline{a}_\ell$ is generally oblique to $S_1$(i.e., $\underline{a}_\ell \cdot \underline{n}_{S_1} \neq 0$ or $\underline{a}_\ell \cdot \underline{n}_{S_1} \neq 1$) but there exists at least a point where $\underline{a}_\ell \cdot \underline{n}_S = 1$ or $\underline{a}_\ell \cdot \underline{n}_{S_1} = 0$, then

$$U_{S_1} = U_{S_2} \quad \text{implies} \quad S_1 = S_2 \ .$$

Under the same hypotheses of Theorem 2.4 the uniqueness result holds also for closed wavefront i.e. if $S_1$ and $S_2$ are $C^2$ closed surfaces one of them inside the others. These first results and others suggest the conjecture that the uniqueness result should hold under the same hypotheses of Theorem 2.4 in a more general admissible wavefront class W and for $m_\ell \neq m_t$ variables in H. Finally we outline the following open problems: the generalization of the results to an extracellular anisotropic medium and to a body volume with piecewise constant conductivity and the numerical approaches for solving the inverse problem.

## 3. Inverse Problem in Terms of Potential Alone

### 3.1 Mathematical Formulation of the Problem

In relation to the feasibility of the inverse problem in terms of potential alone the following question arises: to what extent can the potential distribution on the "heart" surface be computed from BSM? We discuss the relevant features concerning this question summarizing the results of the works [8], [9], [10].

We shall now describe in more detail the mathematical setup of the direct and inverse problems in terms of epicardial potential.

The human body $\Omega_1$ can be considered as an isotropic linear resistive conducting medium excluding a region $\Omega_0$. At any time instant of the heart beat the electric field can be considered quasi-static (see [18]) and the volume conductor $\Omega_1$ is imbedded into an insulating medium (the air) that is the normal derivative of the potential is zero on $\Gamma_1$. We set $\Omega = \Omega_1 - \Omega_0$, K(x) the electrical conductivity and $\Gamma_0 = \partial\Omega_0$ represents a fixed surface surrounding the heart and lying in proximity of the heart surface; in the following $\Gamma_0$ is referred to as "epicardial" surface. At any time t let V(x) be the electric potential distribution in $\Omega$. If $u(x) = V(x)|_{\Gamma_0}$ is known then V(x) in $\Omega$ is characterized by the following mixed boundary problem:

$$\begin{cases} \text{div } K(x) \text{ grad } V(x) = 0 & \text{in } \Omega \\ V(x)=u(x) \text{ on } \Gamma_0, \ \dfrac{\partial V(x)}{\partial n} = 0 & \text{on } \Gamma_1 \end{cases}$$

and the direct problem consists in <u>evaluating $V(x)$ on $\Gamma_1$</u>. If no information is available about $V(x)$ on $\Gamma_0$ but it is possible to measure $V(x)=z(x)$ on $\Sigma\subset\Gamma_1$ then we have:

$$\begin{cases} \text{div } K(x) \text{ grad } V(x)=0 & \text{in } \Omega \\ V(x)=z(x) \text{ on } \Sigma \ , \ \dfrac{\partial V(x)}{\partial n} = 0 & \text{on } \Gamma_1 \end{cases} \tag{3.1}$$

and the inverse problem consists in <u>estimating $V(x)$ on $\Gamma_0$</u>. If the observed potential $z$ on $\Sigma$ were measured error-free, the Cauchy problem for the elliptic operator (3.1) would define a unique solution $V(x)$ but in a highly unstable fashion since, as it is well known, the Cauchy problem for elliptic operators is an ill-posed problem.Let $v\in L^2(\Gamma_0)$ define the state $y(x;v)=y(v)$ as the unique solution in $H^{\frac{1}{2}}(\Omega)$ of

$$\begin{cases} \text{div } K(x) \text{ grad } y(v)=0 & \text{in } \Omega \\ y(v)=v \text{ on } \Gamma_0 \ , \ \dfrac{\partial y(v)}{\partial n} = 0 & \text{on } \Gamma_1 \end{cases}$$

we introduce the following operator

$$Av = y(v)|_{\Sigma}$$

and for $z\in L^2(\Sigma)$ the cost function

$$J(v)=\int_{\Sigma} |y(v)-z|^2 d\sigma = \|Av-z\|^2_{0,\Sigma}$$

We consider the minimization problem:

$$\text{find } u\in L^2(\Gamma_0):J(u)= \inf_{v\in L^2(\Gamma_0)} J(v) \tag{3.2}$$

If $z\in A(L^2(\Gamma_0))$ i.e. there exists a unique solution of the Cauchy problem then $u=V(x)|_{\Gamma_0}\in L^2(\Gamma_0)$ satisfies $Au=z$ hence $u$ is the unique minimizer of $J$. But the problem is still unstable i.e. the operator $A$ admits an unbounded inverse operator in the spaces $H^s$, $\forall s\in R$. In order to solve the problem one must approximate it with a family of stable problems. Many methods are available to perform this stabilization. We investigate the regularization techniques imposing smoothing constraints justified by the physical problem. We consider the following cases:

| Space of Admissible Controls U | | Regularization Operator B |
|---|---|---|
| $U=H^1(\Gamma_0)=\{v\in L^2(\Gamma_0):Bv\in L^2(\Gamma_0)\}$ | with | $B = \text{grad}\,|_{\Gamma_0}$ |
| $U=H^2(\Gamma_0)=\{v\in L^2(\Gamma_0):Bv\in L^2(\Gamma_0)\}$ | with | $B = \Delta|_{\Gamma_0}$ |

We consider the regularization cost function:

$$J_\varepsilon(v)=J(v)+\varepsilon\int_{\Gamma_0}|Bv|^2 d\sigma \ , \quad v\in U \quad , \quad \varepsilon>0 \text{ "small"}$$

Then the inverse problem (3.2) is approximated by the following family of stable problems dependent on $\varepsilon$:

$$\text{Find } u_\varepsilon\in U : J_\varepsilon(u_\varepsilon)=\min_{v\in U} J_\varepsilon(v) \tag{3.3}$$

We refer to [8] for the existence, the uniqueness of the minimizer u of (3.3) and for the convergence of $u_\varepsilon$ to u. For the numerical approximation of the regularized problem (3.3) it is necessary to introduce a finite dimension approximation $U_h$ of the control space U and an approximate observation operator $A_h$ of A. If we consider a parametrization of the control u (having n degrees of freedom):

$$u(x)=\sum_{i=1}^n u_i w_i^o(x) \quad \text{then} \quad y(x;u)=\sum_{i=1}^n u_i y_i(x)$$

where $y_i(x)$ is the solution of:

$$\begin{cases} \text{div } K(x) \text{ grad } y_i(x)=0 & \text{in } \Omega \\ y_i(x)|_{\Gamma_0}=w_i^o(x), \quad \dfrac{\partial y_i(x)}{\partial n}\Big|_{\Gamma_1}=0 & i=1,\dots,n \\ Au = \sum_{i=1}^n u_i y_i(x)\Big|_\Sigma \end{cases} \tag{3.4}$$

Hence a way of building an approximate operator of A consists in the numerical solution of the previous n mixed boundary elliptic problems. The approximate solution $y_i^h(x)$ of $y_i(x)$, solution of (3.4), can be obtained by means of the finite element method applied to the three dimensional variational formulation of problem (3.4). The numerical computation were performed using isoparametric hexaedral elements of first order (for details see [8]). Denoting by $\Sigma^h$, $\Gamma_0^h$ the discretized surfaces respectively of $\Sigma$ and $\Gamma_0$ and by $\{x_j^\Sigma\}_{j=1,m}$,$\{x_k^{\Gamma_0}\}_{k=1,n}$ the mesh points on $\Sigma^h$ and $\Gamma_0^h$ we define:

$$A_h u_h = y^h(x;u_h)|_{\Sigma^h} = \sum_{k=1}^{n} u_k y_k^h(x)|_{\Sigma^h} \tag{3.5}$$

where $u_h = \sum_{k=1}^{n} u_k w_k^o(x)$, $\{w_k^o\}_{k=1,n}$ is a basis of the finite element space used on $\Gamma_o^h$ and $\underline{u}=(u_1,\dots,u_n)^T$ is the vector of the nodal values of $u_h$ on $\Gamma_o^h$. Setting $\underline{z}=\{(A_h u_h)(x_j^\Sigma)\}_{j=1,m}$ the vector of nodal values of $A_h u_h$ on $\Sigma^h$ then in terms of $\underline{z}$ and $\underline{u}$ the relation (3.5) implies

$$\underline{z} = T\underline{u}$$

where

$$T = \{y_i^h(x_j^\Sigma) \quad i=1,n \quad , \; j=1,m\}$$

is the so called "transfer" matrix between $\Gamma_o^h$ and $\Sigma^h$. Moreover the matrix T depends only on the geometric data $\Omega$. The first step of the numerical procedure consists in the computation of T; T is obtained solving the n mixed boundary value problems (3.4), which differ only for the Dirichlet data on $\Gamma_o$, by implementing a suitable version of the frontal method which performs a triangular factorization only once for all the n problems. The matrix T is a very ill-conditioned matrix reflecting the ill-posedness of the inverse problem (3.2). We remark that the transfer matrix can also be computed by means of methods based on boundary integral representations of the potential; these methods have been investigated in [11]. Since, as we shall say later, the accuracy of the transfer matrix is not a relevant factor in the experimental inverse problem all the following results will be related to the transfer matrix computed by the three dimensional variational method outlined above, which is convenient in order to reduce computational time and storage requirement. It is easy to verify that the discretized form of the regularized problem (3.3) can be restated in the following least square form:

$$\text{find } \underline{u}_\varepsilon \in R^n : \min_{v \in R^n} \|\mathcal{T}\underline{v}-\underline{b}\|^2_{R^m} + \varepsilon\|R\underline{v}\|^2_{R^n} \tag{3.6}$$

where $\mathcal{T}$ is a suitable matrix related to the transfer matrix T,$\underline{b}$ is related to the observation z(x) on $\Sigma$ and R is obtained by discretizing the operator B (see [8] for details). For results about convergence and error estimates of the solution of the approximate regularized problem to the solution of the continuous one (3.3) we refer again to [8]. We shall limit the discussion about the numerical results obtained using $\mathcal{T}$ =T i.e. instead of the residual $\int_{\Sigma^h}|A_h u_h - z|^2 d$ consider the discrete residual $\|T\underline{u}-\underline{z}_m\|$ where $\underline{z}_m$ is the vector of the measured potential values at the nodes of $\Sigma^h$. In the application no improvement in accuracy was achieved using a distributed observation z(x) obtained from the measured vector $\underline{z}_m$ by suitable interpolation. We remark that problem (3.6) must be solved for a sequence of $\underline{z}_m$ data i.e. the measured surface potential at different time instants of the heart beat and several values of the smoothing parameter $\varepsilon$ must be considered. Hence in order to reduce the computational load it is important to apply a very efficient procedure. This has been done by computing, only once, the generalized singular value decomposition of the matrix T and R (see [15], [20]) after that the solution $\underline{u}_\varepsilon$ can be written explicitly as a function of $\varepsilon$.

## 3.2 Numerical Results

Numerical experiments have been performed with test functions and experimental data using two different geometries. The geometry G1 was related to the isolated dog heart experiment and the other, G2, to a human torso. In the first G1 we have a cylindrical tank containing the isolated dog heart imbedded in a physiological solution. On the entire normal heart beat the potential values were measured by means of electrodes of which 156 on the lateral surface $\Sigma$ of the tank and 122 lying on ideal surfaces $\Gamma_0$ at an average distance from the heart surface of 0.5 and 1. cm. The transfer matrix T between 122 nodes on $\Gamma_0$ and 156 nodes on $\Sigma$ was computed using a three dimensional mesh made up of 1200 hexaedral elements and 1446 nodes. In the human geometry G2 we built a transfer matrix between 158 nodes on a surface lying in proximity of the epicardial surface and 230 nodes on the chest surface using a three dimensional mesh of 1500 hexaedral elements and 1900 nodes. The 230 nodes were in correspondence to the sites of body surface measurements. We consider the following test functions:

$$\phi(x,y,z) = \begin{cases} F1=a_1\tilde{x}\tilde{z}+b_1\tilde{y}/r_1^3 \\ \\ F2=a_2\tilde{y}\tilde{z}+b_2\tilde{x}\tilde{y}/r \end{cases}, \quad \text{with } r_i^2=(x-x_0^i)^2+(y-y_0^i)^2+(z-z_0^i)^2,\ i=1,2 \quad \text{and } \tilde{x}=x-x_0^i,\ \tilde{y}=y-y_0^i,\ \tilde{z}=z-z_0^i$$

where in the G1 geometry we choose:

$$a_1=a_2=1.,\ b_1=10^5,\ b_2=0.5\times10^4,\ x_0^i=1.,\ y_0^i=1.5,\ z_0^i=9.,\ i=1,2$$

and in the G2 geometry we set: $a_1=1.$, $a_2=0.25$, $b_1=5\text{x}10^5$, $b_2=10^5$, $x_0^1=2.$, $y_0^1=2.5$, $z_0^1=25.5$, $x_0^2=1$, $y_0^2=1.$, $z_0^2=21.5$.

For the two geometries we have: $\Delta\phi=0$ in $\Omega, \phi|_{\Gamma_0}=u_e$, $\frac{\partial\phi}{\partial n}\Big|_{\Gamma_1}=g\neq0$ and introducing:

$$\Delta u=\Delta w=0 \text{ in } \Omega,\ u|_{\Gamma_0}=u_e,\ w|_{\Gamma_0}=0,\ \frac{\partial u}{\partial n}\Big|_{\Gamma_1}=0,\ \frac{\partial w}{\partial n}\Big|_{\Gamma_1}=g$$

it results $\phi=u+w$; denoting by z the trace of $\phi$ on $\Sigma$ it follows:

$$z = Au_e+w|_\Sigma$$

Hence, consider the following discrete form of the regularized problem:

$$\underline{u}_\varepsilon \in R^n :\ \min_{v \in R^n} \frac{1}{m}\|T_0\underline{v}-\underline{z}_m\|^2+\varepsilon\|R\underline{v}\|^2 \qquad (3.7)$$

where $T_0\underline{v}=T\underline{v}+\underline{w}$, $\underline{w}$ the vector of the nodal values on $\Sigma$ of the finite element approximation of w and n,m are the mesh point numbers on $\Gamma_0^h$ and $\Sigma^h$, respectively. As a first step, we solved the direct problem i.e. $T_0\underline{u}_e=\underline{z}_c$ with $\underline{u}_e$ vector of the nodal values of $\phi$ on $\Gamma_0^h$ and we compared $\underline{z}_c$ with the vector $\underline{z}_a$ or the nodal values of $\phi$ on $\Sigma^h$. In tables 1,2 we report the relative error RE between $\underline{z}_c$ and $\underline{z}_a$ and the mean surface residual $MSR=\|\underline{z}_c-\underline{z}_a\|/\sqrt{m}$ for the two geometries G1 and G2, respectively.

| G1 | RE | MSR |
|---|---|---|
| F1 | 0.05 | 2.6 |
| F2 | 0.25 | 10.2 |

Table 1

| G2 | RE | MSR |
|---|---|---|
| F1 | 0.02 | 1.9 |
| F2 | 0.08 | 2.7 |

Table 2

Then we solved problem (3.7) using a matrix R related to an approximation of the gradient regularization operator on $\Gamma_0$ and we applied the precedure taking as surface data $\underline{z}_m$:

A) the simulated surface data with a noise vector $\underline{e}$ of independent normally distributed zero mean pseudo random variables with common variance $\sigma^2$, added to it, i.e., $z_m=T_0\underline{u}_e+\underline{e}$

B) the analytical surface data with the pseudo random noise vector $\underline{e}$ added to it, as above, i.e., $\underline{z}_m=\underline{z}_a+\underline{e}$

We remark that in A) the approximation errors of the model are not present while in B) these errors are included. We introduce the correlation coefficient $CC=(u_e-\bar{u}_e,u_\varepsilon-\bar{u}_\varepsilon)/\|u_e-\bar{u}_e\|\cdot\|\underline{u}_\varepsilon-\bar{u}_\varepsilon\|$ where $\bar{u}=\sum_{i=1}^n u_i/n$ is the mean value; in Table 3,4 we report the RE and CC between $\underline{u}_e$ and $\underline{u}_{\varepsilon_0}$ obtained using the surface data A) and B) respectively. These results are referred to the optimal regularized solution $\underline{u}_{\varepsilon_0}$ where $\varepsilon_0$ is the optimal choice of the smoothing parameter $\varepsilon$, i.e., the minimizer of $\|\underline{u}_e-\underline{u}_\varepsilon\|$. In these Tables we report also the signal to noise ratio $S/N=\|\underline{z}\|/\sigma\sqrt{m}$ and the root mean square surface residual $MSR=\|T_0\underline{u}_{\varepsilon_0}-\underline{z}_m\|/\sqrt{m}$.

| G1 | | φ=F1 | | | | φ=F2 | | | |
|---|---|---|---|---|---|---|---|---|---|
| | σ | RE | CC | MSR | S/N | RE | CC | MSR | S/N |
| (A) From | 0. | 1.5E-3 | 0.999 | 1.3E-6 | - | 1.8E-3 | 0.999 | 1.2E-6 | - |
| $\underline{z}_m$ = | 0.25 | 0.14 | 0.99 | 0.23 | 163. | 0.16 | 0.99 | 0.22 | 164. |
| $T_o \underline{u}_e + \underline{e}$ | 7. | 0.25 | 0.96 | 6.8 | 6.2 | 0.39 | 0.92 | 6.55 | 6.1 |
| (B) From $\underline{z}_m$ = | 0. | 0.13 | 0.98 | 0.15 | - | 0.35 | 0.95 | 3.1 | - |
| $\underline{z}_a + \underline{e}$ | 7. | 0.24 | 0.97 | 7.68 | 5.5 | 0.37 | 0.93 | 7.36 | 5.7 |

Table 3

| G2 | | φ=F1 | | | | φ=F2 | | | |
|---|---|---|---|---|---|---|---|---|---|
| | σ | RE | CC | MSR | S/N | RE | CC | MSR | S/N |
| (A) From | 0. | 1.2E-3 | 0.999 | 9.E-7 | - | 3.5E-4 | 0.999 | 6.E-7 | - |
| $\underline{z}_m$ = | 0.25 | 0.10 | 0.99 | 0.23 | 343. | 0.20 | 0.98 | 0.22 | 134. |
| $T_o \underline{u}_e + \underline{e}$ | 5. | 0.24 | 0.97 | 4.73 | 17.2 | 0.40 | 0.91 | 4.96 | 6.3 |
| (B) From | 0. | 0.19 | 0.98 | 1.02 | - | 0.28 | 0.96 | 2.46 | - |
| $\underline{z}_m$ = | 5. | 0.26 | 0.96 | 4.98 | 17.2 | 0.36 | 0.93 | 4.71 | 6.4 |
| $\underline{z}_a + \underline{e}$ | 7. | 0.27 | 0.96 | 7.1 | 11.9 | 0.46 | 0.88 | 6.36 | 4.8 |

Table 4

These results indicate that, increasing the noise level, the RE increases rapidly. Comparing the results in terms of RE obtained in the cases (A) and (B), it can be seen that in presence of a "realistic" instrumental noise level of $\sigma=5\div7$ μV (microvolt), the approximation errors of the model do not influence appreciably the accuracy of the results. The pattern match between $\underline{u}_{\varepsilon_0}$ and the analytical data $\underline{u}_e$ on $\Gamma_o$, measured by CC, shows that the pattern of the reconstructed distribution is always in very good agreement with the analytical one i.e,, the comparison is qualitatively good. Moreover, the behaviour of CC is quite different from that showed by RE since, increasing the noise level, the CC does not decrease as rapidly as RE increases. From these results, related to the inverse problem in the two geometries, it follows that the regularization algorithm in presence of a realistic instrumental noise level on the data correctly reproduces only the pattern of the potential on $\Gamma_0$ i.e., those features like the number and approximate locations of maxima and minima and the qualitative behaviour of the equilevel lines which are the features relevant to the understanding of the information content related to the epicardial potential maps in the applications. The inverse procedure was applied to the data collected in experiments on isolated dog hearts. The "epicardial" maps were inversely computed and compared to the corresponding measured maps for a set of 62 time instants from a normal heart beat.

In Fig.2 we report the correlation coefficient between measured and computed maps at 122 sites lying on an ideal surface $\Gamma_0$ enclosing the heart at an average distance from the epicardium of 0.5 cm (solid line) and 1. cm (dashed line), respectively. Best pattern match was obtained in the time interval of the P wave, of the initial QRS complex and of the T wave.

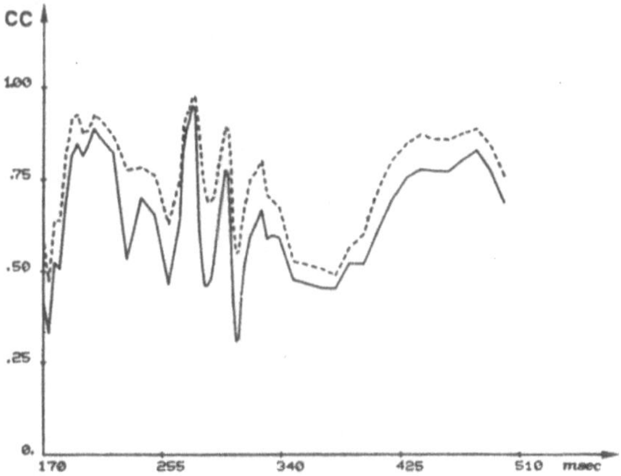


Fig. 2

An extensive numerical investigation was carried out to evaluate the influence of different factors on the accuracy of inversely computed maps (see [8], [9]). We briefly summarize some concluding remarks. Comparing the regularization methods with the Laplace, the gradient and the Euclidean norm operator and the truncated singular values method we found that the gradient regularization operator yielded the best accuracy. Shortening the distance between the inner surface $\Gamma_0$ and the outer surface $\Gamma_1$ of only 0.5 cm, the average RE decreased of 13% and the CC increased of 12%. Reducing the number of the three dimensional mesh elements of $\Omega$ from 1200 to about 400, i.e. reducing the accuracy of the transfer matrix T between the same number of nodes on $\Gamma_0^h$ and $\Sigma^h$, the average RE increased of 1% and the CC decreased of only 1%. The results showed that, since, at present, the overall noise on the data is not easily reducible, the accessible accuracy can be obtained with a coarser mesh allowing a corresponding reduction of computational time and memory occupation in the construction of the transfer matrix.

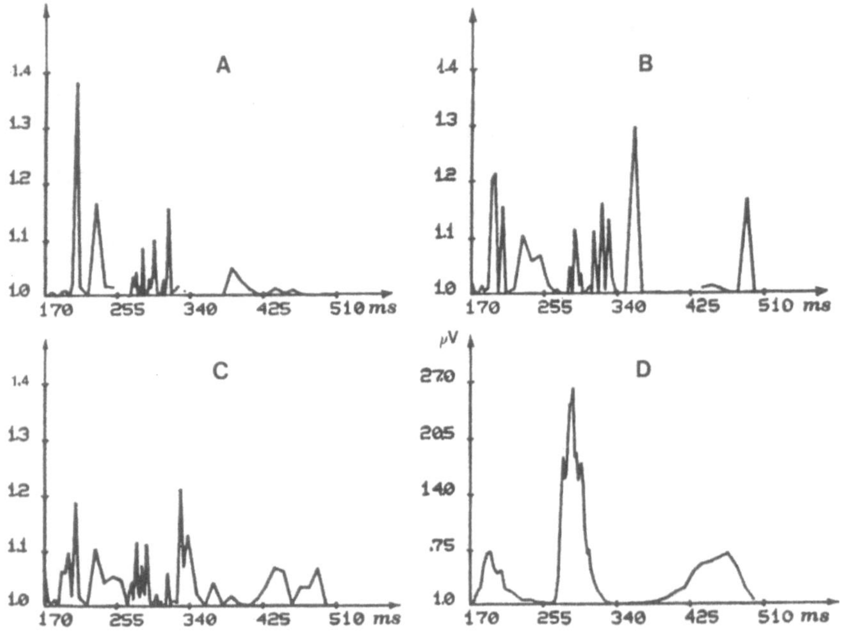


Fig. 3

We discuss now the problem of choosing a quasi optimal smoothing parameter $\varepsilon$. Since in the experimental applications a good estimate of the variance of the overall noise on the data, at any time instant of the heart beat, is not known, for estimating $\varepsilon$ from the data we considered methods not requiring a knowledge of an estimate of the variance. We compared the performance of the maximum likelihood estimate (ML)(see [1]), of the generalized cross validation estimate (GCV) (see [21],[14]) and of an empirical estimator which we shall call Composite Residual and Smoothing Operator (CRESO). The CRESO estimate consists in finding the greatest relative maximizer of $C(\varepsilon)= \|Ru_{\varepsilon}\|^2+2\varepsilon\frac{d}{d\varepsilon}\|Ru_{\varepsilon}\|^2$. The performance of these estimators is shown in Fig.3, (A) for CRESO, (B) for GCV and (C) for ML, respectively, where we report the ratio between the RE of the computed epicardial maps choosing $\varepsilon$ according to the estimators and the optimal RE; discontinuities in the graphs denote the failure of the estimator in the wide range of $\varepsilon$ explored. These results were obtained from the simulated surface data adding a noise vector $\underline{e}$ of $\sigma$=5. Fig.3 (D) shows the signal to noise ratio S/N over the 62 time instants of the heart beat. We remark the average low value of S/N and its high variability. In general all the three estimators were successfully applied and succeeded in recognizing a quasi optimal estimate of $\varepsilon$ when a normally distributed pseudo random noise level was added to the simulated surface data. The ML estimator yields the average best results. Moreover these results showed a good performance of the CRESO estimator. We analyzed the empirical estimator CRESO because the statistical estimators (ML), (GCV) did not work when the regularization procedure was applied to the experimental surface data. In this case we consider the CRESO estimator and other two empirical estimators:the sensitivity (SE) and the quotient (QE) estimators (see [19], [16]). The comparison of the performance over the entire heart beat, when using experimental data, revealed that CRESO worked better than SE and QE. Since the SE estimator did frequently not work, in Fig.4, we compare the performance of CRESO (solid line) and QE (dashed line) reporting the ratio between the estimated RE and the optimal RE. We concluded that CRESO criterion proved to be more effective than QE except for 30 msec. at about the end of QRS complex. We report here one example of the results (for

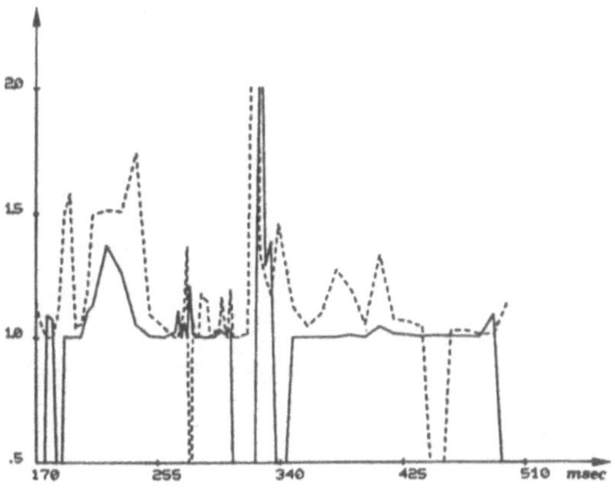


Fig. 4

others see [8] ) concerning the time instant 8 msec. after the beginning of the QRS complex. In the following Fig.5 we denote by: MRE the relative error, CC the correlation coefficient, M+, M- the absolute

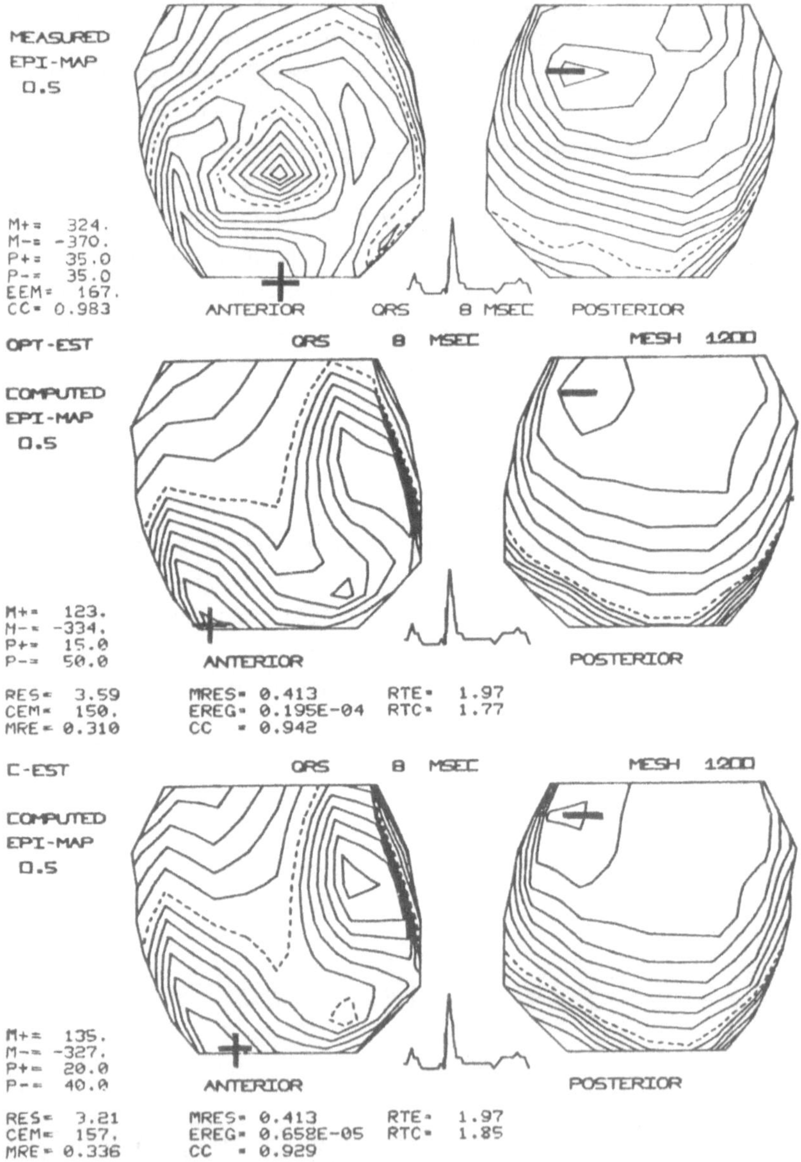


Fig. 5

maximum and minimum in $\mu V$ (microvolt) and P+, P-, are the voltage step between positive and negative equipotential lines specified in $\mu V$, respectively. Moreover, $RES = \|T\underline{u}_{\varepsilon} - \underline{z}_m\| / \sqrt{m}$, $MRES = \min_v \|Tv - \underline{z}_m\| / \sqrt{m}$, $EEM = \|\underline{u}_e\| / \sqrt{n}$, $CEM = \|\underline{u}_{\varepsilon}\| / \sqrt{n}$, $EREG = \varepsilon$, $RTE = \|\underline{u}_e\| / \|\underline{z}_m\|$, $RCC = \|\underline{u}_{\varepsilon}\| / \|\underline{z}_m\|$.

Fig.5 shows the maps at 0.5 cm from the epicardium respectively, measured, computed with optimal $\varepsilon$ and computed with $\varepsilon$ estimated by CRESO.

Finally, the regularization procedure with CRESO estimator was applied to the experimental surface maps measured at 230 sites on the chest surface of a normal human subject. In building the transfer matrix the inhomogeneities of the thorax were not taken into account.

In general the activation sequence of the myocardium which could be inferred from the estimated "epicardial" maps (see [9], [10]) was compatible with the excitation pathways described in literature.

Acknowledgement: The author wishes to thank Mrs. I.Heitz for her careful typing of the camera-ready manuscript.

## References

[1] Anderssen R., Bloomfield P.: A time series approach to numerical differentiation. Technometrics 16, 69-75, 1974.

[2] Baruffi S., Spaggiari S., Stilli D., Musso E. Taccardi B.: The importance of fiber orientation in determining the features of cardiac electric field. In Modern Electrocardiology, Z. Antaloczy ed., Excerpta Medica, pp. 89-92, Amsterdam 1978.

[3] Colli Franzone P., Guerri L., Viganotti C., Macchi E., Baruffi S., Spaggiari S., Taccardi B.: Potential fields generated by oblique dipole layers modeling excitation wavefront in the anisotropic myocardium. Comparison with potential fields elicited by paced dog hearts in a volume conductor. To appear on Circulation Research.

[4] Colli Franzone P., Guerri L., Viganotti C., Macchi E., Baruffi S., Spaggiari S., Taccardi B.: A mathematical model simulating the anisotropy of cardiac electric sources.In Electrocardiology 81, Z. Antaloczy, I.Preda eds., pp. 113-118, Akdémiai Kiado, Budapest 1981.

[5] Colli Franzone P., Guerri L., Taccardi B., Tentoni S., Viganotti C.: Cardiac fibers orientation and potential fields: model studies. In Proceeding of the Int. Cong. on Electrocardiography, Tokyo, 1982.

[6] Colli Franzone P., Guerri L., Viganotti C.: Oblique dipole layer potentials applied to electrocardiology. To appear on Journal of Math. Biol.

[7] Colli Franzone P., Guerri L., Magenes E.: Oblique double layer potentials for the direct and inverse problems of electrocardiology. Forthcoming paper.

[8] Colli Franzone P., Guerri L., Taccardi B., Viganotti C.: The direct and inverse potential problems in electrocardiology. Numerical aspects of some regularization methods and application to data collected in isolated dog heart experiments. Pubbl. N. 222, of the Institute of Numerical Analysis, I.A.N.-C.N.R., Pavia 1979.

[9] Colli Franzone P., Guerri L., Viganotti C.: The inverse potential problem applied to the human case. In Models and Measurements of cardiac electric field, E. Schubert ed., pp. 19-33, Plenum Pub. Corp., 1982.

[10] Colli Franzone P., Guerri L., Taccardi B., Viganotti C.: A numerical procedure to inversely compute epicardial potentials from body surface maps applied to a normal human subject. In Computer Cardiology, pp. 187-190, IEEE Computer Society Press, 1981.

[11] Colli Franzone P., Magenes E.: On the inverse potential problem of electrocardiology. Calcolo XVI, IV, pp. 459-538, 1979.

[12] Corbin L.V.II, Scher A.M.: The canine heart as an electrocardiographic generator: Dependence on cardiac cell orientation. Circ. Res. 41, pp. 58-67, 1977.

[13] Cottini C., Dotti D., Gatti E., Taccardi B.: A 240-probe instrument for mapping cardiac potentials. In the electrical field of the heart P. Rijlant ed., pp. 99-102, Presses Académiques Européennes, 1976.

[14] Golub G., Heath M., Wahba G.:Generalized cross validation as a method for choosing a good ridge parameter. Technometrics 21, 2,pp. 215-223, 1979.

[15] Golub G., Reinsch C.: Singular value decomposition and least squares solutions. Num. Math., 14, pp. 403-420, 1970.

[16] Leonov A.S.: Justification of the choice of the regularization parameter according to quasi-optimality and quotient criteria. U.S.S.R. Comput.Math. Phys., 18, pp. 1-15, 1979.

[17] Lions J.L., Magenes E.: Non homogeneous boundary value problems and applications. vol. 1,Grundlehren 181, Springer Verlag, Berlin, 1971.

[18] Plonsey R., Fleming D.: Bioelectric phenomena. Mac Graw-Hill, New York 1969.

[19] Tykhonov A.N., Arsenine V.:Méthodes de résolution de problémes mal posée. Ed. Mir, Moscow, 1976.

[20] Van Loan C.: Generalizing the singular value decomposition. SIAM J. Numer. Anal., 13, 1, pp. 76-83, 1976.

[21] Wahba G.: Practical approximate solution to linear operator equation when the data are noisy. SIAM J. Numer. Anal., 14, 4, 1977.

# DETERMINATION OF COEFFICIENTS IN RESERVOIR SIMULATION

Richard E. Ewing

## 1. Introduction

The process of determining unknown parameter values, such as porosity and permeability, which are used in a mathematical reservoir model to give the best fit to measured well production history is commonly called "history matching." Theoretically, one would like to have an automatic routine for history matching which is applicable to simulators of varying complexity and which can determine a set of parameters giving a good history match with a reasonable amount of computational time and effort. In this paper we shall survey and compare some of the methods proposed in the literature for automated history matching, emphasizing the inherent difficulties which must be addressed, and indicating some possible future directions for research in this area.

In recent years a number of authors have investigated the subject of history matching. Although an inhomogeneous reservoir physically requires an infinite number of parameters for complete property description, a computational reservoir simulator can allow specification of only a finite number of parameters. The most straightforward approach for specifying general properties in a finite difference or finite element simulator which utilizes a spatial grid network is to allow porosity and permeability to vary independently within each grid block or element. In field scale simulations using up to 25,000 grid blocks, this could require an algorithm for determining 50,000 or more unknowns simultaneously. While minimizing the modeling error, this technique generates an extremely difficult and ill-conditioned optimization problem. Therefore, the feature of the history-matching problem in reservoir simulation that distinguishes it from parameter estimation problems in other fields of science and engineering is the large dimensionality of both the system state and the unknown parameters. To reduce both the statistical uncertainty and the computational complexity, one must decrease the number of unknowns and if possible, utilize any additional available information.

One commonly used approach for reducing the number of unknowns is to divide the reservoir into a relatively small number of zones and to assume uniform reservoir properties within each zone. As the number of zones is decreased for computational tractability, the modeling error increases. The specification of an optimal level of zoning has been discussed by several authors [16,17,34,43].

An alternative to parametrization is to use prior geological information in the form of an assumed probability distribution for the reservoir properties considered as random variables. This information need not be limited to measurements on the reservoir under study but can be based on geological information about property variability in reservoirs of the same type. A form of Bayesian estimation then can be employed to determine the unknown properties. In applying this technique one must consider the effects of the error in the prior statistics employed upon the results. The Bayesian techniques have been compared with zonation methods in the literature [29,43].

The reliability of the parameter estimates from history matching has been studied by several authors [14,16,17,34,43,44]. Nonlinear regression analysis is used to obtain standard deviations and correlation coefficients for reliability estimates. The assumptions required for this nonlinear regression analysis are; (1) parameter estimates are unique, (2) the model can be approximated by linearization around the parameter estimates, and (3) measurement errors in the pressure are normally distributed with zero mean and some standard deviation. Under these assumptions confidence limits for the unknown parameters can be determined.

Each of the history matching methods mentioned earlier involve various mathematical programming (finite-dimensional) optimization techniques. More recently variational or optimal control (infinite-dimensional) methods have been applied to history matching problems [8,10,11,16,17,29,43,48,50]. These techniques have realized an important saving in computing time in automatic history matching of single-phase flow regimes. However, these methods are currently limited to first-order gradient iterative methods which generally converge more slowly than those of higher order. Nevertheless, the amount of computation required per iteration is significantly less than that required for higher-order optimization methods [50] and first-order methods are attractive for automatic history matching. The variational methods allow determination of the gradient of the performance index with respect to the parameters in a manner that circumvents calculation of the sensitivity coefficients. If one wishes to use a higher order iterative method like the Gauss-Newton or Marquardt method, a more complex variational approach allows the efficient calculation of the necessary sensitivity coefficients [20]. The optimal control approach has been applied to determination of relative permeability curves in multi-phase conditions recently [10,50].

In Section 2, we shall describe some representative cases of reservoir simulation formulations and the basic concepts of history matching. We shall discuss both single-phase and multi-phase problems. In Section 3, we shall present some error estimates on a much simpler, but closely related, problem and discuss the origin and importance of various sources of error. Section 4 will contain a brief survey and comparison of some of the techniques which have appeared in the literature. Finally, in Section 5 we shall discuss some of the important difficulties which still complicate the history matching process in reservoir simulation and indicate critical directions for additional research.

## 2. Formulation of History-Matching Problems

We shall first consider nonsteady-state flow of a slightly compressible oil with viscosity $\mu$ in a thin horizontal uniform radial reservoir of thickness $h$, wellbore radius $r_w$, outer boundary radius $r_e$, permeability $k$ and porosity $\phi$. The fluid properties, viscosity $\mu$ and compressibility $c$, are assumed to be known while the properties of the reservoir, porosity $\phi$ and permeability $k$ are assumed to be unknown (constants at first). The porosity is the fraction of void space in the reservoir through which the fluid can flow while the permeability is the measure of the "ability of the fluid to flow in the reservoir" defined by Darcy's Law:

$$\vec{u} = -\frac{k}{\mu}\left\{\frac{\partial p}{\partial r} - \rho\vec{g}\right\} \tag{2.1}$$

where $\vec{u}$ is the superficial or Darcy velocity of the fluid, $p$ and

$\rho$ are the pressure and density of the fluid, $\vec{g}$ is the gravity vector and $r$ is the radial distance measure in the reservoir. Since the reservoir is assumed horizontal for exposition purposes, we shall omit the gravity terms in (2.1).

The mathematical model which determines the pressure distribution as a function of time and space in the reservoir during primary recovery is:

$$
\begin{aligned}
&\text{a)} \quad \phi c \frac{\partial p}{\partial t} = \frac{1}{r}\frac{\partial}{\partial r}\left(\frac{k}{\mu} r \frac{\partial p}{\partial r}\right) &&, \ 0 < r_w < r < r_e, \ 0 < t \le T, \\
&\text{b)} \quad p(r,0) = p_o &&, \ 0 < r_w < r < r_e, \\
&\text{c)} \quad \frac{\partial p}{\partial r}(r_e,t) = 0 &&, \ 0 < t \le T, \\
&\text{d)} \quad \frac{2\pi h k r_w}{\mu}\frac{\partial p}{\partial r}(r_w, t) = q, &&, \ 0 < t \le T,
\end{aligned}
\qquad (2.2)
$$

where $p_o$ is an initial pressure distribution throughout the reservoir and $q$ is the volumetric flow rate which is assumed to be uniform on the well bore $r_w$. The only place that data from the flow process can be observed is at the wells. However either the pressure $p$ or the flow rate $q$ (but not both) is observable at the well. One of these quantities must be controlled or specified while the other is measured.

In the history matching process a quantity, say pressure, is measured at the well bore. Then choices of the unknown parameters are chosen so that the results of a simulator run using the particular choice of parameters in the model will be as close to the measured values in a specified sense throughout the production history. Since our choice of parameters is based upon this comparison, we must have as accurate values of pressures from our simulator as possible. In Section 5 we shall see that use of standard numerical discretization schemes will obtain fairly accurate pressures at the well, while approximations to the flow rate at the well do not even converge! Thus, for effective history matching, one should try to see that the well is produced by controlling the flow rates and measuring the pressures if possible.

However, the person who does the history matching via a simulator is rarely able to influence how the well is produced, and often finds that bottom hole pressures have been specified and flow rates measured.

In this situation, one should use alternate discretization procedures in order to obtain accurate computed flow rates at the well with which to compare in the history-matching algorithm. Until this problem is addressed in Section 5, we shall assume that flow rates have been controlled and the pressures at the wells are the observables:

$$p(r_w, t_k) = \tilde{p}_k \qquad , \; k = 1,2,\ldots, N_o, \tag{2.3}$$

where $N_o$ is the number of observations taken at the well through time.

In the general case of primary recovery, or recovery of oil using the resident reservoir pressures, usually several production wells are utilized on a larger scale. If $\Omega$ denotes a general reservoir of normalized uniform thickness $h=1$ and outer boundary $\partial\Omega$, our mathematical model for the pressure $p$ can be given by

$$\begin{aligned}
&\text{a)} \quad \phi c \frac{\partial p}{\partial t} = \nabla \cdot \frac{k}{\mu}\left\{\nabla p - \rho \vec{g}\right\} + \sum_{i=1}^{N_w} q_i \delta(x - x_i, y - y_i), \\
&\qquad\qquad (x,y) \in \Omega, \; 0 < t \leq T, \\
&\text{b)} \quad p(x,y,0) = p_o(x,y), \quad (x,y) \in \Omega \\
&\text{c)} \quad \frac{\partial p}{\partial \nu} = 0, \quad (x,y) \in \partial\Omega, \quad 0 < t \leq T,
\end{aligned} \tag{2.4}$$

where $N_w$ is the number of producing wells with flow rates $q_i$ at the location $(x_i, y_i)$, $i = 1,2,\cdots,N_w$ and $\frac{\partial p}{\partial \nu}$ is the outward normal derivative of $p$ on the boundary $\partial\Omega$.

We note that due to the small size of the well bore in comparison to the dimensions of the reservoir, the model for the flow at the wells has been changed from the Neumann boundary condition given in (2.2.d) to a model using a Dirac delta function $\delta(x - x_i, y - y_i)$ as a source term in (2.4.a). In the process of switching models, we have lost the explicit notion of a well bore radius in (2.4). However, since the observables are pressures at the well bores, as in (2.3), one must reintroduce the size of the well bore into the system (2.4) by using a special well model which describes radial flow at the wells. Typical well models have been described by Matthews and Russell [36], Odeh [37], and Peaceman [38,39] and will not be discussed in detail here except as a possible source of modeling error.

The Observables from (2.4) are of the form

$$p(x_i,y_i,t_k) = p_{ik}, \quad i = 1,\cdots,N_w, \; k = 1,\cdots,N_o. \tag{2.5}$$

We note here that if the fluid is incompressible, $c = 0$ and (2.4.a) is an elliptic partial differential equation while, for compressible fluids, (2.4.a) is a parabolic partial differential equation. Therefore, since $c$ is often small, the numerical techniques for approximating the solution of (2.4) must be able to treat both types of equations. Also for the incompressible or time independent case, the initial pressure $p_o$ is not specified and pressure observations are made at the wells only once in time for the steady-state problem. This decrease in data makes the incompressible case significantly more difficult as a history matching problem for single-phase flow.

Simulation of secondary and tertiary recovery processes are inherently more complex than for single-phase flow. Enhanced recovery processes include waterflooding, miscible and immiscible displacement, chemical flooding, and various thermal processes (steamflooding, in-situ combustion, etc.). Simple models for both miscible and immiscible displacement can be put in the form (see [9,24,38])

$$\begin{aligned}
&\text{a)}\quad \phi c \frac{\partial p}{\partial t} = \nabla \cdot k\lambda \nabla p + \sum_{i=1}^{N_w} q_i \delta(x - x_i, y - y_i) \\
&\qquad\qquad \equiv -\nabla \cdot u + Q, \quad (x,y) \in \Omega, \quad 0 < t \leq T, \\
&\text{b)}\quad u \cdot \nu = 0, \quad (x,y) \in \partial\Omega, \quad 0 < t \leq T, \\
&\text{c)}\quad \phi \frac{\partial s}{\partial t} + \nabla \cdot f(s)u - \nabla \cdot D\nabla s = \sum_{i=1}^{N_w} q_i \delta(x - x_i, y - y_i)\tilde{s}_i, \\
&\qquad\qquad (x,y) \in \Omega, \quad 0 < t \leq T, \\
&\text{d)}\quad f(s)u \cdot \nu - D \frac{\partial s}{\partial \nu} = 0, \quad (x,y) \in \partial\Omega, \quad 0 < t \leq T,
\end{aligned} \tag{2.6}$$

where $k\lambda$ is the total mobility, $p$ and $u$ are the pressure and Darcy velocity of the total fluid, respectively, $s$ is the saturation or fraction of the available pore space occupied by the invading fluid, and $\tilde{s}_i$ is the saturation of the invading fluid at the $i^{th}$ well, specified at injection wells and measured at production wells. For immiscible displacement, $f(s)$ is defined by [15]

$$f(s) = \frac{1}{2}\left(\frac{k_{rr}}{\mu_r} - \frac{k_{ri}}{\mu_i}\right) \tag{2.7}$$

where $\mu_i$ and $\mu_r$ are the viscosities of the invading and resident fluids and $k_{ri}$ and $k_{rr}$ are the respective relative permeabilities for the fluids while for the miscible displacement process $f(s) \equiv s$. In the immiscible displacement regime we have [15]

$$D(s) = -k \frac{k_{ri}}{\mu_i} \frac{k_{rr}}{\mu_r} \left(\frac{k_{ri}}{\mu_i} + \frac{k_{rr}}{\mu_r}\right) \frac{\partial pc}{\partial s} \geq 0 \tag{2.8}$$

which degenerates to zero for $s = 0$ (or the irreducible saturation of the invading fluid) or $s = 1$ (or the irreducible saturation of the resident fluid), while for miscible displacement, we have [15,24,25] the diffusion-dispersion tensor

$$\left[D_{ij}(x,u)\right] = \phi d_m I + \frac{d_\ell}{|\underset{\sim}{u}|}\begin{pmatrix} u_1^2 & u_1u_2 \\ u_1u_2 & u_2^2 \end{pmatrix} + \frac{d_t}{|\underset{\sim}{u}|}\begin{pmatrix} u_2^2 & -u_1u_2 \\ -u_1u_2 & u_1^2 \end{pmatrix} \tag{2.9}$$

where $u = \left(u_1, u_2\right)$ is the total Darcy velocity from (2.6.a) and $d_m, d_\ell$, and $d_t$ are the molecular diffussion and longitudinal and transverse dispersion coefficients, respectively. Since $D$ from (2.9) does not degenerate as in (2.8), we have a transport dominated diffusion process in the miscible displacement regime. Finite difference and finite element error estimates for (2.6) in both the miscible and immiscible regimes appear in [15,23-25].

The observables for (2.6) for history matching purposes are a) pressures at the well bore as in (2.5), b) relative flow rates of the fluids at the wells, and c) the time it takes for the invading fluid to reach the production well (known as breakthrough times) for each well. Thermal and compositional simulators are also based upon systems such as (2.6) but with the added complexities of phase packages, thermal constraints, and mass exchange between phases and components. Thus history matching with these simulators will be even more difficult.

In order to get a good initial approximation to the porosity and permeability in the vicinity of the wells, local measurement techniques are used. First cores taken from the wells during the drilling process are subjected to a variety of experiments in a controlled laboratory environment to determine local porosities and permeabilities within the core samples [35]. Then various logging tools are lowered down the bore hole and electrical, acoustical, radiation, or electromagnetic logs are taken to determine rock and fluid properties in the neighborhood of the wells.

Once good local information around the wells is known, the reservoir engineer uses whatever seismic or geological information is available as well as correlations with information from similar reservoirs or sands [28,41] to extrapolate the porosity and permeability data at the wells throughout the reservoir. This serves as an initial guess of the unknown parameters which is then updated through simulation using these values to obtain the best history match possible.

The history matching problem can then be viewed as a nonlinear, constrained, optimization (least-squares) problem of minimizing the sum of the squares of differences between observed and calculated values. For example, if pressures at $N_w$ wells are the observables, at times $t_1, t_2, \cdots, t_{N_o}$ and $N$ unknown parameters $\alpha_i$ are to be determined, the following problem is formulated:

Problem: Find a vector $\vec{\alpha} = (\hat{\alpha}_1, \cdots, \hat{\alpha}_N)$ that minimizes

$$J = \sum_{i=1}^{N_w} \sum_{k=1}^{N_o} w_{ik} \left[ p^*(x_i, y_i, t_k) - P^*(x_i, y_i, t_k; \vec{\alpha}) \right]^2 \tag{2.10}$$

where $p^*(x_i, y_i, t_k)$ is the measured pressure at well $i$ and time $t_k$, $P^*(x_i, y_i, t_k; \vec{\alpha})$ is the computed pressure from the simulator at well $i$ and time $t_k$ using the vector $\vec{\alpha}$ as parameters, and $w_{ik}$ are specified weights. The weights $w_{ik}$ are used to give a bias to certain data which may be believed to be more or less accurate than other data. If certain a priori knowledge is assumed about the mean values or correlations of the unknown parameters, the objective functional $J$ in (2.10) could be modified to include a Bayesian term penalizing the weighted deviations of the parameters from their mean values (see [29]). Alternately penalty terms biasing assumed correlations could be added to $J$.

For gradient-type optimization procedures, the value of the simulator at any computational point $\ell$ can be adjusted by a suitable choice of the vector $\delta\vec{\alpha}$ in the following manner:

$$P^*_\ell\left(\vec{\alpha}_o + \vec{\delta\alpha}\right) = P^*_\ell\left(\vec{\alpha}_o\right) + \sum_{j=1}^{M} \frac{\partial P^*_\ell}{\partial \alpha_j}\,\delta\alpha_j, \qquad (2.11)$$

where the direction and magnitude of $\vec{\delta\alpha}$ is chosen via the optimization process. The matrix $\left[\frac{\partial P^*_\ell}{\partial \alpha_j}\right]$ for $\ell = 1,\cdots,M$ (M is the number of cells in the discretization) and $j = 1,\cdots,N$ is called the <u>sensitivity</u> matrix ; $\frac{\partial P^*_\ell}{\partial \alpha_j}$ measures the sensitivity of $P^*_\ell$ to changes in parameter $\alpha_j$. The sensitivity coefficients are strongly related to the accuracy with which we can solve the optimization problem, as is indicated in the next section. The method of calculation of these coefficients is the distinguishing factor in several of the automatic history-matching algorithms in the literature.

## 3. Sources of Error in History Matching

There are many different sources of error in the history matching process. In this section we shall illustrate means by which error is introduced into the process via an error analysis on a much simpler problem. Complete error bounds are usually impossible to obtain because the problem does not admit an analytic solution from which to obtain accurate sensitivity coefficients. Even when error bounds are attainable, they are usually overly pessimistic for realistic error estimation. However, they are valuable in understanding the sources and relative magnitudes of errors. Assume for the purposes of illustration that a constant permeability $k$ is the only unknown parameter in the problem (2.2), and only one observation time $t_o$ is used to determine the one unknown.

We shall first introduce some notation. Let $p$ be the actual pressure at the well at observation time $t_o$ which is measured as $p^*$, subject to a measurement error. The data for the mathematics model (2.2), $\phi, c, \mu, p_o$, and $q$, are also subject of measurement error and are given as $\phi^*, c^*, \mu^*, p_o^*$, and $q^*$. Let $p(k)$ and $p^*(k)$ be the solutions of the mathematical models with parameter $k$ using the data in (2.2) and the measured data, respectively. Let $P^*(\tilde{k})$ be the solution of the simulator with measured data $(\phi^*, c^*, \text{etc.})$ and input parameter $\tilde{k}$. Let $\frac{\partial p}{\partial k}(x,y,t,k)$ denote the sensitivity of the mathematical model (2.2) to the parameter $k$.

For the simple problem (2.2) with constant coefficients, one can write down an analytic solution for $p$ at any point given any parameter $k$. This solution defines a function $p:k \to p(r_w,t;k)$ whose derivative with respect to $k$ is the sensitivity coefficient described above. If $k^*$ is the approximate solution of the minimization problem from (2.10), we can use the Mean Value Theorem and the triangle inequality to see that:

$$
\begin{aligned}
|k - k^*| &\leq \sup_{x,t,k} \left|\frac{\partial p}{\partial k}\right|^{-1} |p(k) - p(k^*)| \\
&\leq \sup_{x,t,k} \left|\frac{\partial p}{\partial k}\right|^{-1} \Big\{ |p(k) - p| + |p - p^*| \\
&\quad + |p^* - P^*(k^*)| + |P^*(k^*) - p^*(k^*)| + |p^*(k^*) - p(k)| \Big\} \qquad (3.1) \\
&\equiv \sup_{x,t,k} \left|\frac{\partial p}{\partial k}\right|^{-1} \Big\{ T_1 + T_2 + T_3 + T_4 + T_5 \Big\}.
\end{aligned}
$$

Error is introduced into the process in the following ways: $T_1$ indicates modeling errors, $T_2$ indicates measurement errors, $T_3$ denotes errors introduced by approximately solving the optimization problem, $T_4$ encompasses all the errors, roundoff and discretization, introduced by the numerical simulator, and $T_5$ describes the errors introduced into the model by using measured data. Clearly each of these errors is magnified or diminished depending upon how sensitive the model is to the unknown parameter, measured by $\partial p/\partial k$.

These error sources were estimated for the problem of determining a constant coefficient in a one-dimensional heat flow in [22] together with their effects upon the tolerance bounds for the unknown parameter. The concept of designing an experiment to produce a more favorable sensitivity coefficient and thus better error estimates was also discussed. In this way, if one understands the sources of error in the process and can estimate their sizes, he can develop better experiments or better models to increase the sensitivity to the important parameters and reduce modeling error, or improve the numerical procedures or optimization techniques to decrease the respective errors.

Techniques for determining both $\phi$ and $k$ simultaneously appear in [3,4,22]. See bibliographies of these papers for other work in this area by Cannon, Douglas, DuChateau, and Jones. Techniques for determining unknown coefficients for nonlinear heat conduction problems appeared in [5,6]. Error estimates were also obtained for determination of spatially varying coefficients in two space dimensions for groundwater flow problems by Falk in [27]. This corresponds to the incompressible case of (2.4). Falk's estimates pointed out problems with stagnation and the difficulties for time-independent problems when the data is accessible only at the wells which are widely separated.

## 4. Literature Survey

The mathematical programming problem arising in history matching has received considerable attention. Jacquard [32,33] first addressed the problem. Essentially all the early studies [7,12,33,34,44,46,47] proposed finite dimensional techniques using either one unknown parameter per grid block or zonation, the assumption that large zones with constant properties should be used to match the reservoir performance. There is a clear trade-off between the number of zones with respect to computational effort versus accuracy. In this context an optimum level of description was discussed by several authors [16,17,19,29,34,43]. A nice survey of much of this early work appears in Dogru, et al. [17].

A different approach for reducing the number of parameters is the use of additional information about the means, variances or correlations of the unknown parameters obtained from previous geological data. The prior geological information is incorporated in the optimization problem through a formulation akin to Bayesian estimation. The objective functional is modified to bias the convergence toward expected mean values or other statistical knowledge of the paramaters as discussed in Section 2. Gavalas, et.al. [29], studied various practical aspects of Bayesian estimation concerning the reservoir problem and compared the Bayesian method with the zonation method. Shah, et.al. [43], derived the error covariance of the Bayesian estimates, studied the effects of errors in the assumed prior statistics, and compared the estimates with those obtained by parametrization by zonation.

As we mentioned in Section 3, error estimates are not usually practical for reservoir problems. However the reliability of the parameter estimates were studied quantitatively via statistical methods by several authors [3,14,16-18]. From the residual of the history match, the statistical variance of the match was obtained:

$$\sigma_p^2 \cong \frac{J \min}{N_w N_o - N}, \tag{4.1}$$

where J is from (2.10), $\sigma_p^2$ is the variance of the pressures, $N_w N_o$ is the number of data points and N is the number of parameters to be determined. $\sigma_p$ is the standard deviation of the pressure. Let $a_j$ be the estimates of $\alpha_j$ obtained from the minimization procedure. The variance of each estimated parameter $a_j, j = 1,\cdots,N$, is calculated from the nonlinear least-squares theory [17]

$$\sigma_{a_j}^2 = C_{jj}\sigma_p^2, \quad j = 1,\cdots,N, \tag{4.2}$$

where $C_{jj}$ are the diagonal elements of the matrix $C = (S^T WS)^{-1}$, S is the sensitivity matrix defined in Section 2 and W is a selected weighting matrix (usually the identity matrix). Then individual confidence limits for each parameter $\alpha_j$ are given by [17]

$$a_j - t_{1-\beta/2}\sigma_{a_j} \leq \alpha_j \leq a_j + t_{1-\beta/2}\sigma_{a_j}, \quad j = 1,\cdots,N, \tag{4.3}$$

where t is obtained from the student t test with β risk level. Confidence limits for the pressures are obtained in a similar manner [17].

One point that distinguishes several of the history matching algorithms in the literature is the method by which the sensitivity coefficients, defined in Section 2, are calculated. Dogru and Seinfeld [20] give a survey and comparison of various techniques for computing sensitivity coefficients. In certain special cases when an analytic solution for the problem can be found, the sensitivity coefficients can be computed analytically [19,22]. However, in general, they must be computed via the simulator or special code. Usually the calculation of the sensitivity coefficients comprises the most time-consuming aspect of the history match.

The most straightforward method of numerically calculating the coefficients is through finite differences, using the simulation code [14,16-18]. A full simulation run is made with a fixed set of parameters, then N more runs are made, incrementing each unknown parameter, one at a time. Therefore this technique requires $N + 1$ full simulations or solution of $(N + 1)K$ differential equations (if there are $K$ cells in the reservoir) in order to obtain the sensitivity matrix. The advantage of this technique is that no additional code is required.

Another direct calculation method can be derived by first considering the set of $K$ ordinary differential equations produced by a spatial discretization of the problem (2.4). We obtain the system

$$\frac{d\vec{P}}{dt} = A\vec{P} + f, \quad \vec{P}(0) = p_o, \tag{4.4}$$

where the matrix $A$ depends upon the choice of finite difference or finite element spatial discretization used. Since the sensitivity coefficients are

$$s_{\ell j} = \frac{\partial P_\ell}{\partial \alpha_j}, \quad \ell = 1,\cdots,N, \tag{4.5}$$

we can obtain a set of equations governing the sensitivity coefficients by differentiating (4.4) above with respect to $\alpha_j$ to obtain

$$\begin{aligned} &\text{a)} \quad \frac{d\vec{s}_j}{\partial t} = A\vec{s}_j + \frac{\partial A}{\partial \alpha_j}\vec{P}, \quad j = 1,\cdots,N \\ &\text{b)} \quad \vec{s}_j(0) = 0, \quad j = 1,\cdots,N. \end{aligned} \tag{4.6}$$

If we then solve (4.4) to obtain $\vec{P}$ and then solve (4.6) $N$ times, we can obtain the sensitivity matrix. This direct approach again requires the solution of $(N + 1)$ simulations or $(N + 1)K$ ordinary differential equations. This technique requires only a modification of the code for the right hand side of (4.4) and by emphasizing the time-dependence of the sensitivity coefficients gives a better understanding of the sensitivity through time.

We next consider a variational approach to coefficient calculation. Define the adjoint to (4.4) as

$$\begin{aligned} &\text{a)} \quad \frac{d\psi}{dt} = -\psi A \\ &\text{b)} \quad \psi(t,t) = I. \end{aligned} \tag{4.7}$$

Using the solutions $P$ and $\psi$ to (4.4) and (4.7) respectively, we can, analogously to (4.6), obtain a set of integral equations for the sensitivity coefficients [20]

$$\vec{s}_j(t) = \int_0^t \psi(t,\tau)\frac{\partial A}{\partial \alpha_j}\left[\psi(t,0)p_o + \int_0^\tau \psi(t,s)\vec{f}(s)ds\right]d\tau. \tag{4.8}$$

The variational method requires the solution of $K^2$ adjoint equations plus the $K$ equations in (4.4) followed by $N$ integral equations. As described in [20] for small $N$ compared to $K$, the direct method requires less work, but as $N$ increases the variational method becomes more economical.

If the sensitivity matrix is computed, then higher than first order gradient methods can be utilized in the optimization procedure. However, we have seen that this is computationally expensive. First-order gradient methods generally converge more slowly than those of higher-order but the amount of computation required per iteration is significantly less. Thus first order methods are also useful for history matching. For an iterative first-order method, the sensitivity coefficents are not required as long as the gradient of the performance index with respect to the parameters can be computed [11]. Variational or optimal control methods [8,10,11,16,17,29,43,48-50] can produce the necessary information through the solution of only $2K$ ordinary differential equations via one standard simulation run and a run using the relevant adjoint equations. Recently, optimal control techniques have been extended to address the much more difficult problem of determination of relative permeability curves in multi-phase immiscible problems described by (2.6). The forward simulation of these problems is considerably more complicated and must be significantly improved before routine field scale history matching on these problems are reasonable.

Uniqueness and identification of spatially varying properties have recently received increased attention in groundwater hydrology. Falk [27] and Richter [42] addressed the incompressible flow problem. Cannon and DuChateau considered nonlinear equations with additional boundary data [6] and proposed an algorithm to determine properties from a two-parameter family [21]. Hornung and Messing [31] pointed out uniqueness problems in the identification of soil parameters from infiltration data.

## 5. Difficulties Remaining for History Matching

There are many difficulties still affecting the history matching process. In this section we shall point out a few of these difficulties and suggest some directions for future research. As discussed in Section 3 the major sources of error come from measurement errors, modeling errors, and the inability to solve the nonlinear least-squares problem adequately. Also, there are serious problems with uniqueness, observability, and the general question of well-posedness of the parameter identification problem.

Observability plays a crucial role in the history matching process. Depending upon the locations of the wells and the number of data points available, it may not be possible to uniquely determine all of the unknown parameters. Clearly, one cannot determine more parameters than he has data points. Also a major problem is that observations can only be made at the wells. For example, assume a reservoir were square with an injection well at one corner and a production well at the opposite corner. There is a line of symmetry between the two wells. For an arbitrary distribution of parameters throughout the reservoir, the complete production history for all time would be exactly the same if the original distribution were mirrored about the symmetry line. Thus we have a uniqueness problem even if infinite data is available if we have symmetries of any form in the reservoir geometry.

Also, at a given well, the observed data might reflect the values only in specific zones of the reservoir due to stagnation regions or other geometries. Thus certain parameters may not be observable at a well, or the well data might be very insensitive to certain parameters (the sensitivity coefficients are zero or near zero). One should try to determine these insensitive observations and weight them with a small or zero weight $w_{ij}$ in the objective functional $J$.

The basic problem is the lack of sufficient useful data. In a control theory setting, having observations only at a few wells is akin to trying to control a multi-dimensional surface with a few point controls and is thus an extremely difficult problem especially due to the large size of the reservoir problems. One idea for trying to increase the data in a significant way and greatly aid in the observability problem is to try to augment the standard data with seismic data during production to try to get a better feel for the location of the interface between the injected and resident fluids during an enhanced recovery problem. Due to the inherent difficulty of flooding problems and the economics involved in developing an understanding of the complex flooding mechanisms, seismic augmentation or other techniques to get better observations is definitely feasible.

A very important source of error in the simulation process is measurement error. Techniques for obtaining more accurate, more continuous, and more reliable data are under development. At this point the effects of this measurement error upon the history matching problem through estimates of continuous dependence upon data of the models and the objective functional must be studied until they are better understood.

Also more research must be directed toward developing better, more robust and more computationally efficient optimization procedures. History matching algorithms produce extremely large, highly ill-conditioned, constrained, nonlinear least-squares optimization problems. Often the surface defined by the objective functional will have local relative minima or large flat areas caused by lack of sensitivity which will slow convergence or cause the algorithm to terminate too soon, before a good approximation to the global minimum is reached. Thus robust algorithms that will test for relative minima and insensitivities must be developed.

In the opinion of this author, more and better a priori and a posteriori statistical techniques must be brought to bear in automatic history matching in order to both aid in the convergence to geologically expected values and correlations and to understand the relative magnitudes of the errors in the parameter estimates. Better statistical information could be used in the weighting of the observations in the objective functional, in filtering the observed data, and in setting up the optimization problem in general. Correlations of the determined parameters should be tested against a priori correlations. As more history is obtained, the statistics should be continually updated to obtain a better total history match.

Finally, probably the major source of error in the more complex enhanced recovery processes is modeling error. The partial differential equations are only approximations for the laws governing flow through porous media and should be continually updated to model more of the important physics of the flow. There are regions around the wells and along the chemical or thermal interfaces where the standard Darcy flow models do not adequately describe the flow. The phenomena of viscous fingering which often dominates the flow behavior is not understood nor modeled adequately. The complex flow patterns at the production wells in the presence of fingering flow is not adequately modeled, and it is this type of model that produces the computed pressures with which we compare the observables. More research in these areas is imperative for enhanced recovery processes.

One of the major observables in enhanced recovery simulation is the relative flow rates of the various fluids at the production wells. As noted in [13,26], in all simulators presently in use today the Darcy velocities from which these flow rates do not converge in an asymptotic sense at the wells and these calculations are subject to severe errors. By using concepts of mixed finite elements combined with techniques of subtracting out the leading terms in these velocity singularities at the wells, one can obtain convergent and accurate approximations of the velocities at the wells, [13,26] which can be used effectively in history matching. Extensions of these techniques to a finite difference setting for routine simulator use is necessary.

Another observable in flooding simulation is the breakthrough time, when the injected fluid reaches the production well. As noted by Yanosik and McCracken [51], the standard five-point finite difference procedures used in two space dimensions by most present simulators are subject to severe grid-orientation problems. The production curves and the shape and position of the fluid interphases are radically different, depending on whether the grid lines in the simulator are oriented in a parallel or diagonal relation to the wells. These different orientations predict extremely different breakthrough times and relative flow rates throughout time after breakthrough. Since these are both observables in the history matching process, the results from the simulator must be independent of grid orientation. The finite element simulations of model miscible displacement processes presented by Wheeler and the author [13,26] do not have these grid orientation problems and are thus amenable to history matching. More research in this and other numerical discretization problems is crucial. The enhanced recovery problems are often characterized by sharp, moving interfaces or fronts moving across the reservoir. Present discretization techniques result in numerical dispersion which smear the sharp fronts and distort the physics and the results of the simulation. Again more research is needed.

In summary, one must have accurate, robust, computationally efficient simulators which are based on models which adequately describe the physics. If one is not confident of the accuracy of the forward solution process given input parameters, he has no business using a simulator based upon these processes in a complex history matching procedure for enhanced recovery.

## References


[1] Banks, H.T., "A Survey of Some Problems and Recent Results for Parameter Estimation and Optimal Control in Delay and Distributed Parameter Systems", I CASE Report #81-26, (1981).

[2] Breit, V.S., Bishop, K.A., Green, D.W., and Trompeter, E.E., "A Technique for Assessing and Improving the Quality of Reservoir Parameter Estimates Used in Numerical Simulators", SPE 4546 presented at the 48th Annual Meeting, Las Vegas, Sep. 30-Oct. 3, 1973.

[3] Cannon, J.R., and Dogru, A.H., "Estimation of Permeability and Porosity from Well Test Data", J. Pet. Tech. Forum (1980) pp. 1323-1324.

[4] Cannon, J.R., and Du Chateau, P., "Determination of Unknown Physical Properties in Heat Conduction Problems", Int. J. Eng. Sci., 11 (1973) pp. 783-794.

[5] Cannon, J.R., and Du Chateau, P.D., "Determining Unknown Coefficients in a Non-linear Heat Conduction Problem", SIAM J. Appl. Math. 24 (1973) pp. 298-314.

[6] Cannon, J.R., and Du Chateau P., "An Inverse Problem for a Nonlinear Diffusion Equation", SIAM J. Appl. Math, 39 (1980) pp. 272-289.

[7] Carter, R.D., Kemp, L.F., and Williams, D., "Performance Matching with Constraints", Soc. Pet. Eng. J. (April 1974) pp. 187-196.

[8] Chavent, G., Dupuy, M., and Lemonnier, P., "History Matching by Use of Optimal Control Theory", Soc. Pet. Eng. J. (Feb. 1975) pp. 74-86; Trans., AIME, Vol. 259.

[9] Chavent, G., "A New Formulation of Diphasic Incompressible Flows in Porous Media", Lecture Notes in Mathematics No. 503, Springer-Verlag, 1976.

[10] Chavent, G., Cohen, G., and Espy, M., "Determination of Relative Permeabilities and Capillary Pressures by an Automatic Adjustment Method", paper SPE 9237 presented at 55th Annual Fall Tech. Conf. and Exhib. of SPE of AIME in Dallas, Sept. 21-24, 1980.

[11] Chen, W.H., Gavalas, G.R., Seinfeld, J.H., and Wasserman, M.L., "A New Algorithm for Automatic History Matching", Soc. Pet. Eng. J. (Dec. 1974) pp. 593-608; Trans., AIME, Vol. 257.

[12] Coats, K.H., Dempsey, J.R., and Henderson, J.H., "A New Technique for Determining Reservoir Description from Field Performance Data", Soc. Pet. Eng. J. (March 1970) pp. 66-74; Trans., AIME, Vol. 249.

[13] Darlow, B.L., Ewing, R.E., and Wheeler, M.F., "Mixed Finite Element Methods for Miscible Displacement Problems in Porous Media", Proc. 6th Symp. on Reservoir Simulation, New Orleans, Louisiana, 1982, pp. 137-146.

[14] Dixon, T.N., Seinfeld, J.H., Startzman, R.A., and Chen, W.H., "Reliability of Reservoir Parameters from History Matched Drillstem Tests", SPE 4282 presented at SPE-AIME 3rd Symp. on Numer. Simulation of Res. Performance, Houston, Jan. 10-12, 1973.

[15] Douglas, Jim, Jr., "Finite Difference Methods for Two-Phase, Incompressible Flow in Porous Media", SIAM J. Numer. Anal. (to appear).

[16] Dogru, A.H., "Confidence Limits on the Parameters and Predictions of One-Dimensional, Slightly Compressible, Single-Phase Reservoirs", Ph.D. thesis U. of Texas, Austin, (May 1974).

[17] Dogru, A.H., Dixon, T.N., and Edgar, T.F., "Confidence Limits on the Parameters and Predictions of Slightly Compressible, Single-Phase Reservoirs", paper SPE 4983 presented at the SPE-AIME 49th Annual Fall Meeting, Houston, Oct. 6-9, 1974.

[18] Dogru, A.H., and Knapp, R.M., "The Reliability of the Predicted Performance of Natural Gas Reservoirs Using Reservoir Parameters from Well Test Data Containing Errors", Proceedings of Rocky Mtn. Regional SPE Meeting, Denver, Colo., April 7-9, 1975.

[19] Dogru, A.H., and Seinfeld, J.H., "Design of Well Tests to Determine the Properties of Stratified Reservoirs", Proceedings of 5th SPE Symposium on Reservoir Simulation, Denver, Colorado, February 1-2, 1979.

[20] Dogru, A.H., and Seinfeld, J.H., "Comparison of Sensitivity Coefficient Calculation Methods in Automatic History Matching", Soc. Pet. Eng. J. (October 1981) pp. 551-557.

[21] DuChateau, P., "Monotonicity and Uniqueness Results in Identifying an Unknown Coefficient in a Nonlinear Diffusion Equation", SIAM J. Appl. Math. 41 (1981) pp. 310-323.

[22] Ewing, R.E., Falk. R.S., Bolzan, J.F., and Whillans, I.M., "Techniques for Thermal Conductivity Measurements in Antarctica", Ann. Glaciology 3 (1982) pp. 96-102.

[23] Ewing, R.E., and Russell, T.F., "Efficient Time-stepping Methods for Miscible Displacement Problems in Porous Media", SIAM J. Numer. Anal. 19 (1982) pp. 1-66.

[24] Ewing, R.E., and Wheeler, M.F., "Galerkin Methods for Miscible Displacement Problems in Porous Media", SIAM J. Numer. Anal. 17 (1980) pp. 351-365.

[25] Ewing, R.E., and Wheeler, M.F., "Galerkin Methods for Miscible Displacement Problems with Point Sources and Sinks - Unit Mobility Ratio Case", Proceedings Special Year in Numerical Analysis, Univ. of Maryland, College Park, 1981, pp. 151-174.

[26] Ewing, R.E., and Wheeler, M.F., "Computational Aspects of Mixed Finite Element Methods", Numerical Methods for Scientific Computing, R.S. Stepleman, editor, North Holland Publ. Co., (to appear).

[27] Falk, R.S., "Error Estimates for the Numerical Identification of a Variable Coefficient", (to appear).

[28] Frick, T.C., Petroleum Production Handbook, Vol. II., McGraw-Hill, New York (1962).

[29] Gavalas, G.R., Shah, P.C., and Seinfeld, J.H., "Reservoir History Matching by Bayesian Estimation", Soc. Pet. Eng. J. (Dec. 1976) pp. 337-350; Trans., AIME, Vol. 261.

[30] Hirasaki, G.J., "Estimation of Reservoir Parameters by History Matching Oil Displacement by Water or Gas", SPE 4283 presented at the 3rd Symp. on Numer. Simulation of Res. Performance, Houston, Jan. 10-12, 1973.

[31] Hornung, U., and Messing, W., "Identification of Soil Parameters for an Infiltration Problem", Finite Elements in Water Resources, Proceedings of the 4th Int. Conf., Hannover, 1982, Springer-Verlag, Berlin.

[32] Jacquard, P., "Theory de L'Interpretation des Mesures de Pression", Revue de L'Institute France du Petrole (March 1964).

[33] Jacquard, P., and Jain, C., "Permeability Distribution from Field Pressure Data", Soc. Pet. Eng. J. (Dec. 1965) pp. 281-294; Trans., AIME, Vol. 234.

[34] Jahns, H.O., "A Rapid Method for Obtaining a Two-dimensional Reservoir Description from Well Response Data", Soc. Pet. Eng. J. (Dec. 1966) pp. 315-327; Trans., AIME, Vol. 237.

[35] Keelan, D.K., "A Critical Review of Core Analysis Techniques", J. Can. Pet. Tech. (April, 1972) pp. 1-14.

[36] Matthews, C.S., and Russell D.G., Pressure Buildup and Flow Tests in Wells, Monograph Vol. 1, Henry L. Doherty Series, SPE of AIME, 1967.

[37] Odeh, A.S., "Steady-State Flow Capacity of Wells with Limited Entry to Flow", Trans. AIME. Vol. 243, p. 43, 1968.

[38] Peaceman, D.W., Fundamentals of Reservoir Simulation, Developments in Petroleum Science 6, Elsevier Publishing Co, Amsterdam, 1977.

[39] Peaceman, D.W., "Interpretation of Well-Block Pressures in Numerical Reservoir Simulation", Soc. Pet. Eng. J. (June 1978) pp. 183-194.

[40] Peaceman, D.W., "Interpretation of Well-Block Pressures in Numerical Reservoir Simulation with Nonsquare Grid Blocks and Anisotropic Permeability", Proceedings 6th SPE Symposium on Reservoir Simulation, New Orleans, Louisiana, Feb. 1-3, 1982.

[41] Pryor, W.A., "Reservoir Inhomogeneities in Some Recent Sand Bodies", Soc. Pet. Eng. J. (June 1972) pp. 229-245; Trans. AIME, Vol. 253.

[42] Richter, G.R., "An Inverse Problem for the Steady-State Diffusion Equation", SIAM J. Appl. Math 41 (1981) pp. 210-221.

[43] Shah, P.C., Gavalas, G.R., and Seinfeld, J.H., "Error Analysis in History Matching: The Optimum Level of Parametrization", Soc. Pet. Eng. J. (June 1978) pp. 219-228.

[44] Slater, G.E., and Durrer, E.J., "Adjustment of Reservoir Simulation Models to Match Field Performance, "Soc. Pet. Eng. J. (Sept. 1971) pp. 295-305).

[45] Solorzano, L.N. and Arrendondo, S.E., "Method for Automatic History Matching of Reservoir Simulation Models", SPE 4594 presented at 48th Annual Meeting, Las Vegas, Sept. 30-Oct. 3, 1973.

[46] Thomas, L.K., Hellums, L.J., and Rehais, G.M., "A Nonlinear Automatic History Matching Technique for Reservoir Simulation Models", Soc. Pet. Eng. J. (Dec. 1972) pp. 508-514.

[47] Veatch, R.W., Jr., and Thomas, G.W., "A Direct Approach for History Matching", paper SPE 3515 presented at the SPE 46th Annual Meeting, New Orleans, Oct. 3-6, 1971.

[48] Wasserman, M.L., Emanuel, A.S., and Seinfeld, J.H., "Practical Applications of Optimal Control Theory to History Matching Multiphase Simulator Models", Soc. Pet. Eng. J. (Aug. 1975) pp. 347-335; Trans., AIME, Vol. 259.

[49] Watson, A.T., "Estimation of Two-phase Petroleum Reservoir Properties", Ph.D. thesis, Calif. Inst. of Tech., Pasadena (1979).

[50] Watson, A.T., Seinfeld, J.H., Gavalas, G.R., and Woo, P.T., "History Matching in Two-phase Petroleum Reservoirs", Soc. Pet. Eng. J. (Dec. 1980) pp. 521-532.

[51] Yanosik, J.L., and McCracken, T.A., "A Nine-Point Finite Difference Reservoir Simulator for Realistic Prediction of Unfavourable Mobility Ratio Displacements", SPE 5734, Proc. 4th Symp. on Numerical Simulation of Reservoir Performance, Los Angeles (1976).

# IDENTIFICATION OF NONLINEAR SOIL PHYSICAL PARAMETERS FROM AN INPUT-OUTPUT EXPERIMENT

Ulrich Hornung

## 1. Introduction

In the field of parameter identification in parabolic differential equations some theoretical and numerical results have been obtained in recent years. The linear heat equation with linear or nonlinear boundary conditions was studied by Cannon/Zachmann [1982]. Here the fact that the heat equation can be solved explicitely was used for various problems with overspecified boundary data to show existence and uniqueness of some unknown parameters.Nonlinear problems of the type

$$\partial_t u = \partial_x (D(u) \partial_x u)$$

were treated by Cannon/DuChateau [1980], who showed that in some class of functions $D(\cdot)$ there exists at least one that fits overspecified boundary data best in a certain sense. DuChateau [1981] proved uniqueness of this function in a slightly different class.

Applications of parabolic equations are heat flow, transport of solutes, and flow problems in porous media. In the latter field some experience has been made to identify parameters for practical problems. Tait [1980] studied anisotropic flow in a circular domain. Zachmann et al. [1981] compared different methods to estimate the hydraulic properties of a draining column. In Hornung/Messing [1982] an infiltration experiment was considered.

Since measurements in hydrology are time-consuming it is highly desirable to have simple means to determine the retention curve and the conductivity of a soil.

## 2. The Direct Problem

A vertical column of a homogeneous porous medium is considered. Unsteady water flow through this column is modelled by

$$\partial_t \Theta = \partial_z (K \partial_z (\psi - z)) \quad , \tag{2.1}$$

where $\Theta$ is the water content, $\psi$ the hydraulic pressure head, K the hydraulic conductivity, z the vertical coordinate, and t the time. It is assumed that at the bottom the column is open and there $\psi$ is always held at atmospheric pressure, i.e. we have

$$\psi = 0 \qquad \text{for} \quad z = L \quad . \tag{2.2}$$

At the beginning the probe is at equilibrium, i.e. we have

$$\psi = L - z \qquad \text{for} \quad t = 0 \quad , \tag{2.3}$$

see Figure 1.

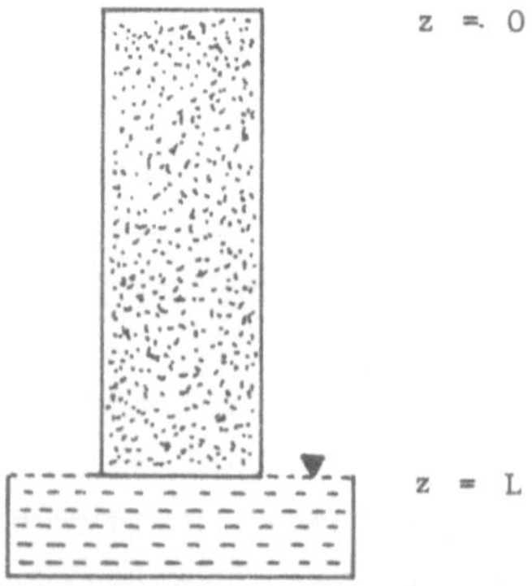


Figure 1

Now the experiment consists in applying some flow $q_o$ at the top of the column and in measuring the flow $q_1$ at its bottom. Hence we have

$$- K \partial_z(\psi - z) = q_o \qquad \text{for} \quad z = 0 \tag{2.4}$$

$$- K \partial_z(\psi - z) = q_1 \qquad \text{for} \quad z = L \ . \tag{2.5}$$

If we knew the functional relationships between $\psi$, $\Theta$ and K, the direct problem would be given by (2.1) through (2.4). This is a degenerate parabolic problem, an existence proof for which was recently given by Alt/Luckhaus [1982]. Standard techniques can be applied to obtain numerical solutions, cf. Haverkamp/Vauclin [1981] and Hornung/Messing [1981].

The identification problem consists in estimating the nonlinear functions $\Theta$ and K from the information given by (2.1) through (2.5). In the sequel we assume that the saturated water content, the residual water content, and the saturated hydraulic conductivity are known. Therefore after renormalization and replacing $\psi$ by $h=-\psi$ we obtain the following problem in dimensionless formulation:

$$\left.\begin{array}{lll}
\partial_t\Theta = - \partial_z(K\,\partial_z(h + z)) & , \ 0 < z < 1, & t > 0 \ , \\
h = z - 1 & , \ 0 < z < 1, & t = 0 \ , \\
h = 0 & , \ z = 1 & , \ t > 0 \ , \\
K\,\partial_z(h + z) = q_o & , \ z = 0 & , \ t > 0 \ , \\
K\,\partial_z(h + z) = q_1 & , \ z = 1 & , \ t > 0 \ .
\end{array}\right\} \tag{2.6}$$

## 3. Representation of Nonlinear Functions

Though theoretically the class of possible functions $\Theta$ and K is very large, for practical purposes a special form using a small number of parameters has to be chosen. The retention curve, i.e. the $\Theta$-h-relationship, has been represented as

$$\Theta = 1/(1+\alpha h)^{\gamma}$$

by Brooks/Corey (1966), cf. Hillel [1980], as

$$\Theta = 1-(1+1/(\alpha h)^2)\cdot \exp(-1/(\alpha h)^2)\cdot (1-C/(\alpha h)^{1/6})$$

by Kovács [1981], or as

$$\Theta = 1/(1+\alpha h)^2$$

by Zachmann et al. [1981]. We have chosen the formula

$$\Theta = 1/(1+(\alpha h)^{\beta})^{1-1/\beta} \tag{3.1}$$

which was proposed by van Genuchten [1980]. The conductivity was given as

$$K = 1/(1+(\alpha h)^{\beta})$$

or as

$$K = \Theta^{\delta}$$

by Gardner (1960), cf. Hillel [1980], and as

$$K = e^{-\delta h}$$

by Rijtema (1965), cf. Feddes et al. [1978]. In this paper the formula

$$K = \Theta^{1/2}\cdot\left(\int_0^{\Theta}\frac{d\tau}{h(\tau)}\Big/\int_0^{1}\frac{d\tau}{h(\tau)}\right)$$

by Mualem [1976], which in connection with (3.1) leads to

$$K = \Theta^{1/2}\cdot(1-(1-\Theta^{1/\gamma})^{\gamma})^2 \quad , \quad \gamma = 1 - 1/\beta \qquad (3.2)$$

is used. There is experimental evidence that formulas (3.1) and (3.2) can be used to describe a large variety of soils approximately. According to van Genuchten [1982] one obtains as a thumb rule

$1.1 < \beta < 1.8$ for fine-textured soils,

$1.8 < \beta < 2.5$ for medium-textured soils,

$2.5 < \beta < 10$ for coarse-textured soils.

Since (2.6) has dimensionless form, the value of $\alpha$ in (3.1) depends not only on the soil, but also on the depth of the column.

## 4. Results of Computer Simulation

The main point of this paper is the question wether or not the identification problem can be solved. In other words: does the infiltration experiment give enough information to decide on the hydraulic properties of the soil? To this end a systematic study of computer simulations was done to investigate the problem of uniqueness and the sensitivity of parameters. Holding the input value $q_o = 0.1$ fixed for a large number of pairs $(\alpha,\beta)$ the output function $q_1(t)$ was calculated from (2.6). This curve is S-shaped with $q_1(0) = 0$ and the limit value $q_1(t) \to q_o$ for large $t$.

The first result of the simulations was that there are different pairs of parameters, i.e. different types of soils, yielding practically the same response curves $q_1(\cdot)$. By this we mean that the curves cannot be distinguished from each other since their difference is too small, see Figure 2.

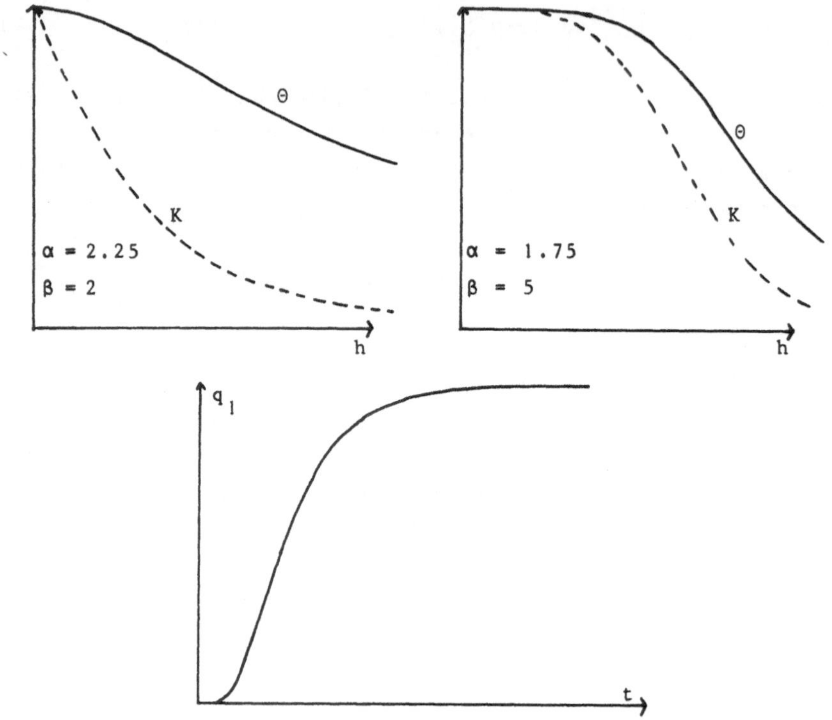


Figure 2

In order to study this result further, the following crude but efficient description of the output curve $q_1(\cdot)$ was selected: The values $t_{10}$ and $t_{90}$ are defined by

$$q_1(t_{10}) = \frac{10}{100}\, q_o \ , \quad q_1(t_{90}) = \frac{90}{100}\, q_o \quad .$$

This may be done since those curves having the same pairs $(t_{10}, t_{90})$ are almost identical on their whole range. Using the two values $t_{10}$ and $t_{90}$ the identification problem amounts in the construction of the inverse of the mapping

$$T : (\alpha,\beta) \longmapsto (t_{10}, t_{90}) \quad .$$

Some insight into the structure of this mapping can be obtained by the diagram of the $t_{10}$- and $t_{90}$-isolines in the $\alpha$-$\beta$-plane, see Figure 3.

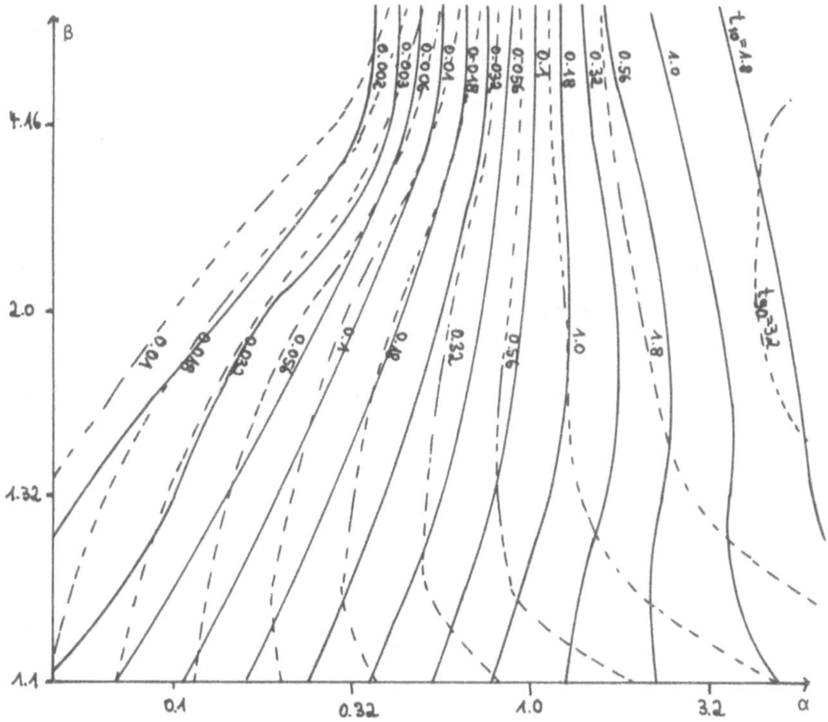


Figure 3

This diagram shows that there is a large region where the isolines are almost parallel. Therefore in this region the Jacobian of the mapping T is almost degenerate. This region is called the region of local nonuniqueness, since in each neighbourhood of a point (α,β) there are other points having the same response.

The second result of the simulations was that after excluding the region of local nonuniqueness in the remaining part there are pairs of points $(\alpha_1, \beta_1)$ and $(\alpha_2, \beta_2)$ having the same response curve, see Figure 4.

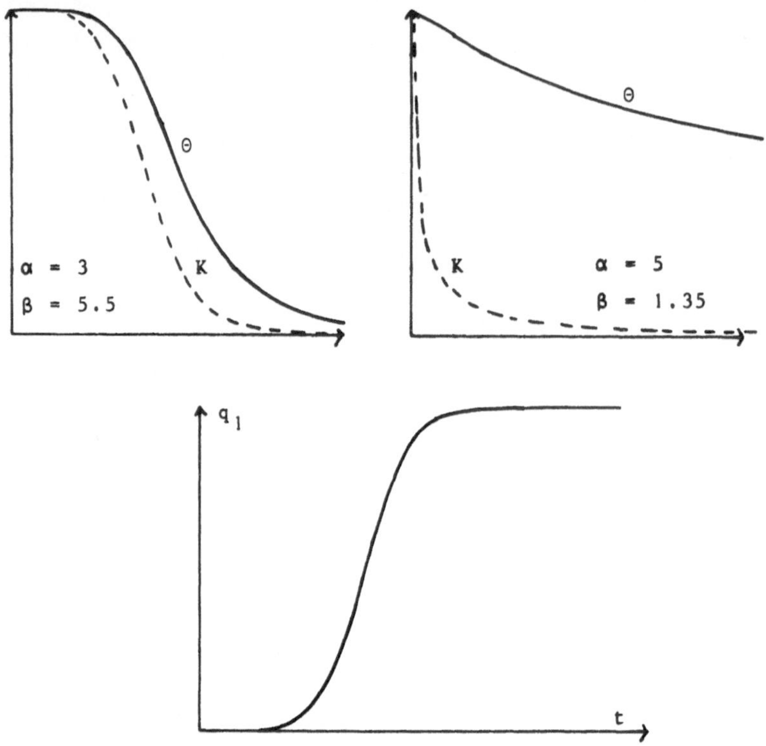


Figure 4

Here the mapping is locally nondegenerate, i.e. the isolines intersect transversally, but nevertheless there are different soils having the same input-output-relation. This phenomenon is called <u>global nonuniqueness</u>.

Summing up it can be said that the experiment described so far does not give enough information to solve the identification problem.

## 5. Additional Measurements

In order to make the identification problem solvable some more information is necessary. The limit value $h_{80}$ of h at depth 1/80 for large t is assumed to be measured using a tensiometer. Now the mathematical problem is to construct the inverse of the augmented mapping

$$\overline{T} : (\alpha,\beta) \longmapsto (t_{10}, t_{90}, h_{80}) \quad .$$

A study of the isolines of the three components $t_{10}$, $t_{90}$, and $h_{80}$, see figure 5, shows that the nonuniqueness problem has partially been resolved. If for some $\beta$ the value $\alpha$ is sufficiently large at least two of the isolines intersect transversally. In this region of the $\alpha$-$\beta$-plane the pair of parameters $(\alpha,\beta)$ is uniquely determined by their $\overline{T}$ - image.

## 6. Conclusion

Under the assumption that an unknown probe of a soil can be adequately represented by formulas (3.1) and (3.2), and if the soil column is large enough, one infiltration

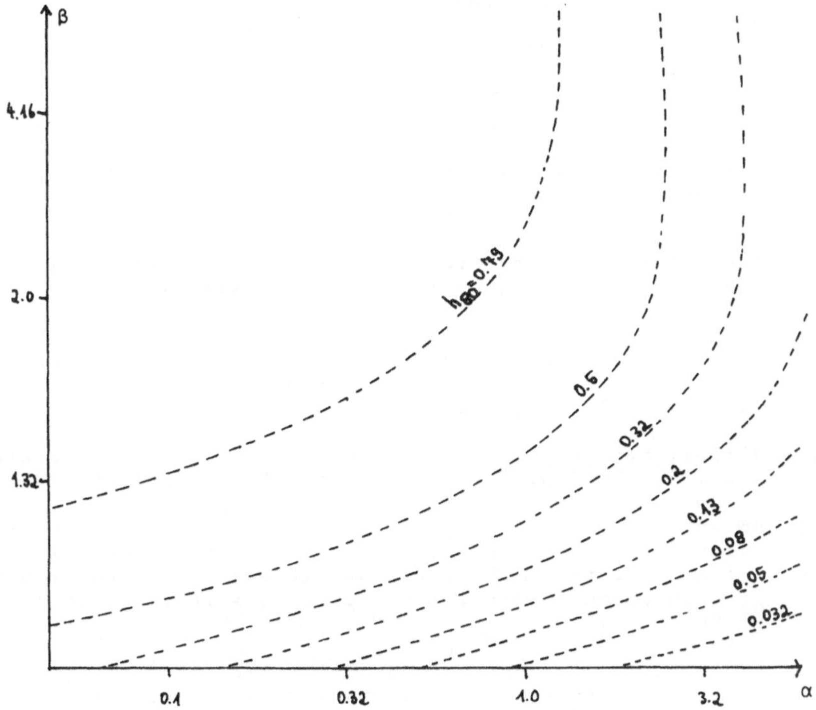


Figure 5

experiment is sufficient to identify its hydraulic properties. Only the registration of the flow $q_1$ and the measurement of the pressure head $h_{80}$ at the end of the experiment are necessary.

Acknowledgement: This work was supported by the Deutsche Forschungsgemeinschaft, Grant Ho 782/2.

## References

[1] Alt, H.W., S. Luckhaus: Quasilinear Elliptic-Parabolic Differential Equations. Sonderforschungsbereich 123, Heidelberg, Preprint 136 (1982).

[2] Cannon, J.R., P. DuChateau: An inverse problem for a nonlinear diffusion equation. SIAM J. Appl. Math. 39 (1980) 272-289.

[3] Cannon, J.R., D. Zachmann: Parameter determination in parabolic partial differential equations from overspecified boundary data. Int. J. Engng. Sci. 20 (1982) 779-788.

[4] DuChateau, P.: Monotonicity and uniqueness results in identifying an unknown coefficient in a nonlinear diffusion equation. SIAM J. Appl. Math. 41 (1981) 310-323.

[5] Feddes, R.A., P.J. Kowalik, H. Zaradny: Simulation of Field Water Use and Crop Yield. Centre for Agricultural Publishing and Documentation, Wageningen, 1978.

[6] Haverkamp, R., M. Vauclin: A comparative study of three forms of the Richard equation used for predicting one-dimensional infiltration in unsaturated soil. Soil Sci. Soc. Amer. J. 45 (1981) 13-20.

[7] Hillel, D.: Fundamentals of Soil Physic. Academic Press, New York, London, Toronto, Sydney, San Francisco, 1980.

[8] Hornung, U., W. Messing: Simulation of two-dimensional saturated/unsaturated flows with an exact water balance. In: A. Verruijt, F.B.J. Barends (Eds.) "Flow and Transport in Porous Media", A.A. Balkema, Rotterdam (1981) 91-96.

[9] Hornung, U., W. Messing: Identification of soil parameters for an infiltration problem. In: K.P. Holz, U. Meissner, W. Zielke, C.A. Brebbia, G. Pinder, W. Gray (Eds.) "Finite Elements in Water Resources", Springer-Verlag, Berlin, Heidelberg, New York (1982) 18.15-18.24.

[10] Kovács, G.: Seepage Hydraulics. Elsevier, Amsterdam, Oxford, New York (1981).

[11] Mualem, Y.: A new model for predicting the hydraulic conductivity of unsaturated porous media. Water Res. Res. 12 (1976) 513-522.

[12] Tait, R.J.: An inverse problem arising in fluid flow in a porous medium. Int. J. Engng. Sci. 18 (1980) 161-167.

[13] van Genuchten, R.: A closed-form equation for predicting the hydraulic conductivity of unsaturated soils. Soil Sci. Soc. Amer. J. 44 (1980) 892-898.

[14] van Genuchten, R.: Private Communication (1982).

[15] Zachmann, D.W., P.C. DuChateau, A. Klute: The calibration of the Richards flow equation for a draining column by parameter identification. Soil Sci. Soc. Amer. J. 45 (1981) 1012-1015.

# ON AN INVERSE NON-LINEAR DIFFUSION PROBLEM

Gerd Eriksson

and

Germund Dahlquist

## 1. Statement of Problems and Basic Facts

Let $u = u_E(x,t)$ be the solution of the autonomous non-linear diffusion equation,

$$\frac{\partial u}{\partial t} = \frac{\partial}{\partial x}\left(D(u)\frac{\partial u}{\partial x}\right), \quad t > 0, \ 0 < x < 1, \tag{1.1'}$$

with initial condition $u_E(x,0) = 1$, (1.1")

and boundary conditions $u_E(0,t) = 0, \ \frac{\partial u_E}{\partial x}(1,t) = 0.$ (1.1"')

If $D(u)$ is a known function, $D(u) \geq a > 0$, the problem will be called the *direct problem*. We shall mainly be concerned with the following inverse problem, where, in addition to the initial and boundary conditions stated above, the function $u_E(1,t)$ is measured, but $D(u)$ is not known. The *inverse problem* is to reconstruct $D(u)$ from the measured estimates of $u_E(1,t)$, which are denoted $\tilde{u}(1,t)$. In practice a point sequence $(\tilde{t}_i, \tilde{u}_i)$, $i = 1,2,\ldots,m$, is given.

In Figure 1.1 $u_E(1,t)$ is plotted for $D(u) = 1$, $D(u) = 2$ and $D(u) = 1 + u$. The figure illustrates the useful facts that $0 \leq u_E(1,t) \leq 1$ and $D_1(u) \leq D(u) \leq D_2(u)$, $\forall u \in [0,1] \Rightarrow u_E(1,t;D_1) \geq u_E(1,t;D) \geq u_E(1,t;D_2)$

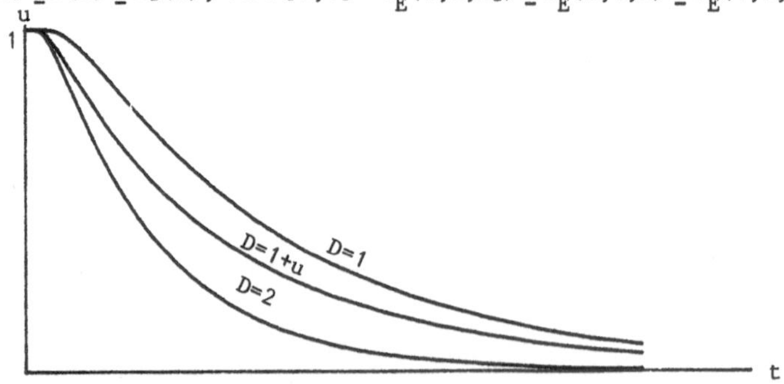


Figure 1.1

We prefer to discuss the functional relationship in terms of the function

$$h(u) = t(u;1)/t(u;D), \tag{1.2}$$

where $t(u;D)$ is the inverse function of $u(1,t;D)$. Note that

$$D := \alpha \cdot D \iff h := \alpha \cdot h.$$

In particular $D(u) \equiv \alpha$ implies $h(u) \equiv \alpha$. The following figures show the relationship between $D(u)$ and $h(u)$. We note that the inverse mapping $h \to D$ is *ill-conditioned*, since the direct mapping $D \to h$ is smoothing. In section 2 we shall try to measure the ill-conditioning. Note also the "boundary layer" at $u = 1$ in Figure 1.5, which indicates that there might exist a transformation of the problem, that is even more advantageous than the one that we have used.

The numerical solution of this non-linear inverse problem is treated in Eriksson [3] and Eriksson [4] . In the latter paper references are given to related works. Cannon and DuChateau [1] perform some theoretical analysis of a similar problem under the assumptions that $D(u)$ and $D'(u)$ are monotone. The set A of permissible

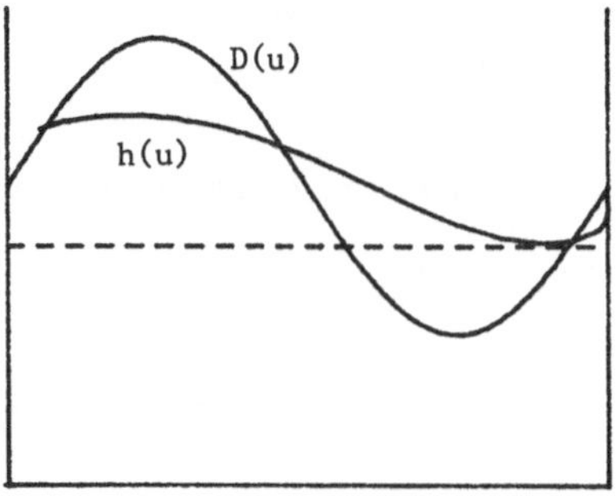


Figure 1.2

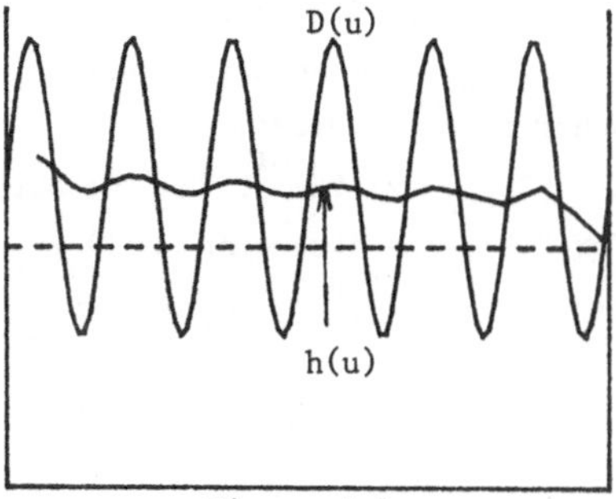


Figure 1.3

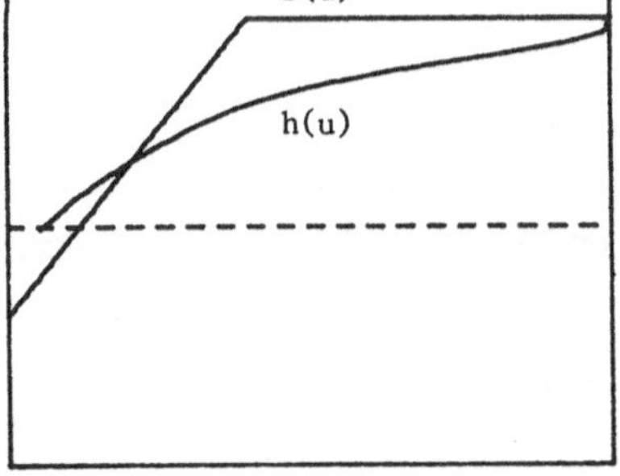


Figure 1.4

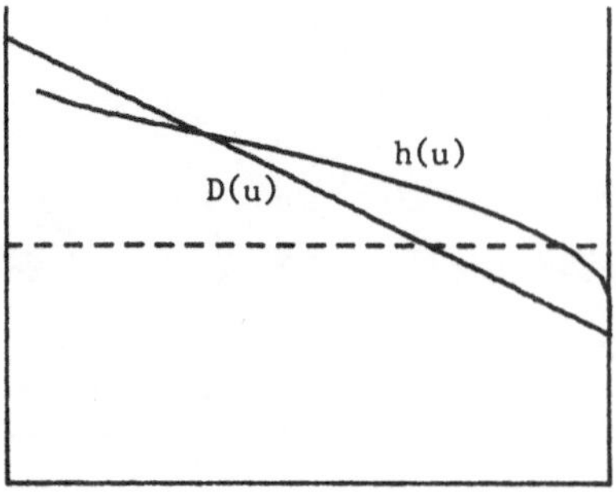


Figure 1.5

functions D(u) is a compact subset of $C^2[0,1]$. They show the existence of a solution of the following "auxiliary problem": find $D \in A$ which gives the infimum of

$$\sup_{0\le t\le T} |u(1,t;D) - \tilde{u}(1,t)|.$$

Usually this infimum is strictly positive, and they don't answer the question when it is equal to zero. Cannon et al. have in several papers after 1964 studied the inverse problem when $D = D(t)$ (instead of $D(u)$), and Stoyan [6] studies the case $D = D(x)$. See also Crank [2] , in particular Ch. 10.

## 2. Boltzmann's Particular Solution and Some Asymptotics

If the boundary condition $\frac{\partial u}{\partial x}(1,t) = 0$ is replaced by, for example, the condition $u(\infty,t) = 1$, then the solution is of the form

$$u_B(x,t) = U(x\cdot(2t)^{-\frac{1}{2}}), \qquad (2.1)$$

B stands for Boltzmann's transformation, see e.g. Crank [2] , section 7.2.

The level curves in the (x,t)-plane of $u_B(x,t)$ are parabolas. In Figure 2.1 we indicate the relationship between the level curves of $u_E$ and $u_B$; they coincide well for small t, but they deviate from each other as t increases.

Set $G = x\cdot(2t)^{-\frac{1}{2}}$. Then we obtain an ordinary differential equation for U(G), see also Crank [2])

$$-G\frac{dU}{dG} = \frac{d}{dG}\left(D\frac{dU}{dG}\right). \qquad (2.2)$$

The *direct* problem becomes a *two-point boundary value* problem, in our

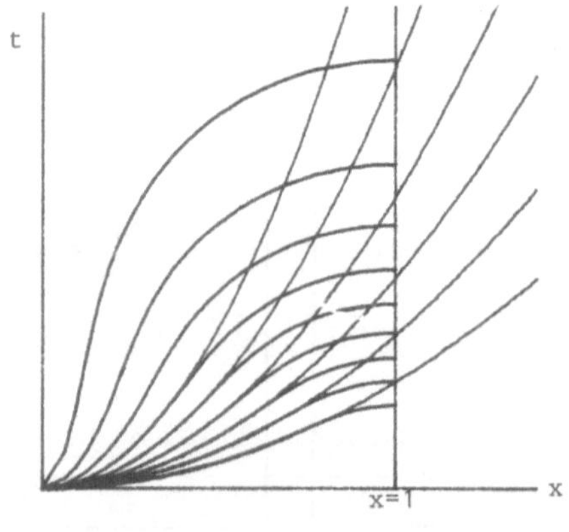


Figure 2.1

case,

$$U(0) = 0, \; U(\infty) = 1. \qquad (2.3)$$

Shampine [5] uses shooting techniques for this problem.

The *inverse* problem has, in the Boltzmann case, an explicit solution, namely

$$D(U) = G'(U) \int_u^1 G(v)dv, \qquad (2.4)$$

where $G(U_i)$ is known for $i = 1,2,\dots,m$. In the numerical computation one must take into account that $G(U)$ is singular at $U = 1$. We see that this inverse problem is ill-conditioned, but no worse than differentiation.

This raises the question of the relation of the solution of Boltzmann's problem $u_B(1,t)$ to $u_E(1,t)$. We don't have many theoretical results, but Figure 2.2 shows that there is a relation of the form $u_E(1,t) \approx \varphi(u_B(1,t))$, which holds fairly independently of the diffusion coefficient function, for which we tried the four cases shown in Figure 2.3.

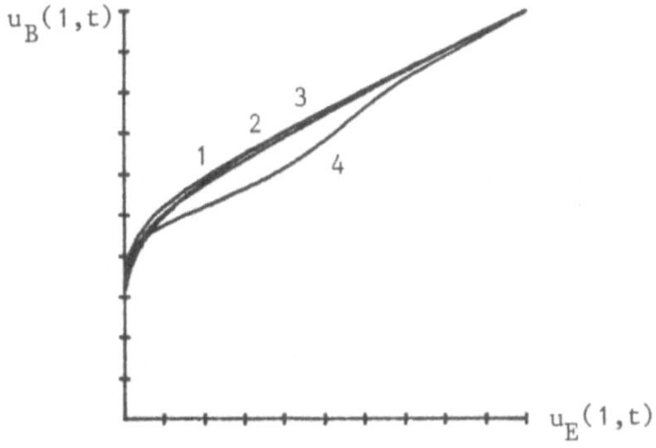


Figure 2.2

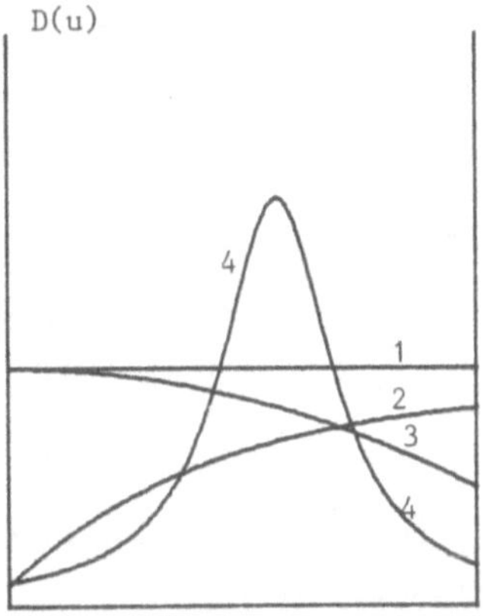


Figure 2.3

This relationship together with (2.4) throws some light on the nature of the ill-conditioning also for our inverse problem.

If $D = \text{const.}$ one can deduce from the explicit solutions of the two problems, in terms of the error function, that (for $x = 1$)

$$1 - u_E(t) \sim 2(1 - u_B(t))$$

$$\text{i.e.} \quad u_B(t) \approx (1 + u_E(t))/2.$$

Figure 2.2 shows that this relation holds with an amazing accuracy for $u_E > 0.2$, not only for $D = \text{const.}$ (case 1), but also for the cases 2 and

3. Even for the difficult case 4 it is obvious that the solution D(U) obtained from (2.4) with $U = (1 + u_E(t))/2$, $G = (2t)^{-\frac{1}{2}}$, should provide a good first approximation to D(u) in our inverse problem. We cannot yet explain this.

We shall now estimate the boundary layer at $u = 1$, see Figure 1.3. Set, without loss af generality, $D(1) = 1$. In view of Figure 2.2 we can instead discuss the Boltzmann problem. Let us accept without proof that $\frac{dU}{dG} \downarrow 0$ when $G \to \infty$, and set

$$H(G) = D(U)\,\frac{dU}{dG} \tag{2.5}$$

Hence $H(G) \downarrow 0$ when $G \to \infty$. Equation (2.2) now can be written

$$dH/dG = -GH/D(U) \tag{2.6}$$

Hence

$$\ln H(G) = -\int \frac{GdG}{D(U)} = \ln C(D) - \frac{G^2}{2} + \int^{\infty} O(1-u)GdG \tag{2.7}$$

where C(D) is an integration constant, the value of which is determined by matching the boundary condition at $G = 0$. Therefore, C(D) depends on the values of D(u) for all $u \in [0,1]$ (not just the vicinity of $u = 1$). After an analysis, the details of which we have to omit, we obtain

$$\begin{aligned} H(G) &= C(D)e^{-G^2/2}(1 + o(1)), \\ 1 - u &\sim H/G. \end{aligned} \tag{2.8}$$

Then, by (2.7)

$$\ln(1-u) + \ln G + \frac{G^2}{2} = \ln C(D) + o(1). \tag{2.9}$$

In the notation of section 1,

$$G^{-2} = 2t(u;D), \quad h(u) = t(u;1)/t(u;D).$$

A straightforward calculation now yields the following estimate of the boundary layer,

$$h(u) - 1 \sim \frac{\ln(C(D)/C(1))}{|\ln(1-u)|} \tag{2.10}$$

## 3. A Working Algorithm

As stated earlier, in the inverse problem m measurements $\tilde{t}_i, \tilde{u}_i$ are given for $x = 1$. A start-approximation D is chosen, and we proceed to solve the non-linear direct problem (1.1) using an iterative Crank-Nicolson method with varying time-steps.

The deviation between the solution at $x = 1$ and the measured values is expressed by the relative residuals

$$g_i = (\tilde{t}_i - t(\tilde{u}_i;D))/\tilde{t}_i \qquad i = 1,2,\dots,m.$$

We now restrict D to the space of cubic splines on some prescribed knots $0 = u_1 < u_2 \dots < u_n = 1$. Thus D is completely specified by its value vector p, where $p_j = D(u_j)$.

We want to choose the parameter vector so as to minimize the residual norm $\|g\|_2$ while keeping D reasonably smooth. For this non-linear optimization problem, a modified Gauss-Newton method is employed where, in each iteration, one has to solve a linear inverse problem. In this process, singular value decomposition is utilized.

The stabilization procedure used to handle the ill-conditioning is a combination of two different kinds of smoothing - on one hand *the choice of the number of spline knots*, n, and on the other hand *the choice of the number of singular values* used, r. The computer program provides the user with the option of making these choices interactively for each iteration.

In each iteration step, we try to find a correction vector, $\delta p$, that minimizes the linearized deviation measure

$$\|g + J\delta p\| \quad \text{where} \quad J = \left(\frac{\partial g_i}{\partial p_j}\right) \quad i \le m,\ j \le n$$

The experimental perturbation necessary for the computation of

$$\frac{\partial g}{\partial p_j} \approx (g(p_j+\varepsilon) - g)/\varepsilon$$

involves solving the direct problem once for each j, with a small perturbation of the parameter in the corresponding knot. SVD is applied to J with singular values $\sigma_1 \ge \sigma_2 \dots \ge \sigma_n$. Now n "candidates" for $\delta p$ are computed: the j:th candidate $(\delta p)_j$ corresponding to $\sigma_k$ being set to zero for $k > j$, i.e.

$(\delta p)_j = -J_j^+ g$ where $J_j^+$ is the pseudo-inverse of rank j.

After each iteration, the user also decides:

- if a new knot grid should be chosen,
- if a new Jacobian J is to be computed,
- if the solution is good enough.

Figures 3.1 and 3.2 show results from the reconstruction of two diffusion functions. Here, we have used a constant D as a first approximation to the unknown diffusion function. The reconstruction was satisfactory after a more than four iteration steps.

| iter. | $\|g\|_2$ | n | new J? | r |
|---|---|---|---|---|
| 0 | 2.05 | 6 | yes | 5 |
| 1 | 0.54 | 6 | no | 5 |
| 2 | 0.29 | 6 | yes | 4 |
| 3 | 0.13 | 6 | no | 5 |
| 4 | 0.015 | | | |

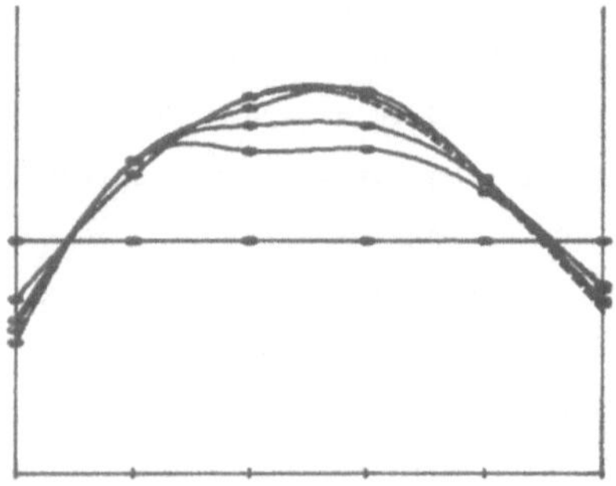

Figure 3.1

| iter. | $\|g\|_2$ | n | new J? | r |
|---|---|---|---|---|
| 0 | 1.99 | 5 | yes | 5 |
| 1 | 0.46 | 5 | no | 4 |
| 2 | 0.28 | 7 | yes | 6 |
| 3 | 0.057 | 7 | no | 6 |
| 4 | 0.030 | | | |

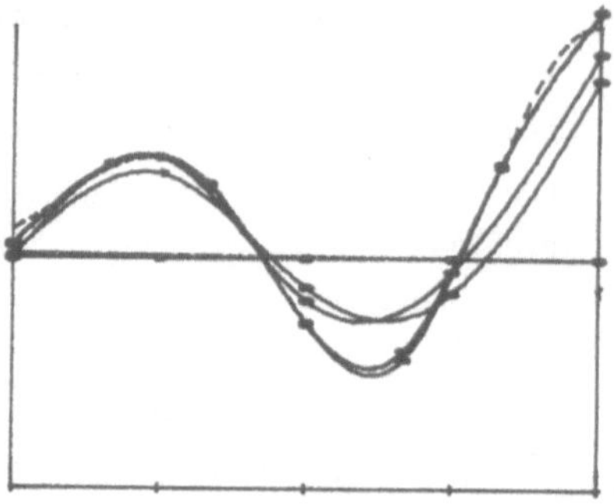

Figure 3.2

In Eriksson, [4] , several numerical experiments are performed with up to 5% noisy data and the number of iterations needed lies, in general, between four and eight. Our algorithm is applicable to a wide range of inverse problems. The question of getting a good start-approximation must, of course, be treated specially in each particular problem. In the diffusion case, a good initial guess is derived from the solution of the much simpler Boltzmann problem, as shown in section 2. Although Eriksson, [4] , reports excellent results with a simple constant start-approximation, a more sophisticated initial guess might be indispensable in cases of very noisy measurements on strongly varying diffusion coefficients.

## References


[1] Cannon, J.R. and DuChateau, P., *An Inverse Problem for a Nonlinear Diffusion Equation,* SIAM J.Appl.Math. 39, Oct. 1980 (272-289).

[2] Crank, J., *The Mathematics of Diffusion,* Clarendon Press, Oxford, 1975.

[3] Eriksson, G., *Numerical Evaluation of a Concentration-dependent Diffusion Coefficient,* Report NADA, KTH, available as TRITA-NA-8220, 1967.

[4] Eriksson, G., *Non-linear Inverse Problems,* Report TRITA-NA-8209, NADA, KTH, 1982.

[5] Shampine, L.F., *Concentration-dependent Diffusion*, Quart.Appl.Math. Jan. 1973 (441-452).

[6] Stoyan, G., *Identification of a Spatially Varying Coefficient in a Parabolic Equation* (in: Inverse and improperly posed problems in differential eq., ed. Anger, G., Akademie-Verlag, Berlin) 1979.

# THE NUMERICAL SOLUTION OF A NON-CHARACTERISTIC CAUCHY PROBELM FOR A PARABOLIC EQUATION

Lars Eldén

Abstract

We study the problem of solving numerically a parabolic partial differential equation in one space dimension, where boundary values are given on one boundary only. For the analysis and numerical solution the problem is reformulated as a Volterra integral equation of the first kind. The problem is ill-posed and we study the nature of the ill-posedness by computing approximately the singular values and functions of the Volterra operator. The integral equation is discretized giving a system of linear equations. Two methods for solving the discrete problem are briefly discussed. The first is the regularization method, and the other is derived using certain properties of the Volterra equation. It is shown that in the case of time-independent coefficients an approximate solution can be obtained in $O(n^2)$ operations, where n is the dimension of the linear system. Numerical examples are given.

## 1. Introduction

Consider the following Cauchy problem for a parabolic partial differential equation

$$\begin{aligned} &u_t = (a(x)u_x)_x, && 0 \leq x \leq 1, \quad 0 \leq t \leq 1, \\ &u(0,t) = g(t), && 0 \leq t \leq 1, \\ &u_x(0,t) = 0, && 0 \leq t \leq 1, \\ &u(x,0) = 0, && 0 \leq x \leq 1, \end{aligned} \tag{1.1}$$

and where the values at the right boundary, $u(1,t) = f(t)$, are unknown and sought for. The coefficient $a(x)$ is assumed to satisfy

$$0 < c_1 \leq a(x) \leq c_2 < +\infty, \qquad 0 \leq x \leq 1,$$

for some constants $c_1$ and $c_2$. For simplicity we may also assume that $a(x)$ is smooth, but this is not necessary (see Section 2).

(1.1) is called a non-characteristic Cauchy problem, since the Cauchy data are given along a line, which is not a characteristic curve for the equation.

The problem (1.1) may arise when one wishes to compute the temperature inside a wall (or a thick tube), where it is impossible to obtain temperature measurements on both sides of the wall. In that case one can (partially) make up for the absence of temperature data at one boundary by measuring also the heat flux at the other boundary:

$$\begin{aligned}
u_t &= (a(x)u_x)_x, && 0 \leq x \leq 1, \quad 0 \leq t \leq 1,\\
u(0,t) &= g_1(t), && 0 \leq t \leq 1,\\
u_x(0,t) &= g_2(t), && 0 \leq t \leq 1,\\
u(x,0) &= \varphi(x), && 0 \leq x \leq 1.
\end{aligned}$$

Using the linearity of this problem, one can easily reduce it to (1.1).

The problem of solving (1.1) is <u>ill-posed</u>: the solution (f) does not depend continuously on the data (g). In order to study the nature of the ill-posedness, and to develop numerical methods, we reformulate the problem as a Volterra integral equation of the first kind with a convolution type kernel

$$\int_0^t k(t-\tau)\, f(\tau)\, d\tau = g(t), \qquad 0 \leq t \leq 1. \tag{1.2}$$

In the special case when the coefficient $a(x)$ is constant,

$$a(x) \equiv \kappa, \tag{1.3}$$

we can find an explicit expression for the kernel function $k(t)$. In Section 2 we derive some properties of the integral equation (1.2), using our knowledge of $k(t)$ in the special case (1.3).

The singular functions [28] of the integral operator (1.2) have a certain symmetry property, which turns out to be useful in the development of a numerical method for (1.2). We derive this property in Section 3, where we also study (for a discretization of (1.2)) how fast the singular values of the operator tend to zero. Note that the ill-posedness of the problems depends on the rate of decay of the singular values.

In Section 4 we consider the discretized version of (1.2), and describe two numerical methods for the solution of the discretized problem. One is the regularization method of Tikhonov-Phillips [31], [24]. The other method is based on the symmetry properties of the singular functions. In both methods the solution is obtained in $O(n^2)$ operations, where $n$ is the dimension of the discretized problem.

Numerical results are given in Section 5, and some extensions of the ideas of this paper are outlined in Section 6.

Surveys of ill-posed problems for differential and integral equations are given in [16], [17], [23], [32], [34]. (1.1) and similar problems have been studied by several authors, [1], [3] - [7], [9], [11] - [15], [19] - [21], [26], [27], [30], [33], [35]. Our approach is mainly practical in the sense that we study (1.2) from the point of view of developing efficient numerical methods. This work was inspired by the papers of Seidman [26], [27], and is also related to the work of Murio [20], [21], and Manselli, Miller [19].

## 2. Reformulation as a Volterra Integral Equation

The basic idea of this paper is to reformulate (1.1) as a Volterra integral equation of the first kind (1.2). This can be done using Laplace transform technique.

Starting out from the initial-boundary value problem

$$\begin{aligned}
&u_t = (a(x)u_x)_x, && 0 \leq x \leq 1, \quad 0 \leq t \leq 1, \\
&u_x(0,t) = 0, && 0 \leq t \leq 1, \\
&u(1,t) = f(t), && 0 \leq t \leq 1, \\
&u(x,0) = 0, && 0 \leq x \leq 1,
\end{aligned} \tag{2.1}$$

taking Laplace transforms, solving (in principle) a two-point boundary value problem for an ordinary differential equation, and taking inverse Laplace transforms, we get

$$g(t) = \int_0^t k(t - \tau)\, f(\tau)\, d\tau, \qquad 0 \leq t \leq 1, \tag{2.2}$$

where $g(t) = u(0,t)$, cf. [29, p. 204-]. Thus, the problem of determing $u(1,t) = f(t)$ from (1.1) is equivalent to that of solving the integral equation (2.2).

It is important to note that (2.2) is a convolution type integral equation.

For a general coefficient $a(x)$ we can not give an explicit expression for the kernel function $k(t)$. However, in the special case when $a(x)$ is constant, $k(t)$ can easily be found (see below). In order to avoid unnecessary complication, we shall in this paper use this special case to derive qualitative properties, which are valid also in the general case. That this is fully justified, even when $a(x)$ has discontinuities, is shown in [2].

A standard procedure to deal with a Volterra equation of the first kind is to transform it to an equation of the second kind, e.g. by differentiation, [18, p. 183]. Here we get

$$k(0)f(t) + \int_0^t k'(t - \tau)\, f(\tau)\, d\tau = g'(t).$$

In our case it is impossible to obtain an equation of the second kind since $k(0) = 0$. Not even repeated differentiations will give the desired result, since all derivatives of $k$ are also zero at $t = 0$. We show this in the constant coefficient case.

When the differential equation is

$$u_t = \kappa\, u_{xx},$$

the kernel function is given by [8,p. 104]

$$k(t) = k_\kappa(t) = \pi\kappa \sum_{n=0}^{\infty} (2n + 1)(-1)^n \exp\left(- \frac{(2n+1)^2\pi^2\kappa t}{4}\right)$$

This is essentially an elliptic theta function [36, p. 86- ], and using Poisson's summation formula, it can be written

$$k_\kappa(t) = \frac{1}{\sqrt{\pi\kappa}\, t^{3/2}} \sum_{n=0}^{\infty} (2n + 1)(-1)^n \exp\left(- \frac{(2n + 1)^2}{4\kappa t}\right) \qquad (2.3)$$

It is shown in [36, p. 89] that $k_\kappa$ is analytic, in the sense that all derivatives $k_\kappa^{(\nu)}(t)$, $\nu = 0,1,2,\ldots$, exist and are continuous for $t > 0$. From (2.3) we also see that

$$\lim_{t\to 0+} k_\kappa^{(\nu)}(t) = 0, \qquad \nu = 0,1,2,\ldots.$$

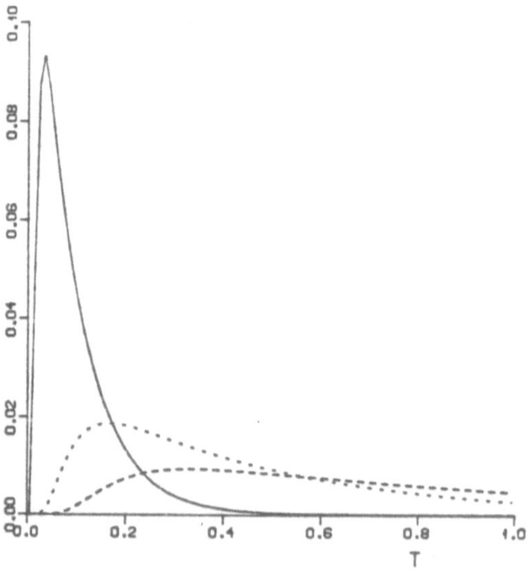


FIGURE 2.1 The function $k_\kappa(t)$ is plotted for $\kappa = 5$ (solid line), $\kappa = 1$ (dotted), $\kappa = 0.5$ (dashed).

The fact that we can not transform (2.2) to an equation of the second kind indicates that the problem of solving (2.2) is as difficult as the problem of solving a Fredholm integral equation of the first kind with a smooth kernel.

In Figure 2.1 we have plotted the function $k_\kappa$ for a few values of $\kappa$.

## 3. Singular Value Analysis

In this section we shall study the equation (2.2) and the integral operator K defined by

$$(Kf)(t) = \int_0^t k(t - \tau)\, f(\tau)\, d\tau, \qquad 0 \leq t \leq 1, \tag{3.1}$$

in an $L^2$-setting. Define the inner product

$$(f_1, f_2) = \int_0^1 f_1(t)\, f_2(t)\, dt.$$

The adjoint operator $K^*$ is easily found to be

$$(K^*g)(\tau) = \int_\tau^1 k(t - \tau)\, g(t)\, dt.$$

A singular system [28, p. 142-] for K,

$$(u_n, v_n, \sigma_n)_{n=1}^{\infty} ,$$

is a sequence of functions $u_n$, $v_n$, called singular functions, and real numbers $\sigma_n$, called singular values, satisfying

(i) $(u_n, u_m) = \delta_{nm}$,
(ii) $(v_n, v_m) = \delta_{nm}$,
(iii) $\sigma_1 \geq \sigma_2 \geq \ldots \geq 0$,
(iv) $Kv_n = \sigma_n u_n$,
(v) $K^*u_n = \sigma_n v_n$.

($\delta_{nm}$ = 1 if n = m, 0 otherwise). Note that our definition differs somewhat from that in [28].

The last two equations, (iv) and (v), imply that $v_n$ and $u_n$ are the eigenfunctions of $K^*K$ and $KK^*$ respectively, with eigenvalues $\sigma_n^2$.

It can be shown, cf. [28, p. 148], that the singular values $\sigma_n$ tend to zero as n tends to infinity (note that since k(t) is continuous, the operator K is bounded).

If the Picard criterion [28, p.164] is satisfied, i.e. if

$$\sum_{n=1}^{\infty} \frac{|(g, u_n)|^2}{\sigma_n^2} < +\infty ,$$

and

$$(g,u) = 0 \text{ for every } u, \text{ such that } K^*u = 0,$$

then the solution of the equation Kf = g is

$$f = \sum_{n=1}^{\infty} \frac{(g, u_n)}{\sigma_n} v_n \tag{3.2}$$

In practice, the Picard criterion is extremely restrictive, because g is generally contaminated with errors.

From (3.2) it is easily seen that the problem is ill-posed: the solution does not depend continuously on the data. Equivalently: the inverse of K is unbounded.

The faster the singular values tend to zero, the more severe becomes the ill-posedness, since the Picard criterion becomes more and more restrictive.

Due to the special structure of the integral operator (3.1), the singular functions have a certain symmetry property.

Lemma 3.1: The singular functions $u_n$, $v_n$ of the operator (3.1) satisfy either

$$u_n(t) = v_n\,(1 - t),$$

or

$$u_n(t) = -\,v_n\,(1 - t).$$

This is a straightforward generalization of a simple property of triangular Toeplitz matrices [22], and here we only sketch the proof.
Proof: Make the change of variables $\tau = 1 - s$; we get

$$(Kf)(t) = \int_{1-t}^{1} k(t + s - 1)\, f(1 - s)\, ds, \qquad 0 \le t \le 1,$$

The integral operator $\tilde{K}$ defined by

$$(\tilde{K}y)(t) = \int_{1-t}^{1} k(t + s - 1)\, y(s)\, ds, \qquad 0 \le t \le 1,$$

is selfadjoint. Let $\Psi_n$ be an eigenfunction of $\tilde{K}$ corresponding to the eigenvalue $\lambda_n$ :

$$\int_{1-t}^{1} k(t + s - 1)\, \Psi_n\,(s)\, ds = \lambda_n\, \Psi_n(t).$$

Now put $s = 1 - \tau$ :

$$\int_{0}^{t} k(t - \tau)\, \Psi_n(1 - \tau)\, d\tau = \lambda_n\, \Psi_n(t)$$

If $\lambda_n \ge 0$ put $u_n = \Psi_n$, otherwise put $u_n = -\,\Psi_n$. □

Now consider the equation

$$\int_{0}^{t} k(t - \tau) v_n(\tau)\, d\tau = \pm\, \sigma_n v_n(1 - t), \qquad 0 \le t \le 1, \tag{3.3}$$

and assume that $\sigma_n \neq 0$. We immediately see that $v_n(1) = 0$. Differentiate (3.3) and use $k(0) = 0$:

$$\int_0^t k'(t - \tau)v_n(\tau)d\tau = \mp \sigma_n v_n' (1 - t).$$

We get $v_n'(1) = 0$.

Thus by repeated differentiation we can prove:

Lemma 3.2: The singular functions $v_n$ of the operator (3.1), corresponding to non-zero singular values $\sigma_n$, satisfy

$$v_n^{(\nu)}(1) = 0, \qquad \nu = 0,1,2,\ldots .$$

Considering the expression (3.2) for the solution of $Kf = g$, and using Lemma 3.2, we now see that formally

$$f^{(\nu)}(1) = 0, \qquad \nu = 0,1,2,\ldots \tag{3.4}$$

This will be used later in the development of a numerical method for (2.2).

We have already seen that the problem of solving (2.2) is ill-posed. It can also be seen that the ill-posedness is due to two different factors:

K is a smoothing operator, (3.5i)

f(t) can not be reconstructed for t close to 1. (3.5ii)

The ill-posedness due to (3.5i) becomes worse for smaller values of the coefficient a(x), cf. Figure 2.1.

The ill-posedness due to (3.5ii) does not depend on the size of a(x). Note that if we put $f(t) = \delta(1 - t)$, a delta function with all its mass concentrated at $t = 1$, then $Kf = 0$.

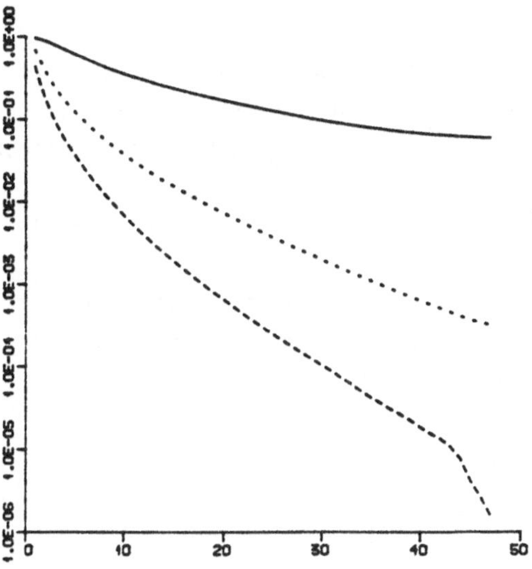


FIGURE 3.1 The 47 largest singular values of the matrix K, for $\kappa = 5$ (solid line), $\kappa = 1$ (dotted), and $\kappa = 0.5$ (dashed).

To study the ill-posedness we computed approximations of the singular values and singular functions, using a discretized version of (2.2)

$$Kf = g \tag{3.6}$$

where K is a 50×50 matrix (see the next section concerning discretization). We considered the constant coefficient case $u_t = \kappa u_{xx}$.

In Figure 3.1 we have plotted the 47 largest singular values of the matrix K for different values of $\kappa$. The rate of decay is approximately exponential (note that the y axis is logarithmic), and depends very much on the value of $\kappa$.

From now on we consider only the case $\kappa = 1$. Here we have 48 singular values in the range 1 to $10^{-4}$, and two of the order of magnitude $10^{-11}$. This pattern, a comparatively slow decay to begin with, and a couple of very small singular values, is even more pronounced for larger values of $\kappa$.

It is interesting to consider the approximations of the corresponding singular functions. The singular vectors corresponding to the singular values $\sigma_1, \sigma_{48}, \sigma_{49}, \sigma_{50}$ are plotted in Figure 3.2

From Figure 3.2 and from plots of the other singular vectors we can see that the ill-conditioning of the linear system (3.6) due to (3.5ii) is associated with the last two singular vectors only. Further, whereas the condition number of the matrix K in (3.6) is $0(10^{11})$, the condition number of the matrix

$$K_{48} = \sum_{i=1}^{48} \sigma_i v_i u_i^T ,$$

is

$$\kappa(K_{48}) = \sigma_1/\sigma_{48} \approx 2 \cdot 10^3,$$

which is quite moderate. Thus, for this value of $\kappa$ the ill-conditioning due to the smoothing properties of K is much less severe than the ill-conditioning due to (3.5ii).

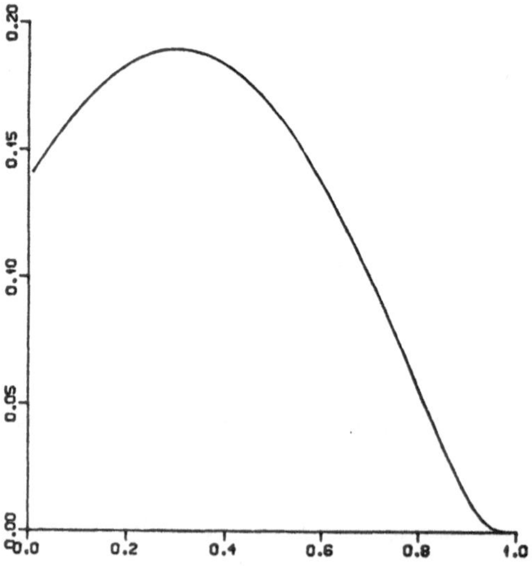


FIGURE 3.2a $v_1$ $(\sigma_1 = 0.69)$

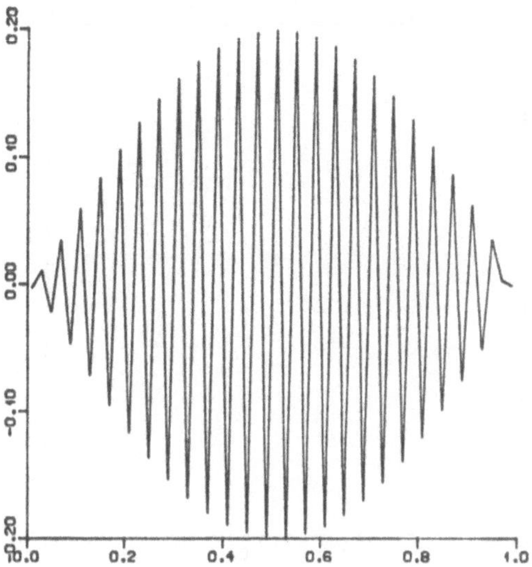


FIGURE 3.2b $v_{48}$ $(\sigma_{48} = 3.1 \cdot 10^{-4})$

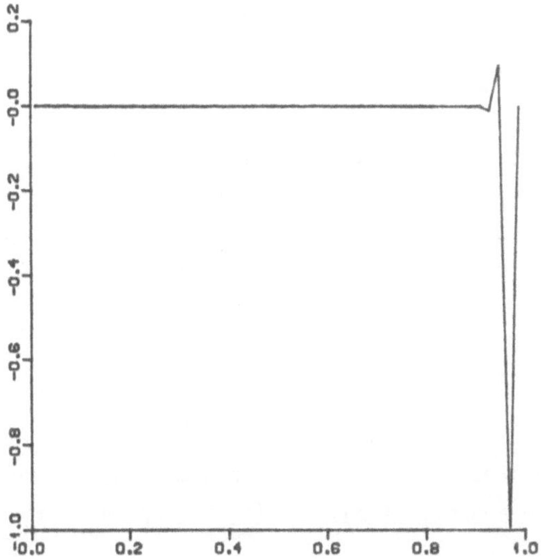


FIGURE 3.2c $v_{49}$ $(\sigma_{49} = 4.5 \cdot 10^{-11})$

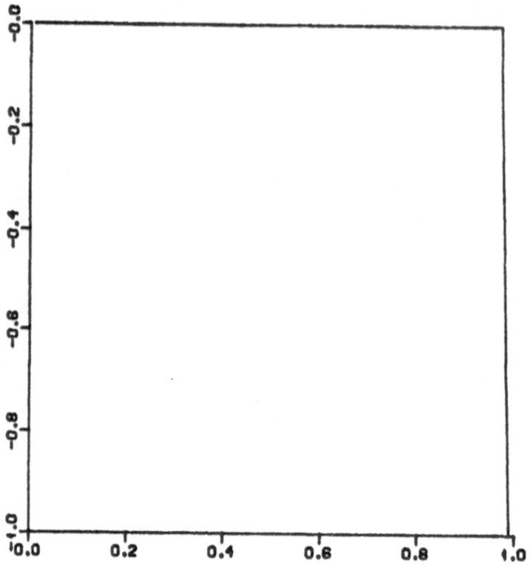


FIGURE 3.2d $v_{50}$ $(\sigma_{50} = 4.1 \cdot 10^{-11})$

## 4. Numerical Methods

There are several ways to discretize the integral equation (2.2). Here we consider one of the simplest, the midpoint method [18, p. 184]. Divide the interval [0,1] into n equal subintervals of length $h = 1/n$, and put $t_j = jh$. The discrete version of (2.2) is the system of linear equations

$$Kf = g, \tag{4.1}$$

where K is a square, lower triangular n×n-matrix. The components $f_j$ of the vector f are approximations of $f(t_{j-\frac{1}{2}})$, and $g_j = g(t_j)$, $j = 1,2,\ldots,n$.

Due to the fact that the integral equation (2.2) has a convolution type kernel, the matrix K becomes a Toeplitz matrix

$$K = \begin{bmatrix} k_1 & & & & & \\ k_2 & k_1 & & & & \\ k_3 & k_2 & k_1 & & & \\ \vdots & & \ddots & \ddots & & \\ k_{n-1} & k_{n-2} & \cdots & k_2 & k_1 & \\ k_n & k_{n-1} & \cdots & k_3 & k_2 & k_1 \end{bmatrix}, \tag{4.2}$$

where $k_j = k(t_{j-\frac{1}{2}})$, $j = 1,2,\ldots,n$, (for convenience we use different notational conventions for K, f and g).

In the constant coefficient case the matrix elements can readily be computed using the explicit expression for the function. For a general coefficient we suggest the following: solve numerically

$$\begin{aligned} u_t &= (a(x)u_x)_x, \\ u_x(0,t) &= 0, \\ u(1,t) &= d(t), \\ u(x,0) &= 0, \end{aligned} \tag{4.3}$$

where

$$d(t) = \begin{cases} 1, & 0 \leq t \leq h/2, \\ 0, & \text{otherwise.} \end{cases}$$

Then take

$$k_j = \hat{u}(0, t_{j-\frac{1}{2}}),$$

where $\hat{u}$ is the numerical solution of (4.3).

If n is taken large enough, the system (4.1) becomes extremely ill-conditioned (cf. Section 3). Such systems can be handled in several ways. However, when choosing a method, it is desirable that we can take advantage of the triangular structure and/or the Toeplitz structure of the matrix K. This means that we are looking for methods with an operation count $O(n^2)$ (or less).

One method, which seems obvious in view of the discussion in Section 3 is the method of truncated singular value decomposition (SVD), see e.g. [34]. Our experiments indicate that in many cases it will be very easy to decide where to truncate the SVD. However, in the computation of the SVD we can not take advantage of the structure of K, and the operation count is $O(n^3)$.

Another standard method for ill-conditioned linear systems is the regularization method of Tikhonov-Phillips [31], [24], [34]. Here we replace (4.1) by

$$\min_f \left\{ ||Kf - g||_2^2 + \mu^2 ||Lf||_2^2 \right\}, \tag{4.4}$$

where the norm is the Euclidean vector norm, and where L is usually chosen as a discretization of a differentiation operator. The value of the regularization parameter $\mu$ determines the smoothness of the solution. (4.4) can be written equivalently

$$\min_f \left|\left| \begin{pmatrix} K \\ \mu L \end{pmatrix} f - \begin{pmatrix} g \\ 0 \end{pmatrix} \right|\right|_2 . \tag{4.5}$$

In many cases the matrix L is a Toeplitz matrix too. E.g. when L corresponds to a first derivative, we have

$$L = \begin{bmatrix} 1 & -1 & & & & \\ & 1 & -1 & & & \\ & & 1 & -1 & & \\ & & & \ddots & \ddots & \\ & & & & 1 & -1 \end{bmatrix}.$$

An efficient and stable algorithm for the solution of (4.5), where both K and L are triangular Toeplitz matrices, is given in [10]. A QR-decomposition of the matrix in (4.5) is computed using plane rotations, and the algorithm utilizes the structure of the problem so that the number of operations is $O(n^2)$.

Some numerical results using the regularization method are reported in the next section.

The analysis of Section 3 showed that singular functions $v_n$ and the solution (of the continuous problem) should satisfy

$$v_n^{(\nu)}(1) = f^{(\nu)}(1) = 0, \qquad \nu = 0,1,2,\ldots \tag{4.6}$$

Also, from the numerical experiments we saw that the smallest singular values of the discretized problem are associated with singular vectors, which do not tend smoothly to zero as t tended to 1, i.e. the last few components of these singular vectors are not small, see Figure 3.2.

Therefore, these singular vectors can be forced out of the solution, and a better conditioned problem can be obtained, if we impose the constraint (4.6). Thus we replace (4.1) by

$$\min_{f \in \mathcal{B}} ||Kf - g||_2 \; ; \qquad \mathcal{B} = \left\{ f : f_j = 0, j = p+1,\ldots,n \right\},$$

for some value of p. This is equivalent to solving a least squares problem

$$\min_{\bar{f}} ||\bar{K}\,\bar{f} - g||_2 \, , \tag{4.7}$$

where

$$\bar{K} = \begin{pmatrix} k_1 & & & \\ k_2 & k_1 & & \\ \vdots & & \ddots & \\ k_p & \cdots & k_2 & k_1 \\ k_{p+1} & \cdots & & k_2 \\ \vdots & & & \vdots \\ k_n & \cdots & & k_{n-p+1} \end{pmatrix}, \qquad \bar{f} = \begin{pmatrix} f_1 \\ f_2 \\ \vdots \\ f_p \end{pmatrix}.$$

This can be solved as follows. Compute the decomposition

$$\bar{K} = Q\binom{0}{S} \, , \tag{4.8}$$

where Q is orthogonal and S is a lower triangular p×p matrix. Then (4.7) is equivalent to

$$\min\left|\left|\binom{0}{S}\bar{f} - \begin{pmatrix}\tilde{g}_1\\ \tilde{g}_2\end{pmatrix}\right|\right|, \qquad \tilde{g} = Q^T g = \begin{pmatrix}\tilde{g}_1\\ \tilde{g}_2\end{pmatrix} \begin{matrix}\}\ n-p\\ \}\ p\end{matrix} ,$$

and the solution is obtained by solving

$$S\bar{f} = \tilde{g}_2.$$

The matrix S in (4.8) can be computed using p(n-p) plane rotations, and thus for p close to n this is a method with operation count $O(n^2)$.

We can not expect this method to work when the ill-conditioning due to the smoothing properties of K is significant. However, for large coefficients a(x) it should be useful. This is verified numerically in the next section.

There are other efficient methods that can be used for our problem, e.g. the one described in [22]. That method is iterative, and is based on the regularization method.

## 5. Numerical Experiments

In this section we report some numerical experiments obtained in the special case of a constant coefficient, $u_t = \kappa u_{xx}$.

The function values $k(t_{j-\frac{1}{2}})$ were computed using the analytical formula. The solution f(t) was selected, and the data vector g was computed by solving an initial-boundary value problem (2.1). Then normally distributed perturbations with mean zero and different standard deviations were added to g. In all cases the dimension of the discrete problem was taken equal to 50.

The computations were performed on a DEC-10 computer, where the relative machine precision is approximately $10^{-8}$.

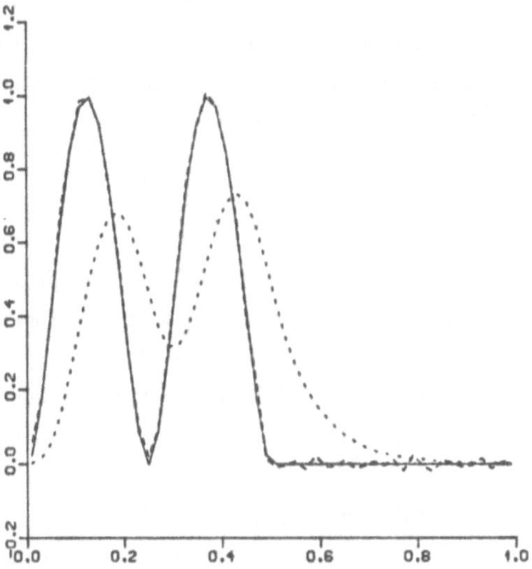


a.

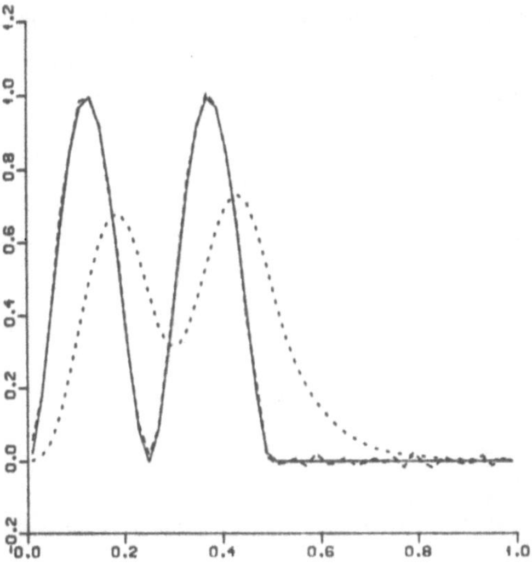


b.

FIGURE 5.1 Results obtained for $\kappa = 5$. Perturbations with standard deviation $10^{-3}$ were added. Exact solution (solid line), data (dotted), and approximate solution (dashed).

a. The regularization method, L corresponding to a first derivative, $\mu = 10^{-8}$.

b. The constraint method, $p = 49$.

In Figure 5.1 we show results obtained for $\kappa = 5$. In this case the condition number of K is equal to $0.9 \cdot 10^8$, but if we remove the smallest singular value the corresponding condition number becomes approximately 16. Therefore we obtain very good results with both methods.

In the case $\kappa = 1$, the condition number of the matrix where we have excluded the two smallest singular values is approximately equal to $2 \cdot 10^3$. When no perturbations were added both methods gave very accurate results. Howevever, for this value of $\kappa$, the problem is much more sensitive to perturbations in the data. When we added perturbations with standard deviation $10^{-3}$, only the regularization method produced acceptable results, see Figure 5.2. Even if we put p as small as 35 in the second method, the solution oscillated wildly. Obviously, for this value of $\kappa$ and for errors of this magnitude, the ill-conditioning due to the smoothing properties of $\kappa$ is so severe, that it is necessary to use a method, which can suppress both kinds of ill-conditioning.

Several other numerical examples were solved. The results reported here are typical.

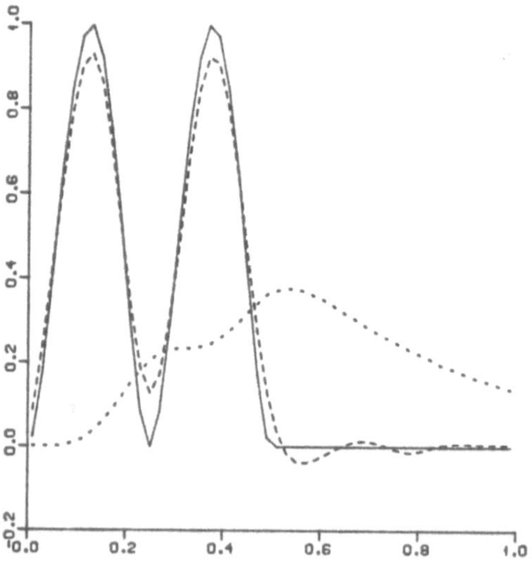


FIGURE 5.2 Result for $\kappa = 1$. Perturbations with standard deviation $10^{-3}$ were added. The regularization method, L corresponding to a first derivative, $\mu = 10^{-3}$.

## 6. Conclusion and Extensions

We have considered two numerical methods for solving a Cauchy problem for a parabolic equation, and described efficient algorithms for the implementation of the methods. One of the methods, the regularization method, was shown to be applicable to a relatively wide class of problem, while the other, the method based on constraining the solution, was useful only when the coefficient in the parabolic equation was large or the perturbations of the data were small.

The methods described in this paper can be generalized straightforwardly to the problem, where the unknown is the flux of heat at the right boundary, i.e. $u_x(1,t)$.

The case when the coefficient depends also on t,

$$u_t = (a(x,t)u_x)_x ,$$

can also be transformed into a Volterra equation of the first kind. In this case, however, it becomes more difficult to determine the kernel function, since we no longer have an equation of convolution type:

$$g(t) = \int_0^t k(t,\tau)f(\tau)\, d\tau , \qquad 0 \le t \le 1.$$

It is also possible to transform other Cauchy problems for parabolic equations into equivalent integral equations. Consider e.g. the problem [11]

$$\begin{aligned} &u_t = (a(x)u_x)_x , && 0 \le x \le 1, \qquad 0 \le t \le 1, \\ &u(0,t) = g_1(t), && 0 \le t \le 1, \\ &u_x(0,t) = g_2(t), && 0 \le t \le 1, \\ &u(1,t) = g_3(t), && 0 \le t \le 1, \end{aligned}$$

where the unknown is the initial function $h(x) = u(x,0)$. Here we get a Fredholm equation of the first kind

$$\int_0^1 k(x,t)h(x)dx = g(t).$$

## 7. Acknowledgements

This work was supported by the Swedish Natural Science Research Council.

During the course of the work I have had many helpful and stimulating discussions with Lars-Erik Andersson, Arne Enqvist and Bengt Winzell.

## References

[1] B.K. Amonov, S.P. Shishatski, An a priori estimate of the solution to the Cauchy problem for a second order parabolic equation with data given on the time surface, and related uniquness theorems, Soviet Math. Dokl. 13(1972), 1153-1154.

[2] L.E. Andersson, A note on the regularity of the solution of a parabolic equation with discontinuous coefficient, Department of Mathematics, Linköping University, 1983, to appear.

[3] J.B. Bell, The noncharacteristic Cauchy problem for a class of equations with time dependence. I. Problem in one space dimension, SIAM J. Math. Anal. 12(1981), 759-777.

[4] J.R. Cannon, A priori estimate for continuation of the solution of the heat equation in the space variable, Ann. Mat. Pura Appl. 65(1964), 377-388.

[5] J.R. Cannon, A Cauchy problem for the heat equation, Ann. Mat. Pura Appl. 66(1964), 155-166.

[6] J.R. Cannon, J. Douglas,Jr., The Cauchy problem for the heat equation, SIAM J. Numer. Anal. 4(1967), 317-336.

[7] J.R. Cannon, R.E. Ewing, A direct numerical procedure for the Cauchy problem for the heat equation, J. Math. Anal. Appl. 56(1976), 7-17.

[8] H.S. Carslaw, J.C. Jaeger, Conduction of heat in solids, Oxford University Press (Clarendon), London, 1948.

[9] J. Douglas, Jr., Approximate continuation of harmonic and parabolic functions, Numerical Solution of Partial Differential Equations, Ed. J.H. Bramble, Academic Press, New York, 1966.

[10] L. Eldén, An efficient algorithm for the regularization of ill-conditioned least squares problems with triangular Toeplitz matrix,SIAM J. Sci. Stat. Comp., to appear.

[11] R.E. Ewing, R.S. Falk, Numerical approximation of a Cauchy problem for a parabolic partial differential equation, Math. Comp. 33(1979), 1125-1144.

[12] R.E. Ewing, The Cauchy problem for a linear parabolic partial differential equation, J. Math. Anal. Appl. 71(1979), 167-186.

[13] L. Garifo, V.E. Schrock, E Spedicato, On the solution of the inverse heat conduction problem by finite differences, Energia Nucleare 22(1975), 448-460.

[14] F. Ginsberg, On the Cauchy problem for the one-dimensional heat equation, Math. Comp. 17(1963), 257-269.

[15] V.M. Isakov, On the uniqueness of the solution of the Cauchy problem, Soviet Math. Dokl. 22(1980), 539-642.

[16] Lavrentiev, Romanov, Vasiliev, Multidimensional inverse problems for differential equations, Lecture Notes in Mathematics 167, Springer, Berlin 1970.

[17] M.M. Lavrentiev, V.G. Romanov, S.P. Shishatskiy, Ill-posed problems in mathematical physics and analysis, Izdatelstvo Nauka, Moscow, 1980.

[18] P. Linz, A survey of methods for the solution of Volterra integral equations of the first kind, in The application and numerical solution of integral equations, ed. R.S. Anderssen et al., Sijthoff & Noordhoff, Alphen an den Rijn, 1980.

[19] P. Manselli, K. Miller, Calculation of the surface temperature and heat flux on one side of a wall from measurements on the opposite side, Ann. Mat. Pura Appl. 123(1980), 161-183.

[20] D.A. Murio, The mollification method and the numerical solution of an inverse heat conduction problem, SIAM J. Sci. Stat. Comput. 2(1981) 17-34.

[21] D.A. Murio, On the estimation of the boundary temperature on a sphere from measurements at its center, J. Comp. Appl. Math. 8(1982), 111-119.

[22] D.P. O'Leary, J.A. Simmons, A bidiagonalization-regularization procedure for large scale discretizations of ill-posed problems, SIAM J. Sci. Stat. Comp. 2(1981), 474-489.

[23] L.E. Payne, Improperly posed problems in partial differential equations, Society for Industrial and Applied Mathematics, Philadelphia, 1975.

[24] D.L. Phillips, A technique for the numerical solution of certain integral equations of the first kind, J. ACM 9(1962), 84-97.

[25] W.W. Schmaedeke, Approximate solutions for Volterra integral equations of the first kind, J. Math. Anal. Appl. 23(1968), 604-613.

[26] T.I. Seidman, Ill-posed problems arising in boundary control and observation for diffusion equations, in Inverse and improperly posed problems in differential equations, ed. Anger, Akademie-Verlag, Berlin, 1979.

[27] T. Seidman, Recovery of a diffused signal, Seminaires IRIA, 1979, (Analyse et controle de systemes), INRIA, Rocquencourt, 1980, pp. 71-82.

[28] F. Smithies, Integral equations, Cambridge University Press, Cambridge, 1965.

[29] I.H. Sneddon, The use of integral transforms, McGraw-Hill, New York, 1972.

[30] A.N. Tikhonov, Theoremes d unicité pour l equation de la chaleur, Matematichesky Sbornik 42(1935), 199-206.

[31] A.N. Tikhonov, Solution of incorrectly formulated problems and the regularization method, Dokl. Akad. Nauk SSSR 151 (1963), 501-504 = Soviet Math. Dokl. 4(1963), 1035-1038.

[32] A.N. Tikhonov , V.Y. Arsenin, Solutions of ill-posed problems, Winston, Wiley, New York, 1977.

[33] A.N. Tikhonov, V.V. Glasko, Methods of determining the surface temperature of a body, USSR Comp. Math. Math. Phys. 7 no 4 (1967), 267-273.

[34] J.M. Varah, A practical examination of some numerical methods for linear discrete ill-posed problems, SIAM Review 21(1979), 100-111.

[35] C.F. Weber, Analysis and solution of the ill-posed inverse heat conduction problem, Int. J. Heat Mass Transfer 24(1981), 1783-1792.

[36] D.V. Widder, The heat equation, Academic Press, New York, 1975.

# THE INVERSE PROBLEM IN GEOELECTRICAL PROSPECTING ASSUMING A HORIZONTALLY LAYERED HALF-SPACE

E. Mundry


## Abstract

The Marquardt method is used for the inversion of geoelectrical measurements for a stratified half-space (determination of layer thicknesses and resistivities). Equivalent models are obtained from eigenvalue analysis of the Gaussian matrix.


## 1. Introduction

The main problem in geophysical prospecting is the determination of the distribution of a physical quantity within the earth from measurements of natural or artificial fields at the earth's surface.

Examples are the determination

-- of the electrical conductivity from geoelectrical or magnetotelluric measurements,

-- of rock densities from gravimetric measurements,

-- of rock magnetization from magnetic measurements.

Generally, a model of the ground must be assumed for the interpretation of the observed data. The inverse problem then consists of the determination of the geometric and physical parameters of the model, which is usually non-linear. In the last several years, the classical least-squares method and its modern extension (ridge regression analysis, generalized matrix inversion) have become important tools for solving such non-linear estimation problems. The papers of BOSUM [1] and HJELT [4] can be mentioned as examples for the interpretation of magnetic data and that of PEDERSEN [14] for magnetic and gravimetric data.

The inverse problem in geoelectrical prospecting assuming a horizontally layered half-space will be discussed in this paper. This model is widely used in hydrogeology, for the geophysical exploration of stratified mineral deposits, and for solving special problems in geological mapping. Attempts have been made to determine a continuous conductivity function $\sigma(z)$ with depth z (OLDENBURG [13] and COEN/YU [2]), but it seems to be difficult to construct a realistic model from the non-unique results. A step-like function of $\sigma$ (where the conductivity in each layer is homogeneous) corresponds better to the geological situation. It should be mentioned that a similar mathematical procedure for direct current (d.c.) measurements (described in the following sections) is also in use for interpreting magnetotelluric soundings (WU [16] and MÜLLER [11]).

## 2. D.C. Resistivity Soundings

A Schlumberger configuration is used in most cases for geoelectrical soundings (Fig. 1). A current I is fed into the ground via two electrodes, A and B. The potential difference U is measured in the center between electrodes M and N. The distance MN is chosen to be sufficiently small in comparison to AB so that, in principle, the gradient of the electrical potential is measured. For fixed positions of M and N, the distance AB between the current electrodes is increased symmetrically so that deeper and deeper parts of the subsurface influence the voltage U.

The data are recalculated in the form of the apparent resistivity

$$\rho_a = K\, U/I,$$

where the geometric factor

$$K = \pi\, (\overline{AB}/2)^2/\overline{MN} \qquad (\overline{MN} \ll \overline{AB})$$

is defined in such a manner that $\rho_a$ is identical with the resistivity $\rho = 1/\sigma$ in the case of a homogeneous half-space. A plot of $\rho_a$ vs. AB/2 (= L/2 in Fig. 2) on a double-logarithmic scale is called a sounding curve (Fig. 2). The inverse problem is to determine the number of layers n, the corresponding thicknesses $h_i$ ($i = 1, \ldots, n-1$), and resistivities $\rho_i$ ($i = 1, \ldots, n$) from this sounding curve.

## 3. The Forward Solution

In a source-free space, the equation for the potential

$$\Delta V_i = 0 \tag{1}$$

can be derived from the equations

$$\underline{j} = \sigma \underline{E}, \quad \underline{E} = -\nabla V, \text{ and } \nabla \cdot \underline{j} = 0 \tag{2}$$

for each layer i ($\underline{j}$ = current density, $\underline{E}$ = electrical field, V = potential).

It is sufficient to solve equation (1) for a point source A with current I placed at the center of a cylindrical coordinate system (r,z). The influence of the other electrode, B, (current = -I) is obtained by superposition. The general solution of equation (1) is

$$V_i = \int_0^\infty [A_i(\lambda)e^{-\lambda z} + B_i(\lambda)e^{\lambda z}]J_0(\lambda r)d\lambda\ , \tag{3}$$

where $J_0$ is the zero order Bessel function of the first kind. The coefficients $A_i$ and $B_i$ can be determined using the following boundary conditions:

$$\begin{aligned} &V_i = V_{i+1}, \quad \sigma_i \partial V_i/\partial z = \sigma_{i+1}\partial V_{i+1}/\partial z \quad (z = z_i) \\ &\partial V_i/\partial z = 0 \quad (z = 0,\ r \neq 0) \\ &V_1 \rightarrow I\rho_1/2\pi R \quad (R = \sqrt{r^2 + z^2} \rightarrow 0) \\ &V_n \rightarrow 0 \quad (R \rightarrow \infty). \end{aligned} \tag{4}$$

The apparent resistivity for a Schlumberger configuration is given by

$$\rho_\alpha(r) := \left.\frac{-\pi r^2}{I}\frac{\partial V_1}{\partial r}\right|_{z=0} = r^2 \int_0^\infty T_1(\lambda)\cdot\lambda\cdot J_1(\lambda r)d\lambda\ , \tag{5}$$

where $r = \overline{AB}/2$.

The function $T_1(\lambda)$ can be calculated by means of the following recurrence relationships:

$$R_{n-1} := k_{n-1}$$

$$R_i = \frac{k_i + R_{i+1}U_{i+1}}{1 + k_iR_{i+1}U_{i+1}} \quad (i = n-2, n-3, \ldots, 1) \tag{6}$$

$$T_1 = \rho_1 \frac{1 + R_1U_1}{1 - R_1U_1}$$

where $k_i = (\rho_{i+1} - \rho_i)/(\rho_{i+1} + \rho_i)$

$U_i = \exp(-2\lambda h_i)$ .

The integration in equation (5) can be carried out using the filter method of GHOSH [3]. For logarithmic equidistant $\lambda$ and $r$ values, the apparent resistivity can be calculated by convolution of $T_1$ with some filter coefficients, which can be stored in advance:

$$\rho_a^{(j)} = \sum_{\ell=\ell_1}^{\ell_2} C_\ell T_1^{(\ell-j)} \tag{7}$$

For details, see [9].

## 4. Inversion Procedures

The function $T_1(\lambda)$ can be calculated from the sounding curve $\rho_a(r)$ by means of a Hankel transformation:

$$T_1(\lambda) = \int_0^\infty \frac{\rho_a(r)}{r} J_1(\lambda r)dr. \tag{8}$$

The recurrence relationship (6) can now be solved for $R_i$ and $R_{i+1}$; and owing to the asymptotic behavior when $\lambda \rightarrow \infty$, it should be possible to determine the parameters $k_i$ and $h_i$ recursively, starting with the top layer (PEKERIS [15], KOEFOED [9]).

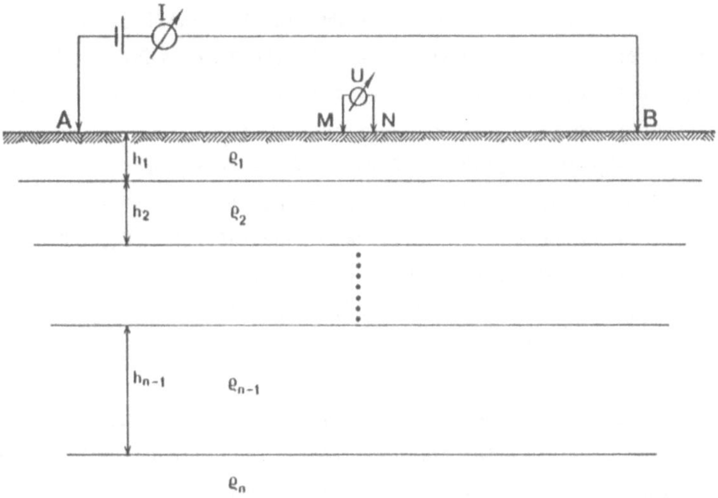


Figure 1: Principle of resistivity sounding using a Schlumberger array over an n-fold stratified half-space with layer thicknesses $h_i$ and resistivities $\rho_i$

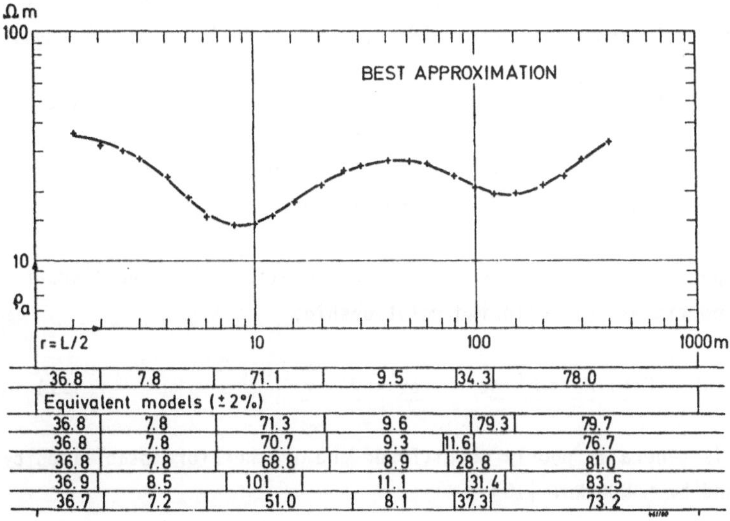


Figure 2: Top: Sounding curve (+) and optimum model curve (——) for geoelectrical sounding no. 46 in the Mahanadi Delta, India; Bottom: Optimum model and equivalent models (z axis = r axis)

In practice this method gives unsatisfactory results: It doesn't work in most cases for 3 or more layers due to the magnification in each step of the errors in the observed data. Furthermore, additional errors are introduced by the numerical Hankel transform of the original data.

Since an analytical solution of the problem does not seem to be possible in practice, the Marquardt or ridge regression method (MARQUARDT [10]) or the generalized matrix inversion method (JACKSON [6], JUPP & VOZOFF [8]) is used for the inversion of d.c. resistivity data (INMAN [5], JOHANSEN [7], MUNDRY & DENNERT [12]). In this case, the number n of the layers must be known.

To avoid negative values for the parameters and to match the values of $\rho_a^{(j)}$ $(j = 1, \ldots, m)$ for the logarithmic sounding curve, it is recommended that logarithmic values be used as follows:

$$\begin{aligned} y_j &= \log \rho_a^{(j)} && (j = 1, \ldots, m) \\ \beta_i &= \log h_i && (i = 1, \ldots, n-1) \\ \beta_{i+n-1} &= \log \rho_i && (i = 1, \ldots, n) \end{aligned} \tag{9}$$

Using a Taylor series, the model is linearized with respect to $\beta_i$ near an initial parameter vector $\underline{\beta}^{(o)}$ ($\underline{\beta}$ has $k = 2n-1 < m$ dimensions):

$$\underline{y} = \underline{\bar{y}} + \underline{F} \cdot \underline{\Delta\beta} + \underline{\varepsilon} \,, \tag{10}$$

where $\underline{y}$ = data vector, $\underline{\bar{y}}$ = vector of the theoretical values for $\underline{\beta}^{(o)}$, $\underline{F}$ = Jacobian matrix, and $\underline{\varepsilon}$ = error vector.

The parameter vector $\underline{\beta}$ is refined by iteration. The least-squares method yields the following relationship:

$$(\underline{F}^T\underline{F}) \cdot \underline{\Delta\beta} = \underline{F}^T \cdot \underline{\Delta y} \qquad (\underline{\Delta y} = \underline{y} - \underline{\bar{y}}) \tag{11}$$

$\underline{F}^T\underline{F}$ is generally poorly conditioned and is therefore stabilized by a method by MARQUARDT [10]:

$$(\underline{F}^T\underline{F} + \mu^2\underline{I}) \cdot \underline{\Delta\beta} = \underline{F}^T \cdot \underline{\Delta y} \,, \tag{12}$$

where $\underline{I}$ is the identity matrix. (In practice, a normalization of $\underline{F}^T\underline{F}$ is done before the calculations.)

A computer programm has been developed for this procedure by MUNDRY & DENNERT [12] which incorporates a special strategy for determining $\mu^2$ so that $\mu^2$ is as small as possible but large enough to ensure the convergence of the iterative refinement procedure.

Once the eigenvalue equation

$$(\underline{F}^T\underline{F})\cdot\underline{V} = \underline{V}\cdot\underline{\Lambda}^2 \tag{13}$$

is solved for each iteration ($\underline{\Lambda}^2$ = diagonal matrix of eigenvalues, $\underline{V}$ = matrix of eigenvector of $\underline{F}^T\underline{F}$), then the solution of equation (12) for any $\mu^2$ value can be obtained by

$$\underline{\Delta\beta} = \underline{V}(\underline{\Lambda}^2 + \mu^2\underline{I})^{-1}\underline{V}^T\cdot\underline{\Delta y}\ , \tag{14}$$

which is a simple series (MARQUARDT [10]).

In most cases, 5 to 10 iterations for $\underline{\beta}$ are needed for sufficient accuracy. Of course it is possible to fix some of the parameters, for example, the resistivity of a specific layer.

To obtain a geologically acceptable model, it is often necessary to establish near the MARQUARDT solution some kind of confidence interval for the parameters. This is done by means of the following:

If an eigenvalue $\lambda_j^2$ of the Gaussian matrix $\underline{F}^T\underline{F}$ is zero, then any multiple of the corresponding eigenvector $\underline{v_j}$ can be added to the optimum solution $\underline{\beta}^{(opt)}$ without changing the sum of squared errors. In most cases, some of the eigenvalues are very small compared to the largest one. An equivalent model is then constructed by adding the product of $c_j$ and the eigenvector $\underline{v_j}$ to the optimum solution:

$$\underline{\beta}^{(eq)} = \underline{\beta}^{(opt)} + c_j\underline{v_j} \tag{15}$$

The factor $c_j$ is determined by regula falsi in such a way that the corresponding model curve $\bar{\rho}_a(\underline{\beta}^{(eq)})$ deviates from $\bar{\rho}_a(\underline{\beta}^{(opt)})$ by a prescribed maximum error. Two equivalent models can be calculated for each of the eigenvectors ($c_j$ positive or negative).

## 5. Example

A geoelectrical sounding in the Indian Mahanadi Delta, where $(AB/2)_{max}$ = 400 m is shown as a practical example. The form of the curve suggests 5 layers, but for geological reasons 6 layers (with a small layer of intermediate resistivity above the low conducting half-space) were assumed (k = 11 parameters).

An initial estimation was made by comparing the curve with 3-layer model curves and by using the so-called auxiliary point diagrams for the successive approximate reduction of the n-layer case to a two-layer model (see KOEFOED [9]).

The optimum solution is shown in Fig. 2, together with 5 equivalent solutions (±2 % deviation for the model curves). As can be seen from the first equivalent model, the resistivity of the next to last layer has nearly the same resistivity as the underlying half-space (79.3 Ωm instead of 79.7 Ωm). Thus, the sounding curve may be interpreted with a 5-layer model.

## 6. References

[1] W. Bosum: Ein automatisches Verfahren zur Interpretation magnetischer Anomalien nach der Methode der kleinsten Quadrate. Geophys. Prospecting 16, 107-126 (1968).

[2] S. Coen, M.W.H. Yu: The inverse problem of the direct current conductivity profile of a layered earth. Geophysics 46, 1702 - 1713 (1981).

[3] D.P. Ghosh: Inverse filter coefficients for the computation of apparent resistivity standard curves for a horizontally stratified earth. Geophys. Prospecting 19, 769 - 775 (1971).

[4] S.-E. Hjelt: Performance comparison of non-linear optimization methods applied to interpretation in magnetic prospecting. Geophysica 13, 143 - 166 (1975).

[5] J.R. Inman: Resistivity inversion with ridge regression. Geophysics 40, 798 - 817 (1975).

[6] D.D. Jackson: Interpretation of inaccurate, insufficient and inconsistent data. Geophys. J.R. astr. Soc. 28, 97 - 109 (1972).

[7] H.K. Johansen: A man/computer interpretation system for resistivity soundings over a horizontally stratified earth. Geophys. Prospecting 25, 667 - 691 (1972).

[8] D.L.B. Jupp, K. Vozoff: Stable iterative methods for the inversion of geophysical data. Geophys. J.R. astr. Soc. 42, 957 - 976 (1975).

[9] O. Koefoed: Geosounding Principles, 1 -- Resistivity Sounding Measurements. Elsevier, Amsterdam (1979).

[10] D.W. Marquardt: An algorithm for least-squares estimation of non-linear parameters. J. Soc. indust. Appl. Math. 11, 431 - 441 (1963).

[11] W. Müller: Inversion by simultaneous fitting of apparent resistivity and phase angle. Acta Geodaet., Geophys. et Montanist. Acad. Sci. Hung. 12, 215 - 222 (1977).

[12] E. Mundry, U. Dennert: Das Umkehrproblem in der Geoelektrik. Geol. Jb. E 19, 19 - 38 (1980).

[13] D.W. Oldenburg: The interpretation of direct current resistivity measurements. Geophysics 43, 610 - 625 (1978).

[14] L.B. Pedersen: Interpretation of potential field data, a generalized inverse approach. Geophys. Prospecting 25, 199 - 230 (1977).

[15] C.L. Pekeris: Direct method of interpretation in resistivity prospecting. Geophysics 5, 31 - 46 (1940).

[16] F.T. Wu: The inverse problem of magnetotelluric sounding. Geophysics 33, 972 - 979 (1968).

# TWO DIMENSIONAL VELOCITY INVERSION FOR ACOUSTIC WAVES WITH INCOMPLETE INFORMATION

A.J. Hermans

## 1. Introduction

We consider a two-dimensional inverse problem. The index of refraction or propagation speed of a medium is known, in some localized region up to small perturbations. Our objective is to construct these small perturbations from observation of the scattered field generated by a known incident field. It follows that in the ideal case the observations have to be collected along a closed curve surrounding the finite support of the perturbations. We shall present a method to solve the incomplete problem, where the data collection took place along a part of the closed curve.

The method we use is based on the theory developped by J.K. Cohen and N. Bleistein [1]. In the one-dimensional case, they applied their theory and obtained rather accurate approximations of the perturbations in the index of refraction. In this case it is very easy to collect data such that a direct inversion procedure can be used. In two- and three-dimensional cases only for ideal situations complete data collection can take place, therefore we study the incomplete case for a two-dimensional situation.

As we stated we consider small perturbations (order $\varepsilon$) of the known index of refraction. The perturbation may be discontinuous although in the reconstruction some smoothing takes place automatically because of the discretization in the Fourier domain. Extreme cases such as trapped modes can not be treated in this way. An important role in the method plays the Born approximation for the scattered field; it is known, however, that this method leads to rather poor results at high frequencies. Therefore we apply the Rytov approximation in that situation.

The results thus obtained are further treated by means of stationary phase and it follows that the forward scattered field is related to the Radon transformation of the perturbation of the index of refraction while the backscattered field may yield additional information.

In the last sections it is shown how for the high frequency problem the incomplete information situation can be handled. In the cases we are interested in, it turns out that a Fredholm integral equation of the second kind with a regular kernel can be obtained. There are several techniques available to solve this equation. We pay no attention to the selection of an appropriate technique. Until now no numerical experiments are carried out.

## 2. The Problem

We consider the scalar Helmholtz equation with variable refraction index $n^2(x) = n_0^2(1+\varepsilon\,\alpha(\underline{x}))$ where $n_0$ is a constant, $\varepsilon$ a small parameter and $\alpha(\underline{x})$ the unknown perturbation.
The Helmholtz operator becomes

$$L = \nabla^2 + n_0^2\omega^2/c^2\ (1 + \varepsilon\ \alpha(\underline{x})) \tag{1}$$

and with $k^2 = n_0^2\omega^2/c^2$
where k is the wave number.
We may write

$$L = L_0 + \varepsilon k^2\ \alpha(\underline{x}) \tag{2}$$

where $L_0 = \nabla^2 + k^2$ is the unperturbed operator

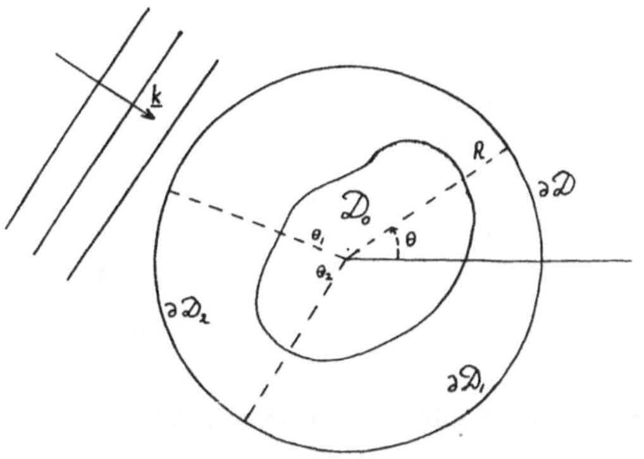


Fig. 1.

The support of $\alpha(\underline{x})$ is $D_0$ which is unknown beforehand, although $\partial D$ can be chosen not intersecting $D_0$. The contour $\partial D$ consists of two parts $\partial D_1$ and $\partial D_2$. Along $\partial D_1$ the scattered field is measured for given incoming fields while along $\partial D_2$ the field is unknown.
For simplicity we take for $\partial D$ a circular contour with radius R and the polar angle $\theta$ where $0 \le \theta < 2\pi$ and $\partial D_2$ with $0 \le \theta_1 < \theta < \theta_2 < 2\pi$.
We consider a function $u(\underline{x},\varepsilon)$ which satisfies the differential equation

$$L\ u = 0 \quad \forall \underline{x} = (x_1,x_2) \text{ in } \tilde{D} \tag{3}$$

where $\tilde{D}$ is infinite. The contour $\partial D$ is not the boundary of $\tilde{D}$, but of a region D containing the support of $\alpha(\underline{x})$. At infinity a proper radiation condition has to be imposed. We use as probe field an incident field

$U_0 = e^{i\underline{k}\cdot\underline{x}}$ which propagates in the direction of $\underline{k}$ where $|\underline{k}| = k = n_0\omega/c$. In this case the scattered field $\tilde{U}_s$ fulfils the radiation condition

$$\lim_{r\to\infty} \sqrt{r}\left(\frac{\partial \tilde{U}_s}{\partial r} - i\,k\,\tilde{U}_s\right) = 0 \tag{4}$$

Because $D_0$ is a finite region we assume that U can be expressed as

$$U = U_0 + \varepsilon\,\tilde{U}_s \tag{5}$$

where $U_s$ can be written as a regular perturbation series

$$\tilde{U}_s = U_s + \varepsilon\,U_2 + \varepsilon^2\,U_3 + \ldots\ldots. \tag{6}$$

This approximation is known as the Born approximation.

Inserting (5) and (6) in (3) and equating like powers in $\varepsilon$ to be zero we obtain

$$L_0\,U_0 = 0 \tag{7}$$

$$L_0\,U_s = -k^2\alpha(\underline{x})\,U_0(\underline{x}) \tag{8}$$

$$L_0\,U_i = -k^2\,\alpha(\underline{x})\,U_{i-1} \quad \text{for } i = 2,3,4 \text{ with } U_1 \equiv U_s \tag{9}$$

The incident field $U_0$ is such that (7) is satisfied.

If we consider $\alpha(\underline{x})$ to be known (8) can be solved easily by means of a source distribution. As a first approximation for the scattered field we find for $\underline{x} \notin D_0$

$$U_s(\underline{x}) = \frac{ik^2}{4}\int_{D_0} \alpha(\underline{x}')U_0(\underline{x}')H_0^{(1)}\{\,k|\underline{x}-\underline{x}'|\}ds' \tag{10}$$

For large values of $|\underline{x}|$, $U_s(\underline{x})$ obeys the radiation condition. For the inverse problem $\alpha(\underline{x})$ has to be determined. For this purpose (10) is not appropriate. We measure $U_s(\underline{x})$ along $\partial D_1$ and (10) may be considered as an ill posed problem. The solution of this Fredholm equation of the first kind does not exist in the general case. Cohen and Bleistein therefore derive a different equation which can be solved uniquely for the complete case.

Introduce v to be a solution of $L_0\,v = 0$. Green's theorem reads

$$\int_D (U_s\,L_0\,v - v\,L_0\,U_s)ds = \int_{\partial D}\left(U_s\,\frac{\partial v}{\partial n} - v\,\frac{\partial U_s}{\partial n}\right) d\,l \tag{11}$$

and leads to

$$\int_D U_0\,v\,\alpha(\underline{x})ds = \theta(\underline{k}) \tag{12}$$

$$\text{with } \theta(\underline{k}) = \frac{1}{k^2}\int_{\partial D}\left(U_s\,\frac{\partial v}{\partial n} - v\,\frac{\partial U_s}{\partial n}\right) d\,l \tag{13}$$

Choosing $v = U_0 = e^{i\,\underline{k}\cdot\underline{x}}$ we obtain for (12)

$$\int_{D_0} e^{2i\,\underline{k}\cdot\underline{x}}\alpha(\underline{x})ds = \theta(\underline{k}) \tag{14}$$

Equation (14) gives the Fourier transform of $\alpha(\underline{x})$ in terms of the observed values of $U_s$ and $\partial U_s/\partial n$ on all of $\partial D$. In order to invert (14) we use the entire family of probes in the incident field $U_0$ which yields the result

$$\alpha(\underline{x}) = \frac{1}{\pi^2}\int \exp\{-2i\,\underline{k}\cdot\underline{x}\}\,\theta(\underline{k})d^2k \tag{15}$$

Because $U_s$ is not measured but $\tilde{U}_s$ $\alpha(\underline{x})$ is determined up to order $\varepsilon$. The accuracy of the result can be improved, which, however, is not done here.

## 3. Incomplete Problem

In this section we shall introduce an iterative procedure to determine $\alpha(\underline{x})$ if $U_s$ and $\partial U_s/\partial n$ are known along a section $\partial D_1$ of $\partial D$. Instead of (13) we write

$$\theta(\underline{k}) = \theta_1(\underline{k}) + \theta_2(\underline{k})$$

$$\text{where } \theta_i(\underline{k}) = \frac{1}{k^2}\int_{\partial D_i}\left(U_s\frac{\partial v}{\partial n} - v\frac{\partial U_s}{\partial n}\right) d\,l \tag{16}$$

Along $\partial D_2$ the unknown functions $U_s$ and $\partial U_s/\partial n$ are related to the unknown $\alpha(\underline{x})$ by means of (10). The iterative scheme works as follows: the ith approximation of $\alpha(\underline{x})$ is given by

$$\alpha^i(\underline{x}) = \frac{1}{\pi^2}\int \exp\{-2i\,\underline{k}\cdot\underline{x}\}[\theta_1(\underline{k})+\theta_2^{\,i-1}(\underline{k})]\,d^2k \quad \text{for } i=1,2,3,4,.. \tag{17}$$

where

$$\theta_2^{\,i-1}(\underline{k}) = \frac{1}{k^2}\int_{\partial D_2}\left(U_s^{\,i-1}\frac{\partial v}{\partial n} - v\frac{\partial U_s^{\,i-1}}{\partial n}\right)d\,l \tag{18}$$

in which $U_s^{\,i-1}$ and its derivative are determined by

$$U_s^{\,i-1}(\underline{x}) = \frac{ik^2}{4}\int_{D_0}\alpha^{i-1}(\underline{x}')U_0(\underline{x}')H_0^{(1)}\{k|\underline{x}-\underline{x}'|\}d\,s' \tag{19}$$

where $D_0$ is the support of $\alpha^{i-1}(\underline{x})$. We take $\alpha^0(\underline{x}) \equiv 0$ in D.

This iterative scheme solves an operator equation which formally can be written as

$$\alpha(\underline{x}) = \alpha'(\underline{x}) + P(\alpha(\underline{x})) \tag{20}$$

where $\alpha'(x) = \frac{1}{\pi^2}\int \exp\{-2\,i\underline{k}\cdot\underline{x}\}\theta_1(\underline{k})\,d^2k$ follows from the measurements. The operator P is defined by (17) - (19) where the iteration is written out as

$$\alpha^i(\underline{x}) = \alpha'(\underline{x}) + P(\alpha^{i-1}(\underline{x})) \tag{21}$$

Convergence in $L_2$ norm depends on $\partial D_2$ as is expected. If the length of $\partial D_2$ is not too large compared with the length of $\partial D_1$ we expect that

$$\|\alpha^{i} - \alpha^{i-1}\|_{L_2} < \lambda \, \|\alpha^{i-1} - \alpha^{i-2}\|_{L_2} \tag{22}$$

where $\lambda \leq 1$. In principle $\lambda$ depends on $\partial D_2$ and can be computed, but a closer view at the equation shows that this is very complicated.

## 4. High Frequency Procedure

In this section we study the incomplete inverse problem for high frequency probe fields. Our procedure consists of an estimation of the scattered field by means of a regular perturbation series (6). We only used the first term to obtain an estimate for $\alpha(\underline{x})$ based on an incomplete information. The question arises whether a high frequency procedure can be obtained by taking the high frequency limit in (10).
A different approach is to use the ray method directly, see H. Schomberg [2].
A closer look at the regular perturbation series shows that the expansion parameter $\varepsilon$ should be replaced by $\varepsilon$ k. Hence the procedure is valid for small values of $\varepsilon$ k or $(\varepsilon k) = O(1)$ for $\varepsilon \to 0$.
Therefore a different approach is needed for large values of k. We shall construct the Rytov approximation of the solution of

$$L\,U = \Delta U + k^2(1 + \varepsilon\alpha(\underline{x}))U = 0 \tag{23}$$

This procedure has been applied for instance by F. Stenger [3].
We shall see that there is a close relation with (10). The Rytov approximation is obtained by insertion of

$$U(\underline{x}) = \exp\{\, i\, kS(\underline{x},\varepsilon)\} \tag{24}$$

with $S(\underline{x}.\varepsilon) = S_0(\underline{x}) + \varepsilon S_1(\underline{x}) + \varepsilon^2 S_2(\underline{x}) + \ldots$ .
Hence the phase function $kS(\underline{x},\varepsilon)$ is written as a regular perturbation series. We shall see, that k $S_0(\underline{x})$ equals the phase of the unperturbed incident wave. The equation for S becomes

$$i\,k\,\Delta S - k^2\,\nabla S.\nabla S + k^2(1 + \varepsilon\alpha(\underline{x})) = 0 \tag{25}$$

Inserting the perturbation series (24) and equating like powers in $\varepsilon$ to zero leads to the following equations for $S_0$ and $S_1$

$$\varepsilon^0:\ i\,k\,\Delta S_0 - k^2\,\nabla S_0\,.\,\nabla S_0 + k^2 = 0 \tag{26}$$

$$\varepsilon^1:\ i\,k\,\Delta S_1 - 2k^2\,\nabla S_0.\,\nabla S_1 + k^2\alpha(\underline{x}) = 0 \tag{27}$$

The phase function of the incoming wave $S_0 = \frac{\underline{k}.\underline{x}}{k}$ is a solution of (26), hence, (27) can be rewritten as

$$\Delta S_1 + 2\,i\,\underline{k}\,.\,\nabla S_1 = i\,k\,\alpha(\underline{x}) \tag{28}$$

The solution of this equation becomes

$$S_1(\underline{x}) = \frac{k}{4}\int_{D_0}\alpha(\underline{x}')\,e^{-i\underline{k}.(\underline{x}-\underline{x}')}H_0^{(1)}\{k|\underline{x}-\underline{x}'|\}d\,S' \tag{29}$$

This expression has great resemblance with the expression for the approximation of the scattered wave field.

## 5. Asymptotic Expansion of (10) and (29)

In this section we investigate the behaviour of the integral

$$J(\underline{x},k) = \frac{i}{4}\int_{D_0}\alpha(\underline{x}')e^{i\,\underline{k}.\underline{x}'}H_0^{(1)}\{k|\underline{x}-\underline{x}'|\}dS' \tag{30}$$

which is a common part of (10) and (29) for large values of $k = |\underline{k}|$. For convenience sake we rotate the coordinate systems, such that the vector $\underline{k}$ is parallel with respect to the horizontal axis.

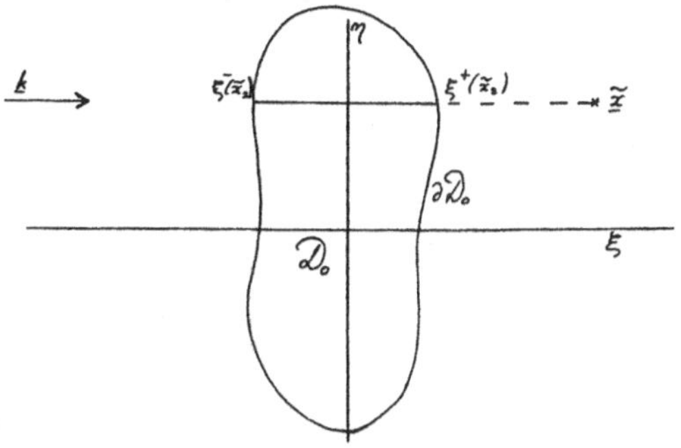


Fig. 2.

We then write:

$$J(\tilde{\underline{x}},k) = \frac{i}{4}\int_{D_0}\alpha(\xi,\eta)e^{ik\xi}H_0\{k\sqrt{(\tilde{x}_1-\xi)^2+(\tilde{x}_2-\eta)^2}\}d\xi d\eta \tag{31}$$

For large values of k and the contour $\partial D$ at finite distance from the region $D_0$, we have

$$J(\tilde{\underline{x}},k) = \frac{e^{\frac{\pi i}{4}}}{2\sqrt{2\pi k}}\int_{D_0}\frac{\alpha(\xi,\eta)e^{ik\xi+ik\sqrt{(\tilde{x}_1-\xi)^2+(x_2-\eta)^2}}}{\{(\tilde{x}_1-\xi)^2+(\tilde{x}_2-\eta)^2\}^{1/4}}d\xi d\eta \tag{32}$$

We first look at the case where $\tilde{x}_1 > \xi^+(x_2)$.
Application of the method of stationary phase for the integral with respect to $\eta$ leads to

$$J(\tilde{\underline{x}},k) \simeq \frac{i\,e^{ik\tilde{x}_1}}{2k}\int_{\xi^-(\tilde{x}_2)}^{\xi^+(\tilde{x}_2)}\alpha(\xi,\tilde{x}_2)d\xi. \tag{33}$$

Inserting this result in (29) leads to an approximation for the correction of the phase function of the forward scattered field

$$S_1^f(\tilde{\underline{x}},k) \approx \frac{1}{2}\int_{\xi^-(\tilde{x}_2)}^{\xi^+(\tilde{x}_2)} \alpha(\xi,\tilde{x}_2)d\xi \tag{34}$$

which is independent of k. This expression can be obtained directly with the aid of the ray method, with the assumption that the curved rays may be approximated by straight lines.
This assumption is reasonable for small values of $\varepsilon$. We see that for the high frequency limit $S_1$ equals the Radon transform of $\alpha$. If we use all directions for the probe fields, we can use techniques as are used for CT scanners. In the next section we give a procedure which solves the incomplete problem. It is of interest to consider the backscattered result just as well. In this situation we have $\tilde{x}_1 < \xi^-(x_2)$ and application of the method of stationary phase in (32) leads to

$$J(\tilde{\underline{x}},k) \approx \frac{i}{2k}\, e^{-ik\tilde{x}_1}\int_{\xi^-(\tilde{x}_2)}^{\xi^+(\tilde{x}_2)} e^{2ik\xi}\alpha(\xi,\tilde{x}_2)d\xi \tag{35}$$

We now find as an approximation for the phase function of the back-scattered field

$$S_1^b(\tilde{\underline{x}},k) \approx \frac{1}{2}\int_{\xi^-(\tilde{x}_2)}^{\xi^+(\tilde{x}_2)} e^{-2ik(\tilde{x}_1-\xi)}\,\alpha(\xi,\tilde{x}_2)d\xi \tag{36}$$

The Born approximation now becomes for the backscattered field

$$U_s^b \approx \frac{ik}{2}\, e^{-ik\tilde{x}_1}\int_{\xi^-(\tilde{x}_2)}^{\xi^+(\tilde{x}_2)} e^{2ik\xi}\alpha(\xi_1,\tilde{x}_2)d\xi \tag{37}$$

This result coincides with the one-dimensional approach of Cohen and Bleistein, where the amplitude of the backscattered field is considered as the Fourier transform of the unknown index of refraction.
We expect that this result can be used for structures resembling the one-dimensional case.

## 6. Incomplete Problems in the Fourier Domain

In the next section we need a procedure to find the original function $f(x_1,x_2)$ if its Fourier Transform $F(p_1,p_2)$ is given in a part of the $\underline{p}$ plane.

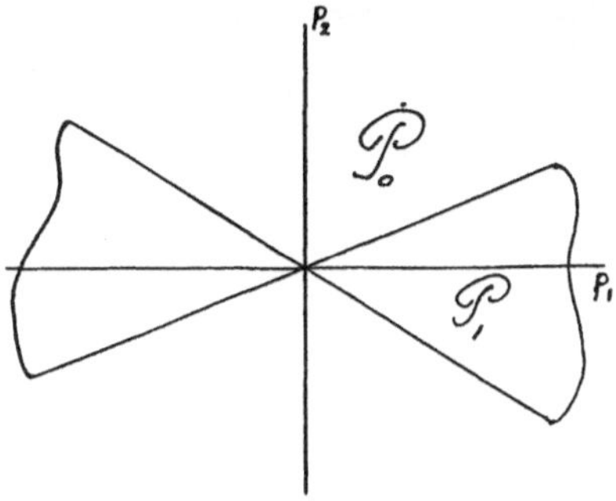


Fig. 3.

We assume F to be given in $B_0$. The Fourier Transform of $f(x_1,x_2)$ is given by

$$F(\underline{p}) = \iint e^{i\underline{p}\cdot\underline{x}} f(\underline{x}) d^2x \qquad (38)$$

and its inverse by

$$f(\underline{x}) = \frac{1}{(2\pi)^2} \iint e^{-i\underline{p}\cdot\underline{x}} F(\underline{p}) d^2p \qquad (39)$$

The inverse transform needs the function F over the whole $\underline{p}$ plane. We derive a regular integral equation of the second kind for the function $f(\underline{x})$ for the situation that $F(\underline{p})$ is only given in $P_0$.

We define the information function $A(\underline{p})$ and its complement $A_c(\underline{p})$ as

$$A(\underline{p}) = \begin{cases} 1 & \forall \underline{p} \in P_0 \\ 0 & \forall \underline{p} \in P_1 \end{cases} \quad \text{and}$$

$A_c(\underline{p}) = 1 - A(\underline{p})$ respectively.

These functions have two-dimensional Fourier transforms $a(\underline{x})$ and $a_c(\underline{x})$ respectively. Then the relation

$$a_c(\underline{x}) = \delta(\underline{x}) - a(\underline{x}) \quad \text{holds.} \qquad (40)$$

It is easily seen that for the convolution of $a(\underline{x})$ and $f(\underline{x})$ we have the expression

$$a(\underline{x}) * f(\underline{x}) = \frac{1}{(2\pi)^2} \iint_{P_0} F(p)\ e^{-i\underline{p}\cdot\underline{x}} d^2p = h(\underline{x}) \qquad (41)$$

The right-hand side of (41) is assumed to be known and with the help of (40) we obtain an integral equation of the second kind for $f(\underline{x})$

$$f(\underline{x}) - a_c(\underline{x}) * f(\underline{x}) = h(\underline{x}) \quad \text{i.e.}$$

$$f(\underline{x}) - \iint a_c(\underline{x}-\underline{x}')\ f(\underline{x}')d^2x' = h(\underline{x}) \qquad (42)$$

For simple regions $P_1$ the kernel function $a_c(\underline{x}-\underline{x}')$ can be evaluated explicitly.

If the support of $f(\underline{x})$ is contained in a ball with finite radius the integral equation is a Fredholm integral equation of the second kind. The solution is unique if the homogeneous equation has no eigen solution. In the case that $f(\underline{x})$ has not a finite support there exists no unique solution of (42). This can be easily seen because the Fourier plane can be completed arbitrarily provided that $F(\underline{p})$ is square integrable.

As an illustration we consider the one-dimensional case when $F(p)$ is not determined for $p_1 < p < p_2$.

The integral equation then becomes

$$f(x) + \int_L a_c(x - x')\, f(x')\, dx' = h(x) \tag{43}$$

where

$$h(x) = \left\{\int_{-\infty}^{p_1} + \int_{p_2}^{\infty}\right\} \frac{1}{2\pi} F(p) e^{-ipx} dp$$

and

$$a_c(x) = \frac{1}{2\pi} \int_{p_1}^{p_2} e^{-ipx}\, dp = \frac{1}{2\pi i x} \left[e^{-ip_1 x} - e^{-ip_2 x}\right]$$

$$= \frac{e^{-i\left(\frac{p_1+p_2}{2}\right)x}}{\pi x} \sin\left[\frac{p_2-p_1}{2}\, x\right] \tag{44}$$

L is the finite support of $f(x)$.

The Fredholm alternative assures us a unique solution of (43) provided that unity is not an eigenvalue of the homogeneous equation

$$f(x) + \lambda \int_L a_c(x-x') f(x') dx' = 0 \ . \tag{45}$$

The trace inequality gives us an upperbound for the largest eigenvalue $\lambda_1$

$$\lambda_1 \le \int_L k(x,x)\, dx$$

where $k(x,x') = a_c(x-x')$. This leads to

$$\lambda_1 \le \frac{L(p_1-p_2)}{2\pi} \quad \text{where L is the length of } L.$$

Hence if $\frac{L(p_1-p_2)}{2\pi} < 1$ we have always a unique solution of (43).

In the case that $L(p_1-p_2)/2\pi \ge 1$ a unique solution exists if one is not an eigenvalue.

In the two-dimensional situation similar properties hold as can be shown.

## References

[1] J.K. Cohen and N. Bleistein: An inverse method for determining small variations in propagation speed. SIAM J. Appl. Math. Vol. 32, 1977.

[2] H. Schomberg: Nonlinear image reconstruction from projections of ultrasonic travel times and electric current densities. Proc. of the conference on Mathematical aspects of computerized tomography, ed. G.T. Herman and F. Natterer, Springer, 1980.

[3] F. Stenger: An algorithm for ultrasonic tomography based on inversion of the Helmholtz equation.

# PART IV

# FREDHOLM INTEGRAL EQUATIONS

# OF THE FIRST KIND

# EXPLOITING THE RANGES OF RADON TRANSFORMSIN TOMOGRAPHY

Frank Natterer

## 1. Radon Transforms and their Ranges

The classical Radon transform of a function f with support in the unit disc $\Omega$ of $\mathbb{R}^2$ is defined as

$$Rf(s,\omega) = \int_{x\cdot\omega=s} f(x)dx = \int f(s\omega + t\omega^{\perp})\,dt$$

Here, $s \in \mathbb{R}^1$ and $\omega = (\cos\varphi, \sin\varphi)$, $\omega^{\perp} = (-\sin\varphi, \cos\varphi)$, $0 \leq \varphi < 2\pi$. Thus, R associates with each f the set of its line integrals.

The range of R is characterized by the Helgason-Ludwig theorem (see e.g. [3]), the trivial part of which is

<u>THEOREM 1:</u> If $g \in$ range (R), then

$$\int s^m g(s,\omega)\,ds$$

is homogeneous polynomial of degree m in $\omega$.

<u>PROOF:</u> Let $g = Rf$ be an element of range (R). Integrating g against the monomials yields

$$\int s^m g\,ds = \int\int s^m f(s\omega + t\omega^{\perp})\,dt\,ds$$

$$= \int (x\cdot\omega)^m f(x)\,dx$$

which is a homogeneous polynomial of degree m in $\omega$.

While R is the relevant integral transform in transmission computed tomography and in positron emission tomography (PET), the attenuated Radon transform

$$R_\mu f(s,\omega) = \int_{x\cdot\omega=s} f(x) e^{-D\mu(x,\omega^\perp)}\, dx \quad ,$$

$$D\mu(x\,,\,\omega) = \int_0^\infty \mu(x + t\omega)\, dt$$

comes up in single particle emission computed tomography (SPECT).

The analog of Theorem 1 for $R_\mu$ is given in [6]:

<u>THEOREM 2:</u> If $g \in$ range $(R_\mu)$, then

$$\int\int e^{1/2(I + iH)R\mu}\, s^m\, g\, e^{ik\varphi}\, ds\, d\varphi = 0$$

for $0 \leq m < k$, where H is the Hilbert transform.

We see that the functions in the ranges of Radon transforms are highly structured, and it is the purpose of the present paper to make use of this structure for problems with incomplete information about the data or the operator.

For the applications of the Radon transforms mentioned in the following, see [4].

## 2. Incomplete Data Problems

While the inversion of R is well understood if the data g are available for $|s| \leq 1$ and $|\varphi| \leq \pi/2$, the problem of recovering f from only some part of g is still difficult. The limited angle problem is one of the well-known incomplete data problems. Here, g is known for $|\varphi| > \Theta$ only, where $0 < \Theta < \pi/2$. Peres [8] and Louis [5] suggested a completion procedure for g based on the structure of g as described by Theorem 1: If $g(\cdot,\omega)$ is expanded in terms of orthogonal polynomials $P_m$ with weight w,

$$g(s\,,\omega) = w(s) \sum_{m=o}^{\infty} P_m(s)\, q_m(\omega)$$

then

$$q_m(\omega) \quad = \quad \int P_m(s)\, g(s\,,\omega)\, ds$$

is a polynomial of degree m in ω. Each $q_m$ can be computed for $|\varphi| > \Theta$ from g, hence $q_m$ can be determined in the missing range $|\varphi| \leq \Theta$ by trigonometric extrapolation. Of course this process is highly unstable, see [2]. Nevertheless, the algorithm given in [5] is highly competitive.

The success of exploiting the range in limited angle tomography encouraged us to look at other incomplete data problems from the same point of view. The exterior Radon problem calls for the reconstruction of f in the annulus $0 < a \leq |x| \leq 1$ from "hollow projections" $g(s,\cdot)$, $a \leq |s| \leq 1$. In principle, this can be done by Cormack's formula [1]: If

$$f(r,\omega) \quad = \quad \sum_k f_k(r)\, e^{ik\varphi}$$

$$g(s,\omega) \quad = \quad \sum_k g_k(s)\, e^{ik\varphi}$$

are the Fourier expansions for f and g, then

$$f_k(r) = -\frac{1}{\pi r} \int_r^1 T_k(s/r)\, ((s/r)^2 - 1)^{-1/2}\, g_k'(s)\, ds \qquad (2.1)$$

where $T_k$ is the Chebyshev polynomial of the first kind. For the evaluation of f(x), (2.1) integrates only from $|x|$ to 1. Hence it is a solution to the exterior problem.

However, (2.1) is very unstable: For $u \geq 1$ we have

$$T_k(u) = \cosh(k \operatorname{arc\,cosh} u) \geq \frac{1}{2}\,(u + \sqrt{u^2 - 1}\,)^k \quad ,$$

i.e. the kernel in (2.1) grows exponentially as $k \to \infty$. Thus, (2.1) is virtually useless for numerical calculations.

A simple application of Theorem 1 turns (2.1) into a stable formula: We observe that, by Theorem 1,

$$0 = \frac{1}{\pi r} \int_0^1 P_{k-1}(s/r) g_k'(s)\, ds$$

for each polynomials $P_{k-1}$ of degree $k - 1$ which has the same parity as $k - 1$. Adding this to (2.1) we obtain a family of new inversion formulae.

In order to get a stable formula, $P_{k-1}$ has to have the same exponential growth as the kernel of (2.1). Since, for $u \geq 1$,

$$T_k(u)(u^2 - 1)^{-1/2} = \frac{\cosh\,(k \cosh^{-1} u)}{\sinh\,(\cosh^{-1} u)}$$

an obvious choice for $P_{k-1}$ is

$$P_{k-1}(u) = \frac{\sinh\,(k \cosh^{-1} u)}{\sinh\,(\cosh^{-1} u)} = U_{k-1}(u)$$

where $U_k$ is the Chebyshev polynomial of the second kind. Then,

$$\begin{aligned} T_k(u)(u^2 - 1)^{-1/2} - P_{k-1}(u) &= \frac{\exp\,(-k \cosh^{-1} u)}{\sinh\,(\cosh^{-1} u)} \\ &= \frac{2\,(u + \sqrt{u^2-1}\,)^{-k}}{(u+\sqrt{u^2-1}\,) - (u+\sqrt{u^2-1}\,)^{-1}} \\ &= \frac{(u+\sqrt{u^2-1}\,)^{-k}}{\sqrt{u^2 - 1}} \end{aligned}$$

Writing u for s/r we obtain

$$f_k(r) = -\frac{1}{\pi r}\int_r^1 \frac{(u+\sqrt{u^2-1})^{-k}}{\sqrt{u^2-1}}\, g_k'(s)\, ds + \frac{1}{\pi r}\int_0^r U_{k-1}(u)\, g_k'(s)\, ds \ . \quad (2.2)$$

This is Perry's formula [9]. It is stable, but it is clearly not a solution to the exterior problem, since it integrates over the entire interval [0,1].

In order to construct a formula which retains at least partially the local character of Cormack's formula (2.1) while being not as unstable, we put

$$P_{k-1} = U_{k-1} - R_{k-1}$$

with $R_{k-1}$ a polynomial of the same type as $P_{k-1}$. Let $0 < a < b < 1$. We determine $R_{k-1}$ such that $R_{k-1}(u)$ is close to $U_{k-1}(u)$ for $u \leq a/b$ and $R_{k-1}(u)$ not too large for $u \geq a/b$. Then,

$$f_k(r) = -\frac{1}{\pi r}\int_r^1 \left(\frac{(u+\sqrt{u^2-1})^{-k}}{\sqrt{u^2-1}} - R_{k-1}(u)\right) g_k'(s)\, ds$$

$$+ \frac{1}{\pi r}\int_0^r (U_{k-1} - R_{k-1})(u) g_k'(s)\, ds$$

is a formula which puts only little weight on the missing data $g(s,\omega)$ for $|s| \leq a$ if $r \geq b$ while being not too unstable.

## 3. Problems with Unknown Operator

In emission computed tomography one has to compute the source distribution f from the integral equations

$$e^{-R\mu} Rf = g \qquad \text{(PET)}$$

$$R_\mu f = g \qquad \text{(SPECT)}$$

without knowing the attenuation distribution μ. Thus we have to invert integral operators which we don't know.

In PET we can try to determine μ from the data g simply by observing that $e^{R\mu} g \in$ range (R). Hence, according to Theorem 1, we have

$$\int s^m (e^{R\mu} g)(s,\omega)\, ds = P_m(\omega) \qquad , \; m = 0,1,\ldots \qquad (3.1)$$

with $P_m$ a homogeneous polynomial of degree m in ω:

$$P_m(\omega) = \sum_{\substack{\ell = -m \\ \ell + m \text{ even}}}^{m} e^{i\ell\varphi} P_{m\ell} \qquad (3.2)$$

We want to know to what extent Rμ can be determined from (3.1). We study the following model situation: f is made up of finitely many sources $a_1, \ldots, a_n$ with intensities $f_1, \ldots, f_n$. In this case it is easy to find a continuous Rμ from g for all rays hitting one of the sources: First one finds the positions $a_k$ of the sources by considering $g(\cdot, \omega)$ for n + 1 judiciously choosen directions ω. Then one scans around each of the sources $a_k$, obtaining $f_k\, e^{-R\mu}$ for all lines through $a_k$ except for the finitely many ones which hit more than one source and which can be determined by continuity. Finally, we determine the activities $f_k$ up to a common factor by considering lines joining pairs of sources.

In fact, (3.1) determines Rμ precisely to the extent we expect:

<u>THEOREM 3:</u> Let $f(x) = \sum_{k=1}^{n} f_k \delta(x - a_k)$ and let $R\mu$ be continuous. If $\mu^*$ is such that $R\mu^*$ is continuous and

$$\int s^m (e^{R\mu^*} g)(s,\omega)\, ds = P_m(\omega) \qquad , m=0,1,\ldots,n-1$$

is a homogeneous polynomial of degree m, then there is a constant c such that

$$R(\mu^* - \mu)(x \cdot \omega, \omega) = c$$

for $x \in \operatorname{supp}(f)$ and all $\omega$.

<u>PROOF:</u> The function $h = e^{R(\mu^* - \mu)}$ satisfies

$$\int s^m h(s,\omega)\, Rf(s,\omega)\, ds = P_m(\omega) \qquad , m=0,\ldots,n-1 .$$

Making use of the adjoint

$$R^* g(x) = g(x \cdot \omega, \omega)$$

of R we obtain

$$\int f(x)(x \cdot \omega)^m\, h(x \cdot \omega, \omega)\, dx = P_m(\omega) \qquad , m=0,\ldots,n-1 .$$

Since f is a finite linear combination of Dirac measures, this can be rewritten as

$$\sum_{k=1}^{n} f_k (a_k \cdot \omega)^m h(a_k \cdot \omega, \omega) = P_m(\omega) \qquad , m=0,\ldots,n-1 .$$

For each $\omega$, this is a $n \times n$ linear system for the unknowns $f_k\, h(a_k \cdot \omega, \omega)$, $k=1,\ldots,n$ with a Van der Monde-matrix. If $(a_k - a_\ell) \cdot \omega \neq 0$ for $k \neq \ell$, it can be solved by Cramer's rule

$$f_k\, h(a_k \cdot \omega, \omega) = \frac{Q_k(\omega)}{Q(\omega)} \tag{3.3}$$

where Q is the Van der Monde determinant

$$Q(\omega) = \begin{vmatrix} 1 & & 1 \\ a_1 \cdot \omega & & a_n \cdot \omega \\ \vdots & & \vdots \\ (a_1 \cdot \omega)^{n-1} & & (a_n \cdot \omega)^{n-1} \end{vmatrix}$$

and $Q_k(\omega)$ is obtained from $Q(\omega)$ by replacing its k-th column by $(P_o(\omega), P_1(\omega), \ldots, P_{n-1}(\omega))^T$. Obviously, Q as well as $Q_k$ are homogeneous polynomials of degree $1 + 2 + \ldots + n - 1 = n(n-1)/2$ in $\omega$.

The zeros of Q are the directions $\omega_{k\ell}$ for which $(a_k - a_\ell) \cdot \omega_{k\ell} = 0$ for a pair $k,\ell$ (i.e. $\omega_{k\ell}$ is orthogonal to the line joining the sources $a_k$, $a_\ell$), and the multiplicity of $\omega_{k\ell}$ is precisely the number of pairs of sources for which this happens. Thus, Q has precisely $n(n-1)/2$ real zeros.

It follows from our assumptions on $R\mu$ and $R\mu^*$ that $h(a_k \cdot \omega, \omega)$ depends continuously on $\omega$, and so does $Q_k(\omega)$ / $Q(\omega)$. Therefore, $Q_k$ has to have the $n(n-1)/2$ zeros $\omega_{k\ell}$ of Q with the right multiplicities. $Q_k(\omega)$ being a homogeneous polynomial with degree $n(n-1)/2$ in $\omega$, it follows that it is a constant multiple of Q. Hence

$$h(a_k \cdot \omega, \omega) = c_k \qquad , \quad h = 1, \ldots, n$$

with some constants $c_k$. For $\ell \neq k$ we have

$$c_k = h(a_k \cdot \omega_{k\ell}, \omega_{k\ell}) = h(a_\ell \cdot \omega_{k\ell}, \omega_{k\ell}) = c_\ell \quad ,$$

hence $c_k$ does not depend on k. The proof is finished.

Theorem 3 is certainly very satisfactory from a theoretical point of view: For each ray, either $g = 0$, in which case we don't care about the value of $R\mu$, or $g \neq 0$, in which case $R\mu$ is determined by the theorem, up to an additive constant which is easily determined by using an exterior reference source.

Unfortunately, Theorem 3 is only of very limited practical value. In practice, only a discretized version of g is available. For simplicity, let us assume that the functions $g(\cdot,\omega)$ are known for $\omega_j = (\cos\varphi_j, \sin\varphi_j)$, $\varphi_j = \pi j/p$, $j = 0,\ldots,p-1$. A natural discretization of (3.1) is

$$\int s^m (e^{R\mu} g)(s,\omega_j)\, ds = P_m(\omega_j), \quad m=0,1,\ldots \qquad (3.4)$$

for $j = 0,\ldots,p-1$. If $R$ is a continuous function satisfying (3.4) in the situation of Theorem 3, the proof of that theorem gives instead of (3.3)

$$f_k\, h\,(a_k \omega_j, \omega_j) = \frac{Q_k(\omega_j)}{Q(\omega_j)}, \quad k = 1,\ldots,n$$

for $j = 0,\ldots,p-1$. This simply says that the quantities we want to show to be constant are rational functions of degree $n(n-1)/2$. Since this is a very large number in practice, $R\mu^*$ is largely undetermined by the discrete conditions.

In order to exploit Theorem 3 numerically, one uses one of the standard approaches for highly ill-posed problems: One models the unknown attenuation distribution by a few parameters and tries to find these parameters by a least squares procedure. This approach seems to be reasonable, as is shown by the following simple example:

Consider two circles $K_1$, $K_2$ with radii 0.45, 0.35 and midpoints (0,0), (0, 0.05), respectively. Let $\mu = 0$ outside $K_1$, $\mu = \mu_1$ in $K_1 - K_2$, and $\mu = \mu_2$ in $K_2$. This is a very rough model for the attenuation distribution in a cross section of the head. The source distribution f is assumed to be 1 in the circle of radius 0.15 around (-0.2, 0), see Figure 1.

We computed $g(s,\omega)$ for these choices of $\mu$ and f, putting $\mu_1 = 4$, $\mu_2 = 2$, for 64 equally spaced directions (i.e. $p = 64$), 64 values each. With these data we did a least squares method on (3.4), using only the 64 equations with $m = 0$. Starting with the initial guess $\mu_1 = 5$, $\mu_2 = 3.5$ we obtained after 3 Gauss-Newton steps $\mu_1 = 4.03$, $\mu_2 = 2.06$.

In Figure 2 we give a 3D plot of the reconstruction of f from g without correcting for attenuation, whereas Figure 3 shows the reconstruction of f after correcting g with the computed attenuation factors. The improvement is obvious. Note that this improvement has been made without knowing the true attenuation coefficients.

In SPECT the situation is similar but more difficult: Rather than determining the unknown factor $\exp(-R\mu)$ we have to find the kernel of the integral equation $R_\mu f = g$. We use Theorem 2 in the same way as we used Theorem 1 in PET, i.e. we try to compute the kernel $D\mu$ of the integral equation $R_\mu f = g$ from the consistency conditions given in Theorem 2. The extent to which this is possible is given in [7]:

<u>THEOREM 4:</u> Let $f = \sum_{k=1}^{n} f_k \delta(x - a_k)$ and let $\mu \in C_o^\infty(\Omega)$. If $\mu^* \in C^\infty(\Omega)$ satisfies

$$\iint e^{1/2(I+iH)R\mu^*} s^m g(s,\omega)\, ds\, d\varphi = 0$$

for $m = 0,1,\dots,n-1$ and $k > m$, then there is a constant c such that

$$D(\mu - \mu^*)(x,\omega) = c$$

for $x \in \operatorname{supp}(f)$ and all $\omega$.

In [7] an attempt has been made to compute $\mu$ in this way.

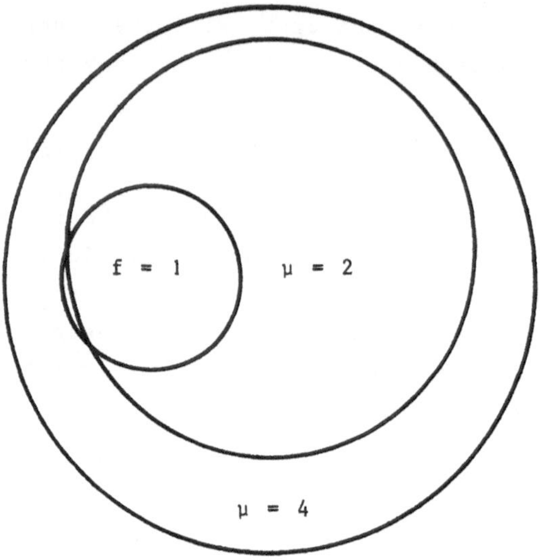


Figure 1: Attenuation and source distribution in head phantom.

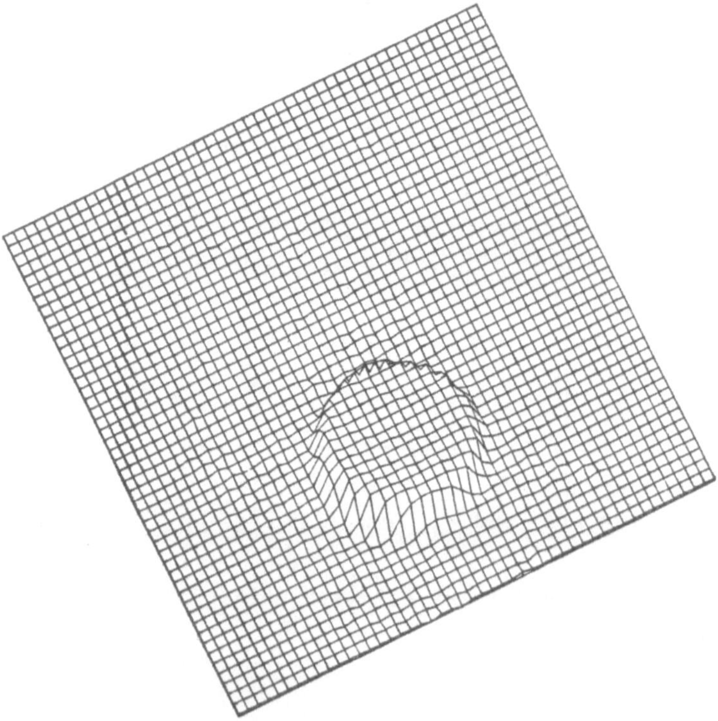

Figure 2: Reconstruction of source distribution without correcting for attenuation.

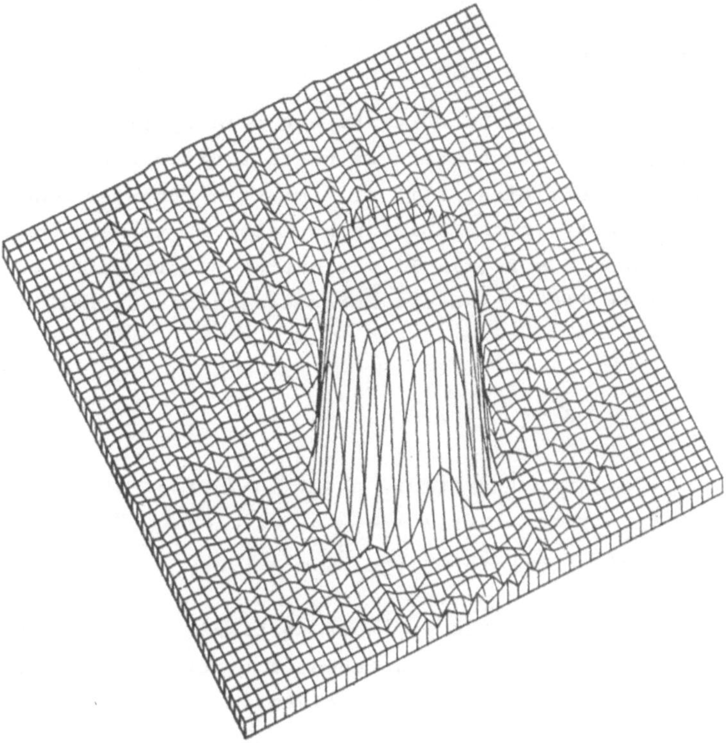

Figure 3: Reconstruction of source distribution after correcting for attenuation. The reconstruction from data without attenuation is almost identical and is not displayed.

## 4. Beam Hardening

In this section we describe one more application of consistency conditions which we haven't tried out yet. The attenuation coefficient f in transmission computed tomography depends not only on the point x but also on the energy E of the x-ray beam $f = f_E(x)$. If the energy distribution in the x-ray beam is T(E), then the measured scanner output yields the quantity

$$g = \int T(E)\, e^{-Rf_E}\, dE \quad .$$

It is assumed (see [10], chapt. 3.2) that there is a low degree polynomial $P(g) = a_o + a_1 g + \dots + a_n g^n$ such that P(g) is equal to $Rf_{\overline{E}}$ for some $\overline{E}$.

In the light of the previous sections, the obvious thing to do is to compute the unknown coefficients $a_o,\dots,a_n$ from the consistency conditions given in Theorem 1, i.e.

$$\sum_{j=o}^{n} a_j \int s^m g^j(s,\omega)\, ds = P_m(\omega) \qquad , \quad m = 0,1,\dots$$

where $P_m$ is a homogeneous polynomial of degree m. Considering that n = 2 seems to be sufficient in practice, a few equations of this system should suffice to determine the $a_j$.

## References


[1] Cormack, A.M.: Representation of a Function by its Line Integrals, with same Radiological Applications I, J. Appl. Physics 34, 2722-2727.

[2] Davison, M.E.: The Ill-Conditioned Nature of the Limited Angle Tomography Problem. To appear in SIAM J. Appl. Math.

[3] Helgason, S.: The Radon Transform. Birkhäuser 1980.

[4] Herman, G.T. (ed.): Image Reconstruction from Projection: Implementation and Applications. Springer 1979.

[5] Louis, A.K.: Picture Reconstruction from Projections in Limited Range, Math. Meth. in the Appl. Sci. 2, 209-220 (1980).

[6] Natterer, F.: The Ill-Posedness of Radon's Integral Equations, in: Proceedings of the Conference on Ill-Posed Problems, Delaware 1979.

[7] Natterer, F.: Computerized Tomography with Unknown Sources. To appear in SIAM J. Appl. Math.

[8] Peres, A.: Tomographic Reconstruction from Limited Angular Data. J. Comput. Assist. Tomogr. 3, 800-803 (1979).

[9] Perry, R.M.: Reconstructing a Function by Circular Harmonic Analysis of its Line Integrals, in: Image Processing for 2D and 3D Reconstruction from Projections, Stanford University, Institute for Electronics in Medicine 1975.

[10] Herman, G.T.: Image Reconstruction from Projections. Academic Press 1980.

# REGULARIZATION TECHNIQUES FOR INVERSE PROBLEMS IN MOLECULAR BIOLOGY

Stephen W. Provencher and Robert H. Vogel

## 1. Introduction

In molecular biology, as in most natural sciences, the number of indirect experiments involving ill-posed inverse problems is rapidly increasing. Three of the most important types of inverse problems involve either (a) severely ill-posed linear problems (e.g., Laplace transforms in relaxation or correlation experiments); (b) very large, and perhaps nonlinear, problems (e.g., estimation of three-dimensional structure from x-ray diffraction or electron microscopy); or (c) parameter estimation involving computationally complex models (e.g., multicomponent subnanosecond fluorescence decay strongly convoluted with the instrument response or excitation function).

In this paper we shall discuss four approaches to these problems. Two of these will only be outlined and two discussed in more detail. Common to all of these approaches are two general strategies, the principle of parsimony and the use of prior knowledge. Prior knowledge (e.g., nonnegativity) can be very useful at eliminating the vast majority of members from the (typically infinite) set of solutions that fit the data to within experimental error.

The principle of parsimony says, of all solutions not eliminated by prior knowledge, choose the simplest one, i.e., the one that reveals the least amount of detail or information that was not already known or expected. This is a standard strategy taken by statisticians and experimentalists building models. It is strictly to protect against artifacts and overinterpretation of the data. While the most parsimonious solution may not have all the detail of the true solution, the detail that it does have is necessary to fit the data and therefore less likely to be artifact. The definition of parsimony obviously depends on the problem and prior knowledge. Often smoothness of the solution in a particular space or minimum number of parameters in a model is an appropriate definition.

All four of these approaches maximize an approximate likelihood function, possibly modified by an additive regularizor term, which imposes parsimony or yields an optimal estimate (in the mean square error sense) using prior statistical knowledge of the mean and covariance matrix of the solution [22]. In this way, only the discrete, statistically weighted observed data points are used. This is important because in many cases the statistics of the noise are fairly well known and the noise is often strongly nonstationary. Furthermore, this eliminates the need to extrapolate or interpolate data to estimate (usually infinite) integrals in formal inversion formulas.

## 2. Constrained Regularization

Imposing prior knowledge of inequality constraints can greatly increase the resolution and stability of the solution. We have found this to be especially important when the solution has significant high frequency content, e.g., sharp edges or isolated peaks [15, 18]. In this section we mention two regularization approaches that can impose inequality constraints. They are described in detail elsewhere and will be only briefly summarized here.

A general-purpose regularization algorithm [16] and portable Fortran package [17] has been developed for linear operator equations subject to any linear equality or inequality constraints imposed by prior knowledge. With numerically stable orthogonal transformations [11] , the general quadratic programming problem is reduced to a ridge regression problem, whose statistical properties have been widely studied. The regularization parameter can then be chosen on the basis of classical confidence regions and F-tests [13, 15].

For the package mentioned above, part of the computation time is proportional to the cube of the number of parameters used to represent the solution. Computations with fewer than about 100 parameters can be done economically. This is usually more than adequate for solutions in one dimension, but not in two or three dimensions. Furthermore the operator equations must be linear. Neutron and x-ray diffraction experiments result in nonlinear operator equations, when the operator produces the absolute value squared of the Fourier transform of the desired electron density. In addition, the data often contains enough information so that $O(10^4)$ to $O(10^5)$ parameters are required to represent the solution. For these large, possibly nonlinear, problems a very efficient optimization algorithm [4] using Frieden's maximum entropy regularizor [7] has been developed. It has been applied to estimating the structure of the Pf1 virion down to a resolution of about 4 Å from x-ray fiber diffraction data [5]. Data from a heavy atom derivative and the native structure were simultaneously analyzed to help reduce the nonuniqueness due to the nonlinear operator (the so-called phase problem). The algorithm has also been applied to three-dimensional reconstruction from electron micrographs and could be applied to a wide variety of other large problems, particularly with missing data.

## 3. Fast Spline-Model Method for Certain Separable Least Squares Problems

Many commonly used methods for experimental data can be put in the form

$$y_k = \sum_{j=1}^{N_\lambda} \alpha_j f_k(\lambda_j) + \varepsilon_k, \quad k=1,\ldots,N_y, \tag{1}$$

where the $y_k$ are experimental data with unknown zero-mean noise components, $\varepsilon_k$, with finite variances and the $\alpha_j$ and $\lambda_j$ are to be estimated. The specified functions, $f_k(\lambda)$, are known, but can be expensive to compute. A common case is a convoluted exponential in fast luminescence decay processes,

$$f_k(\lambda) = \int_0^{t_k} \exp(-\lambda\tau)E(t_k-\tau)\, d\tau, \tag{2}$$

where $E(t)$ describes the impulse response of the instrument or the spread of the excitation.

A properly weighted least squares estimator is maximum likelihood when the $\varepsilon_k$ are normally distributed and approximately so when the $\varepsilon_k$ follow Poisson statistics [14]. However, because of the complexity of eq (2) and the fact that $N_y$, the number of data points, is typically $O(10^3)$, such an analysis can be expensive. Furthermore, several complete analyses starting from different points in parameter space should be performed to have a better chance of finding the global optimum. Because of this, there has been considerable interest in transform methods that apply linear operators to the data and typically reduce the problem to solving a set of nonlinear equations. However, we have shown that this can result in serious losses in Fisher information and corresponding increases in the variances of the parameter estimates [19].

In this section we outline a method in which the computation time for a separable least squares analysis becomes <u>independent</u> of the complexity of the model, $f_k(\lambda)$, and of the number of data points, after some preliminary computations have been performed. The first step is to approximate the functions $f_k(\lambda)$ in an expansion of interpolating functions of small support,

$$\tilde{f}_k(\lambda) = \sum_{i=1}^{N_B} \beta_{ki} B_i(\lambda). \tag{3}$$

We use cubic B-spline interpolation at about 40 knots equally spaced on the $z=\ln\lambda$ axis. This approximates model functions of the type in eq (2) typically to within four significant figures. This is generally more than adequate, considering the fact that neither the data nor the model in eq (1) is so accurate anyway, since the model is only an approximation to the true state of nature. Note that this step is simply an interpolation of an exact analytic function. This is quite different and much faster and more reliable than the more common case of fitting splines to the noisy data.

The weighted least squares analysis involves finding the $\alpha_j$ and $\lambda_j$ that minimize the weighted sum of squared residuals,

$$S \equiv \sum_{k=1}^{N_y} w_k [y_k - \sum_{j=1}^{N_\lambda} \alpha_j \tilde{f}_k(\lambda_j)]^2. \tag{4}$$

Newton and modified Gauss-Newton methods require many evaluations of S, the Hessian of S (or an approximation to it), and the gradient of S with respect to $\alpha_j$ and $\lambda_j$. This requires computation of terms like

$$\sum_{k=1}^{N_y} w_k \tilde{f}_k(\lambda_i)\tilde{f}_k(\lambda_j) = \sum_{k=1}^{N_y} w_k \sum_{n=1}^{N_B} \beta_{kn} B_n(\lambda_i) \sum_{m=1}^{N_B} \beta_{km} B_m(\lambda_j) \tag{5}$$

$$= \sum_{n=1}^{N_B} B_n(\lambda_i) \sum_{m=1}^{N_B} B_m(\lambda_j) C_{nm}, \tag{6}$$

where

$$C_{nm} \equiv \sum_{k=1}^{N_y} w_k \beta_{kn} \beta_{km} \tag{7}$$

Another type of term can be similarly evaluated

$$\sum_{k=1}^{N_y} w_k y_k \tilde{f}_k(\lambda_j) = \sum_{n=1}^{N_B} B_n(\lambda_j) d_n, \tag{8}$$

where

$$d_n = \sum_{k=1}^{N_y} w_k y_k \beta_{kn}. \tag{9}$$

The $\beta_{ki}$ in eq (3) are independent of the data and weights. They depend only upon the model and the experimental design, e.g., the spacing of the $t_k$ in eq (2). Very often for a particular series of experiments these are always the same, and the array $\beta$ can be computed once and for all and stored. The complicated model functions in eq (2) do not have to be evaluated again. Similarly the array C in eq (7) need only be computed once if the $w_k$ do not change. At worst it, together with the vector $\underset{\sim}{d}$ in eq (9), need only be evaluated once at the beginning of the analysis of a set of data.

When cubic B-splines are used for the $B_i(\lambda)$, the second derivatives of $\tilde{f}_k(\lambda)$ are continuous. Terms analogous to eqs (5) and (8), but containing first or second derivatives of $\tilde{f}_k(\lambda)$ can be easily and rapidly evaluated by replacing the corresponding $B_n(\lambda)$ and $B_m(\lambda)$ with their derivatives. The shift invariance of B-splines with equally spaced knots further simplifies the computation. Because of the compact support of the cubic B-splines, at most only 16 of the $N_B^2=O(1600)$ terms in the double sum in eq (7) are nonzero and need be evaluated.

Separable Gauss-Newton algorithms, in which the $\alpha_j$ are effectively treated as implicit functions of the $\lambda_j$ and determined by the linear least squares conditions (e.g., with Algorithm I of [20] ), have been implemented using formulas of the types above, as has the full Newton method. What cannot be straightforwardly implemented is separability using the numerically more stable differentiation of the pseudoinverse [8] . However, key parts have been coded in double precision, and numerous comparisons with conventional analyses without the spline-model method showed excellent agreement of the parameter estimates with both methods. Furthermore the computational priorities and strategies are now completely different. The evaluation of $f_k(\lambda)$ and its derivatives in eq (2), which are ordinarily the major burden, are now practically free. The main burden now is the matrix algebra. Therefore a procedure like differentiation of the pseudoinverse with a computational complexity proportional to $N_y=O(10^3)$ would result in a major increase in computation.

Under typical conditions [25], our implementation of the separable modified Gauss-Newton algorithm using the spline-model method results in a speed increase of a factor of O(100). This permits an elaborate series of analyses to be performed from many different starting points in parameter space. This has been implemented in a portable user-oriented Fortran IV program [24] and will be available on request. It also permits a second term in eq (1),

$$\sum_{i=1}^{N_g} \gamma_i g_{ki}, \tag{10}$$

where the $g_{ki}$ are known and the $\gamma_i$ are to be estimated. This is important in allowing corrections for such things as background, and it can be easily handled using formulas similar to eq (8). There is also provision for simultaneously analyzing several sets of data, each having the same set of $\lambda_j$, but different $\alpha_j$ and $\gamma_i$. This can be very useful, e.g. when spectroscopic measurements at several wavelengths in kinetic studies are made to obtain more reliable estimates of the parameters [10].

The spline-model method is a general approach for very rapidly evaluating the objective function in eq (4), as well as its Hessian and gradient. It may therefore be useful in other optimization or parameter estimation procedures, such as homotopy methods, which can involve a very large number of evaluations of the objective function.

## 4. Three-Dimensional Reconstruction from Projections of Disordered Objects

### 4.1 Introduction

Under proper imaging conditions, an electron micrograph yields an estimate of the projection of the electron density of the object. The estimation of the three-dimensional (3-D) electron density from a series of images with the stage tilted to different angles is then formally the same as the inverse Radon transform problem in computer assisted tomography (CAT). However, there are two important additional difficulties. The first is limited data; the stage can only be tilted over a limited angular range, typically $(-60^{\circ}, 60^{\circ})$ rather than $(-90^{\circ}, 90^{\circ})$. This makes the problem even more ill-posed and seriously diminishes the practical applicability of standard Fourier reconstruction techniques. Second, and most important, is the poor quality data. The objects being studied typically have maximum linear dimensions of $O(10^{-6})$ cm rather than $O(1)$ cm as in CAT. This means that the mass of the object is $O(10^{-18})$ times that in CAT. Thus, in general, by the time enough electrons would have interacted with the object to yield sufficient information, it has long since been completely destroyed.

The most successful strategy to reduce this problem has been to form regular two- (or three-) dimensional arrays of identical objects (particles), reduce the electron dose, and combine the information from the many particles using Fourier methods [23]. However, in general, as the size and complexity of the particle increases so does the difficulty of forming highly ordered regular arrays.

The general problem of combining the information from a number of identical disordered objects with unknown orientations is much more difficult because the relative orientations must be estimated, as well as the electron density. We outline a method for doing this with data from a relatively small number of tilt angles over the limited angular range available in electron microscopy.

## 4.2 Theory

The electron density, $y(r,\theta,\varphi)$, is expressed as a truncated expansion of a complete orthonormal set of functions,

$$y(r,\theta,\varphi) = \sum_{nlm} \gamma_{nlm}\Psi_{nlm}(r,\theta,\varphi),\ n=1,2,\ldots,N, \tag{11}$$

where

$$\Psi_{nlm}(r,\theta,\varphi) = K_{nl}S_{nl}(r)Y_l^m(\theta,\varphi),\ l=n-1,n-3,\ldots,1 \text{ or } 0,\quad m=-l,-l+1,\ldots,l, \tag{12}$$

$$K_{nl} = \{2\Gamma[(n-l+1)/2]/\Gamma[(n+l+2)/2]\}^{1/2}, \tag{13}$$

$$Y_l^m(\theta,\varphi) = N_{lm}P_l^{|m|}(\cos\theta)\exp(im\varphi), \tag{14a}$$

$$N_{lm} = \{(2l+1)(l-|m|)!/[4\pi(l+|m|)!]\}^{1/2} \tag{14b}$$

$$S_{nl}(r) = r^l\exp(-r^2/2)L^{l+1/2}_{(n-l-1)/2}(r^2), \tag{15}$$

$L_j^k(\cdot)$ are the generalized Laguerre polynomials and $P_l^m(\cdot)$ the associated Legendre polynomials defined in eqs (22.3.9) and (8.6.6) of [1], respectively, and the $\gamma_{nlm}$ are to be estimated.

These basis functions in eq (11) are the eigenfunctions of the Schrödinger equation for the spherically symmetric harmonic oscillator (see p. 1663 of [12]). They have the following two useful properties:

(a) They are eigenfunctions of the Fourier transform, i.e.,

$$\int \exp(i\underset{\sim}{r}\cdot\underset{\sim}{R})\Psi_{nlm}(r,\theta,\varphi)d^3r = (2\pi)^{3/2}i^{n-1}\Psi_{nlm}(R,\theta,\varphi). \tag{16}$$

This is most easily evaluated by changing to Cartesian coordinates, in which the variables separate and $\Psi_{nlm}$ is just a product of three one-dimensional harmonic oscillator wavefunctions (see p. 1679 of [12] ). This makes the application of the projection-slice theorem [2], which says that the Fourier transform of a projection is a central slice (planar section) through the 3-D Fourier transform of the density, very easy. Thus we can compare the Fourier transform of the projection data directly with the 3-D electron density in eq (11) using eq (16).

(b) All of the angular dependence is in the spherical harmonics, $Y_l^m(\theta,\varphi)$, whose behavior upon rotation of the coordinate system can be easily expressed and rapidly computed using the rotation operators for spherical harmonics [3,9].

The Fourier transform of the projection data can then be modelled by transforming eq (11) and rotating the coordinate system through the known angle, $\tau$, of tilt about the x-axis (arbitrarily defined to be the tilt axis) and through three (unknown) Euler angles, $\underset{\sim}{\omega}$, that reorient the particle's coordinate system to coincide with a reference system defined below,

$$\hat{F}(\rho,\Phi;\underset{\sim}{\omega},\tau) = (2\pi)^{3/2} \sum_{nlm}^{n=N} \gamma_{nlm} i^{n-1} \sum_{m'=-l}^{l} R^l_{m'm}(\underset{\sim}{\omega}) \times \sum_{m''=-l}^{l} R^l_{m''m'}(-\pi/2,\tau,\ \pi/2)\Psi_{nlm''}(R,\pi/2,\Phi), \quad (17)$$

where the Euler angles and rotation matrices $R^l_{m'm}(\cdot)$ are defined by Brink and Satchler [3], which is the only source we could find that was free of errors or inconsistencies. The variables $\rho$ and $\Phi$ on the left-hand side are just the polar coordinates in the x-y plane and are numerically equal to R and $\Phi$ on the right-hand side.

The term $\Psi_{nlm''}(R,\pi/2,\Phi)$ represents a two-dimensional slice through the x-y plane. Thus this uses the projection-slice theorem assuming that the projection is parallel to the z-axis of the coordinate system used in eq (11). For each particle, the unknown vector, $\underset{\sim}{\omega}$, of Euler angles rotates the coordinate system of that particle so that this is the case. If all particles had the same orientation, then no $\underset{\sim}{\omega}$ would be necessary since with large enough N the $\gamma_{nlm}$ could represent the particle in any orientation. In practice we have always used this reasoning and arbitrarily fixed the coordinate system of one particle to be the reference to which all the others are rotated. Thus with $N_p$ particles there are only $(N_p-1)$ vectors $\underset{\sim}{\omega}$. However, with a relatively small N, it might be better to allow all the particles to rotate so that the limited number of terms in eq (11) can be most efficiently used. This would in any case be necessary if one were using only a subset of the spherical harmonic terms to impose a particular symmetry, e.g., icosahedral symmetry [6], if the exact orientations of the symmetry axes were not known.

Despite the relatively compact and computationally efficient model in eq (17), the computational burden in a straightforward weighted least squares analysis would be overwhelming. This is mainly because of the number of data points and the five-fold sum in eq (17). A typical image is digitized to a 64×64 array and a discrete Fourier transform would yield the same number of values, i.e., $O(10^4)$. For 20 particles and 9 tilt angles, this would amount to a nonlinear least squares analysis with $O(10^6)$ rows. This amount of data can be reduced by a factor O(100) with almost no loss in information by applying an orthogonal transformation based on the orthogonality properties of the basis functions with respect to $\Phi$ and a polar coordinate sampling theorem [21].

All of the $\Phi$ dependence in eq (17) is in the term $\exp(im\Phi)$ in the spherical harmonics in eqs (16) and (12). Because of the orthogonality properties of this term, the circular transform,

$$\hat{F}_{\hat{m}}(\rho;\underset{\sim}{\omega},\tau) \equiv (1/2\pi) \int_0^{2\pi} \exp(-i\hat{m}\Phi)\hat{F}(\rho,\Phi;\underset{\sim}{\omega},\tau)d\Phi, \tag{18}$$

eliminates the innermost sum in eq (17) and reduces to

$$\hat{F}_{\hat{m}}(\rho\ ;\underset{\sim}{\omega},\tau) = (2\pi)^{3/2} \sum_{nlm}^{n=N} \gamma_{nlm} i^{n-1} K_{nl} S_{nl}(\rho) N_{l\hat{m}} P_l^{|\hat{m}|}(0)$$
$$\times \sum_{m'=-l}^{l} R^l_{m'm}(\underset{\sim}{\omega}) R^l_{\hat{m}m'}(-\pi/2,\tau,\pi/2) \tag{19}$$

Furthermore there are only nonzero terms when $|\hat{m}|<N$; i.e., there are only (2N-1) $\hat{m}$ values needed to represent all the information relevant to the model in eq (11).

The radial variable, $\rho$, can also be sampled. Although neither the model in eq (11) nor its Fourier transform in eq (16) are of compact support, they both can be considered to be approximately so. This is because the radial parts of both the function and its Fourier transform are strongly damped toward zero with increasing r or R by the Gaussian factor in eq (15). We denote by $r_{max}$ and $\rho_{max}$, respectively, the values of r and $\rho$ beyond which the model and its transform can be considered to be negligibly small compared to the maximum electron density (in the model) or the noise components (in its transform). These cutoff values depend weakly on the value of N in eq (1) and the signal-to-noise ratio in the data, but $\rho_{max}=4$ and $r_{max}=6$ have been found to be sufficiently large for $N\leq 13$. Space-limiting the model to $r\leq r_{max}$, the polar coordinate sampling theorem [21] says that all the information is obtained by sampling $\hat{F}_{\hat{m}}(\rho;\underset{\sim}{\omega},\tau)$ at $\rho$ values given by

$$\rho_{\hat{m}k} \equiv Z_{\hat{m}k}/r_{max}, \qquad (20)$$

when $Z_{\hat{m}k}$ is the kth zero of the Bessel function $J_{\hat{m}}(\rho)$. Band-limiting the model to Fourier components with $\rho\leq\rho_{max}$ (because the high-frequency Fourier components of the signal in the data become negligible compared with the components of the noise) yields a greatly reduced number of data points for each image, O(100) rather than $O(10^4)$,

$$\hat{F}_{\hat{m}k}(\underset{\sim}{\omega},\tau) \equiv \hat{F}_{\hat{m}}(\rho_{\hat{m}k};\underset{\sim}{\omega},\tau). \qquad (21)$$

In order to fit the data to the model in eqs (21) and (19), the data must be transformed as follows:

$$F_{\hat{m}}(\rho;\underset{\sim}{\omega},\tau) = (1/2\pi)\int_0^{2\pi} d\Phi \exp(-i\hat{m}\Phi)\int d^2r \exp(i\underset{\sim}{r}\cdot\underset{\sim}{\rho})f(x,y), \qquad (22)$$

where f(x,y) is used to represent the data because the images are scanned with a Cartesian grid. The integral over $\Phi$ can be performed analytically (see pp. 1678–1680 of [12]) to yield

$$F_{\hat{m}k}(\underset{\sim}{\omega},\tau) = i^{\hat{m}} \int_0^{r_{max}} dr\, r \int_0^{2\pi} d\varphi J_{\hat{m}}(r\rho_{\hat{m}k})\exp(-i\hat{m}\varphi)f(x,y). \qquad (23)$$

This transform is orthogonal with respect to both of the indices $\hat{m}$ and k. The orthogonality with respect to $\hat{m}$ is clear from the orthogonality of $\exp(-i\hat{m}\varphi)$ over the interval $\varphi\in[0,2\pi]$. The orthogonality with respect to k follows from a change of variable to $t=r/r_{max}$, eq (20), and the standard orthogonality relation for integrals involving zeroes of Bessel functions, eq (11.4.5) of [1],

$$\int_0^1 tJ_{\hat{m}}(Z_{\hat{m}k}t)J_{\hat{m}}(Z_{\hat{m}k'}t)dt = 0, \; k \neq k'. \tag{24}$$

In practice, eq (23) is evaluated by numerical quadrature, and $\underset{\sim}{F}$, the vector of $F_{\hat{m}k}(\underset{\sim}{\omega},\tau)$ values is simply the linear transformation

$$\underset{\sim}{F} = C\underset{\sim}{f}, \tag{25}$$

where $\underset{\sim}{f}$ is the vector of image data points, $f(x,y)$, and the matrix C (typically about $100 \times 64^2$) accounts for the quadrature weights, the Cartesian grid of the data points, and the kernel of the transformation in eq (23). To within quadrature error, the matrix C is Hermitean because of the orthogonality of the integral transforms mentioned above. This is very convenient, because, if the covariance matrix of $\underset{\sim}{f}$ is an identity matrix, the covariance matrix of $\underset{\sim}{F}$ is diagonal. That is, uncorrelated stationary noise in the projection data $\underset{\sim}{f}$ remains uncorrelated in the reduced data $\underset{\sim}{F}$, and one can perform a simple weighted least squares fit of $\underset{\sim}{F}$ in eq (25) to the model in eqs (21) and (19). Otherwise one would have to work with a non-diagonal covariance matrix in the least squares analysis. The assumption of uncorrelated stationary noise is often not bad, even when the total signal is Poisson, because a large background must often be subtracted from the total signal to obtain $\underset{\sim}{f}$.

### 4.3 Practical Aspects

The matrix C in eq (25) typically takes about 20 min of CPU time to compute. However, it depends only on such things as $r_{max}$ and $\rho_{max}$ and the number of data points in the scanning grid for the images. These usually remain the same from one experiment to the next, and C can therefore be computed once and for all and stored. The reduction of a complete image to $\underset{\sim}{F}$ in eq (25) then takes only a few seconds of CPU time. From this point onwards, only the reduced data, $\underset{\sim}{F}$, is used.

All of the images must have the same origin. Fortunately, the center of gravity of a projection is independent of the orientation of the particle, and this is used as the origin. This is also a good choice in that it generally results in relatively rapid convergence of the expansion in eq (11). In practice the estimation of the center of gravity of each image must be done with care, after subtraction of the background.

A more general set of functions than $\Psi_{nlm}(r,\theta,\varphi)$ in eq (12) contains a scale factor multiplying r. We have arbitrarily set this to one. However, from section 4.2 it is clear that the r values in the input data must be scaled so that the maximum extent of the projections in the images is about $r_{max}$.

Regularization is imposed by the upper limit N. We start with a relatively small N, typically 5, and increase N until the decrease in the weighted sum of squared deviations of the fit to $\underset{\sim}{F}$ (in a plot versus N) does not seem to be significant. Classical F-tests might also be helpful. Experience so far indicates that imposing parsimony by truncation is not as bad here as truncating Fourier series or transforms, but slightly smoother solutions could probably be obtained by a more gradual tapering of the expansion in eq (11). However, the number of $\gamma_{nlm}$ parameters is N(N+1)(N+2)/6, and such a gradual taper would be very expensive except for very low resolution solutions. Furthermore the stepwise increase in N results in a natural series of solutions with increasing detail. This can be seen from the natural hierarchy of increasing complexity of the spherical harmonics or the wave functions in eq (12) with increasing l or n. We have performed analyses with N as large as 13 (with 455 $\gamma_{nlm}$ parameters), but N=9 (with 165 $\gamma_{nlm}$) is often sufficient to represent reasonably large structures to within the resolution attainable in electron microscopy.

There are generally far more linear ($\gamma_{nlm}$) parameters than nonlinear ($\underset{\sim}{\omega}$) ones. If $N_p$ is the number of particles being analyzed, then there are either $3(N_p-1)$ or $3N_p$ nonlinear parameters, and $N_p$ seldom exceeds 20. Therefore a least squares analysis exploiting separability should greatly improve the rate and region of convergence.

Because of the large size of the problem and the formulation, nonnegativity was not imposed. This is not as serious as in other cases because of the relatively low resolution attainable in electron microscopy.

The disorder of the particles prevents the problem from being reduced to a series of independent two-dimensional reconstructions of slices through the 3-D structure, as is often possible in CAT. The need for a direct 3-D reconstruction brings with it an unavoidable added computational burden, but it does have the advantage that smoothness of the entire 3-D structure tends to be imposed, and this is generally not the case when a series of two-dimensional slices are reconstructed independently.

One of the main advantages of this formulation in terms of a straightforward problem in parameter estimation by weighted least squares is that the approximate covariance matrix of the parameter estimates is obtained. This gives a very clear indication of the reliability of the estimated structures and a warning when N is getting too large. It also permits general theoretical studies of the effects of such things as the number and range of tilt angles and the disorientation of the particles on the uncertainties in the estimates. It turns out that disordered particles can actually bring a benefit in that a wider range of views are obtained over the limited range of tilt angles available than if all particles had the same orientation. In fact simulations indicate that the Fisher information content for $N_p$ randomly oriented particles can be almost as large as that for $N_p^2$ particles of identical orientation, even including the extra uncertainty due to the extra $\underset{\sim}{\omega}$ parameters for the disoriented case.

For a set of particles with identical orientations and an upper limit of N in eq (11), precisely N different tilt angles are needed; otherwise the parameter covariance becomes singular and the parameters indeterminate. With disoriented particles, this requirement can be relaxed, but it is still recommended. It is important to be able to use as few tilt angles as necessary. This permits the total dose of electrons that the particles can tolerate to be divided into larger doses for each tilt angle. This makes it easier and more reliable to subtract background, to locate the center of gravity, and to perform the rest of the steps in the analysis.

The information content deteriorates as the range of tilt angles is restricted, but more slowly for disordered than for ordered particles. In fact, simulations indicate that a tilt range of $(-45^{\circ},45^{\circ})$ is often sufficient with a set of disordered particles. It would be advantageous if the commonly used extremely oblique tilts in the range $(-60^{\circ},60^{\circ})$ could be avoided.

Another advantage of the statistical treatment of the problem is that the large data reduction in eq (25) is immediately demanded by the analysis. Sampling $\hat{F}(\rho,\Phi;\underset{\sim}{\omega},\tau)$ in eq (17) more finely resulted in a practically singular covariance matrix for the weighted least squares analysis, and it was apparent that a sampling theorem had to be used.

The method has been extensively tested with simulated projections and added noise of the level typically found in electron micrographs, and the results have been very encouraging. However, it is necessary to make tests with real micrographs containing the numerous systematic errors and artifacts that occur in electron microscopy, and this has been started.

## 5. References

[1] M. Abramowitz and I.A. Stegun, Handbook of Mathematical Functions (Dover, New York, 1965).

[2] R.N. Bracewell, Strip integration in radioastronomy, Australian J. Phys. 9, 198 (1956).

[3] D.M. Brink and G.R. Satchler, Angular Momentum (Clarendon Press, Oxford, 1968).

[4] R.K. Bryan, Ph. D. Dissertation, University of Cambridge (1980).

[5] R.K. Bryan, M. Bansal, W. Folkhard, C. Nave, & D.A. Marvin (in preparation).

[6] N.V. Cohan, The spherical harmonics with the symmetry of the icosahedral group, Proc. Cambr. Phil. Soc. 54, 28 (1958).

[7] B.R. Frieden, Restoring with maximum likelihood and maximum entropy, J. Opt. Soc. Amer. 62, 511 (1972).

[8] G.H. Golub and V. Pereyra, The differentiation of pseudo-inverses and nonlinear least squares problems whose variables separate, SIAM J. Numer. Anal. 10, 413 (1973).

[9] Z. Kam, The reconstruction of structure from electron micrographs of randomly oriented particles, J. Theor. Biol. 82, 15 (1980).

[10] H. Lachmann, New methods of parameter identification in kinetics of closed and open reaction systems, this volume.

[11] C.L. Lawson and R.J. Hanson, Solving Least Squares Problems (Prentice-Hall, Englewood Cliffs, 1974).

[12] P.M. Morse and H. Feshbach, Methods Of Theoretical Physics, Part II (McGraw-Hill, New York, 1953).

[13] R.L. Obenchain, Classical F-tests and confidence regions for ridge regression, Technometrics 19, 429 (1977).

[14] P.F. Price, A comparison of least-squares and maximum-likelihood estimators for counts of radiation quanta which follow a Poisson distribution, Acta Cryst. A35, 57 (1979).

[15] S.W. Provencher, Inverse problems in polymer characterization: Direct analysis of polydispersity with photon correlation spectroscopy, Makromol. Chem. 180, 201 (1979).

[16] S.W. Provencher, A constrained regularization method for inverting data represented by linear algebraic or integral equations, Comput. Phys. Commun. 27, 213 (1982).

[17] S.W. Provencher, CONTIN: A general purpose constrained regularization program for inverting noisy linear algebraic and integral equations, Comput. Phys. Commun. 27, 229 (1982).

[18] S.W. Provencher, in Photon Correlation Techniques, edited by E.O. Schulz-Dubois (Springer, Heidelberg, 1983).

[19] S.W. Provencher and R.H. Vogel, Information loss with transform methods in system identification: A new set of transforms with high information content, Math. Biosci. 50, 251 (1980).

[20] A. Ruhe and P.A. Wedin, Algorithms for separable nonlinear least squares problems, SIAM Rev. 22, 318 (1980).

[21] H. Stark, Sampling theorems in polar coordinates, J. Opt. Soc. Amer. 69, 1519 (1979).

[22] O.N. Strand and E.R. Westwater, On the numerical solution of Fredholm integral equations of the first kind by the inversion of the linear system produced by quadrature, J. Assoc. Comput. Mach. 15, 100 (1968).

[23] P.N.T. Unwin and R. Henderson, Molecular structure determination by electron microscopy of unstained crystalline specimens, J. Mol. Biol. 94, 425 (1975).

[24] R.H. Vogel, SPLMOD Users Manual, EMBL Technical Report DA06, European Molecular Biology Laboratory, Heidelberg, 1983.

[25] R.W. Wijnaendts van Resandt, R.H. Vogel, and S.W. Provencher, Double beam fluorescence lifetime spectrometer with subnanosecond resolution: Application to aqueous tryptophan, Rev. Sci. Instr. 53, 1392 (1982).

# A COMPARISON OF STATISTICAL REGULARIZATION AND FOURIER EXTRAPOLATION METHODS FOR NUMERICAL DECONVOLUTION

A. R. Davies, M. Iqbal, K. Maleknejad and T. C. Redshaw

## 1. Introduction

The Fredholm integral equation of the first kind of convolution type:

$$(Kf)(x) \equiv \int_{-\infty}^{\infty} k(x-y)f(y)dy = g(x), \quad -\infty < y < \infty, \tag{1.1}$$

occurs widely in the applied sciences. k and g are known kernel and data functions, respectively, and f is to be found. We shall assume that f, g and k lie in suitable function spaces, such as $L_2(\mathbb{R})$, so that their Fourier transforms (FTs) exist. (^ will denote FTs, and ˇ inverse FTs).

The ill-posedness of (1.1) is well known; the degree of ill-posedness depends on the decay rate of the kernel transform $\hat{k}(\omega)$ and also on the smoothness of the solution f. When g is inexact, therefore, we may seek a stable or filtered approximation to f given by

$$f_\lambda(y) = \frac{1}{2\pi}\int_{-\infty}^{\infty} Z(\omega;\lambda)\,\frac{\hat{g}(\omega)}{\hat{k}(\omega)}\,\exp(i\omega y)\,d\omega, \tag{1.2}$$

where $Z(\omega;\lambda)$ is a filter function dependent on a parameter $\lambda$.

Filters may be constructed in several ways [11], in particular from Tikhonov regularization methods where $\lambda$ in (1.2) becomes the regularization parameter. In the latter case, two essential decisions are needed: (i) the choice of smoothing norm which acts as the stabilizing functional, and (ii) the choice of $\lambda$ in some optimal sense. Problem (i) depends on the nature of k and f, and is in general difficult. (Some papers addressing this problem are referenced in [5]). Several deterministic and statistical methods have been proposed for problem (ii), some of which are described briefly in Section 3.

In many practical applications, the function f has compact support, so that the transform $\hat{f}(z)$, $z = \omega + i\zeta$, is analytic. As an alternative to regularization we could then employ a rectangular filter

$$Z(\omega;\omega_N) = \begin{cases} 1, & |\omega| \leq \omega_N \\ 0, & |\omega| > \omega_N \end{cases}, \tag{1.3}$$

where the cut-off frequency $\omega_N$ is chosen optimally, and an attempt is made to restore high frequency information by numerical analytic continuation of the truncated transform. Such an extrapolation process is itself ill-posed and therefore requires stabilization in some manner. When f is non-negative, several stable Fourier extrapolation methods are available, some of which are described in Section 4.

The performance of statistical regularization and constrained Fourier extrapolation will be compared in Section 5.

## 2. Regularization and Approximation

We shall restrict attention to pth order regularization, where the smoothing functional

$$C(f;\lambda) = \| Kf - g \|_2^2 + \lambda \| f^{(p)} \|_2^2 \tag{2.1}$$

is minimized over the subspace $H^p \subset L_2$. Both norms in (2.1) are $L_2$, and $f^{(p)}$ denotes the pth derivative of f. The regularization parameter $\lambda$ controls the trade-off between smoothness imposed on the filtered solution, and the extent to which (1.1) is satisfied. The minimizer of (2.1) in $H^p$ is given by (1.2), where

$$Z(\omega;\lambda) = \frac{|\hat{k}(\omega)|^2}{|\hat{k}(\omega)|^2 + \lambda\omega^{2p}}. \tag{2.2}$$

We assume throughout this paper that the support of each function f, g and k is essentially finite and contained within the interval [0,1), possibly by a change of variable. It is then convenient to adopt the approximating function space $T_{N-1}$, of trigonometric polynomials of degree at most N-1 and period 1, since then the discretization error in the convolution may be made exactly zero at the grid points, and Fast Fourier Transforms (FFTs) may be employed in the solution procedure [5].

Let g and k be given at N equally spaced points $x_n = nh$, $n = 0,\dots,N-1$, with spacing $h = 1/N$. Then g and k are interpolated by $g_N$ and $k_N \in T_{N-1}$, where

$$g_N(x) = \frac{1}{N}\sum_{q=0}^{N-1} \hat{g}_{N,q}\exp(i\omega_q x), \tag{2.3}$$

$$\hat{g}_{N,q} = \sum_{n=0}^{N-1} g_n \exp(-i\omega_q x_n), \tag{2.4}$$

and $$g(x_n) = g_n = g_N(x_n), \quad \omega_q = 2\pi q, \tag{2.5}$$

with similar expressions for $k_N$.

It is easily established that:

(1) Equation (1.1) is exactly equivalent at $\{x_n\}$ to the $N \times N$ discrete system

$$(K\underline{f})_n = g_n, \tag{2.6}$$

where

$$K = \Psi \operatorname{diag}(h\hat{k}_{N,q})\Psi^H, \tag{2.7}$$

and $\Psi$ is the unitary discrete FT matrix with elements

$$\psi_{rs} = N^{-\frac{1}{2}}\exp\left(\frac{2\pi i}{N}rs\right), \quad r,s = 0,\dots,N-1. \tag{2.8}$$

(2) In $T_{N-1}$, $f_\lambda$ in (1.2) is approximated by

$$f_{N;\lambda}(x) = \sum_{q=0}^{N-1} Z_{q;\lambda}\frac{\hat{g}_{N,q}}{\hat{k}_{N,q}}\exp(i\omega_q x), \tag{2.9}$$

where the discrete pth order filter is

$$Z_{q;\lambda} = \frac{|\hat{k}_{N,q}|^2}{|\hat{k}_{N,q}|^2 + N^2\lambda\tilde{\omega}_q^{2p}}, \quad \tilde{\omega}_q = \begin{cases} \omega_q, & 0 \leq q < N/2 \\ \omega_{N-q}, & N/2 \leq q \leq N-1 \end{cases}. \tag{2.10}$$

## 3. Statistical Methods for Optimizing $\lambda$

A list of these methods may be found in [7]. They fall essentially into two classes:

Class 1: methods which depend on a priori knowledge of the noise level in the data, as depicted by the variance $\sigma^2$. (We assume white noise throughout this paper);

Class 2: methods which do not depend on prior knowledge of $\sigma^2$.

Here we shall consider two representative methods from each class.

### 3.1 Class 1 Methods

3.1.1 Turchin's Method (TM) [12]. Two statistical ensembles of functions in $T_{N-1}$ are introduced: (i) An ensemble which describes the fidelity of the random vector $K\underline{f}$ to the data $\underline{g}$, and which is characterized by an a priori multivariate normal distribution with conditional density function (cdf) $P(\underline{g}|\underline{f})$. (ii) An ensemble of smooth functions consistent with pth order regularization and characterized by an a priori probability density function (pdf) $P_\lambda(\underline{f})$. The filtered solution (2.9) is then the mathematical expectation of the ensemble given by the intersection of (i) and (ii). This intersection has an a posteriori cdf given by Bayes' theorem:

$$P(\underline{f}|\underline{g}) = \text{const}.P(\underline{g}|\underline{f})P_\lambda(\underline{f}).$$

The optimal $\lambda$ is defined as that value which makes the variance of the a posteriori intersection ensemble equal to the variance $\sigma^2$ of the a priori ensemble (i). $\lambda$ is determined by solving the nonlinear equation

$$\frac{\sigma^2}{N}\sum_{q=0}^{N-1} Z_{q;\lambda} + \sum_{q=0}^{N-1} |\hat{g}_{N;q}|^2(1 - Z_{q;\lambda})^2 = \sigma^2. \tag{3.1}$$

3.1.2 Turchin-Klein Method (TK) [8]. The above definition of optimality is somewhat heuristic, and a more satisfactory analysis has been proposed by Klein. Define the a priori cdf

$$P(\underline{g}|\lambda) = \int_{\mathbb{R}^N} P(\underline{g}|\underline{f})P_\lambda(\underline{f})d\underline{f},$$

and consider any a priori pdf $P(\lambda)$ which is assumed to be "narrow" in the sense that $\frac{d}{d\lambda}(\log P(\lambda))$ is negligible for values of $\lambda$ of interest.

Then the optimal $\lambda$ is defined as that which maximizes the a posteriori cdf given by Bayes' theorem:

$$P(\lambda|\underline{g}) = \text{const.}P(\underline{g}|\lambda)P(\lambda).$$

This leads to a nonlinear equation for $\lambda$ of the form

$$\sum_{q=0}^{N-1} Z_{q;\mu} - N^2\mu \sum_{q=0}^{N-1} \tilde{\omega}_q^{2p} Z_{q;\mu}^2 \frac{|\hat{g}_{N,q}|^2}{|\hat{k}_{N,q}|^2} = 0, \text{ where } \mu = \lambda\sigma^2. \quad (3.2)$$

3.2 Class 2 Methods

3.2.1 Wahba's Cross-Validation Method (CV) [13]. If the mth data point $g_m$ is ignored, define the corresponding filtered solution (2.9) by

$$f_{N;\lambda}^{[m]} = \min_{\phi \in T_{N-1}} \frac{1}{N} \sum_{\substack{n=0 \\ n \neq m}}^{N-1} |(K\phi)(x_n) - g_n|^2 + \lambda \| \phi^{(p)} \|_2^2$$

The value $(Kf_{N;\lambda}^{[m]})(x_m)$ then predicts the missing datum $g_m$. The optimal $\lambda$ is defined as that which minimizes the weighted mean-square prediction error for all m:

$$V_{CV}(\lambda) = \frac{1}{N} \sum_{m=0}^{N-1} [(Kf_{N;\lambda}^{[m]})(x_m) - g_m]^2 w_m(\lambda). \quad (3.3)$$

For the circulant kernel (2.7) unit weights are appropriate, and (3.3) may then be expressed as

$$V_{CV}(\lambda) = \frac{\frac{1}{N} \sum_{q=0}^{N-1} (1 - Z_{q;\lambda})^2 |\hat{g}_{N;q}|^2}{\left[1 - \frac{1}{N} \sum_{q=0}^{N-1} Z_{q;\lambda}\right]^2}. \quad (3.4)$$

Although an a priori estimate of $\sigma^2$ is not needed, an a posteriori estimate may be found [7].

3.2.2 Maximum Likelihood Method (ML) [5]. By comparing the filter (2.10) with a suitably constructed minimum-variance (Wiener) filter, an optimal $\lambda$ is defined as the minimizer of

$$V_{ML}(\lambda) = N \log\left[\sum_{q=1}^{N-1} |\hat{g}_{N;q}|^2 (1 - Z_{q;\lambda})\right] - \sum_{q=1}^{N-1} \log(1 - Z_{q;\lambda}). \quad (3.5)$$

Again, an a posteriori estimate of $\sigma^2$ may be found [5].

Theorems which estimate the order of convergence of all the statistical regularization methods described above may be found in [5,7].

## 4. Constrained Fourier Extrapolation

Let $f \in L_2(0,a)$, $a > 0$, be real valued, so that its FT is hermitian: $\hat{f}(-\omega) = \overline{\hat{f}(\omega)}$. The classical theorem of Paley-Wiener [2] states that $\hat{f}(z)$ is entire, $z = \omega + i\zeta$, and of exponential type a, i.e. $|\hat{f}(z)| = O(\exp(a|z|))$. In fact, $|\hat{f}(\omega)| \to 0$ as $\omega \to \infty$ since $\hat{f}(\omega) \in L_2(\mathbb{R})$, but the decay of $|\hat{f}|$ is slower than exponential decay.

Let $S_{\omega_N}\hat{f}$ denote the restriction of $\hat{f}(\omega)$ to the finite interval $[-\omega_N,\omega_N]$, which results from applying the rectangular filter (1.3). When $S_{\omega_N}\hat{f}$ is known, the identity theorem for analytic functions states that $\hat{f}(z)$ is determined uniquely for all z, and in particular for all $\omega \in \mathbb{R}$. Of course, in practice, we know only finitely many discrete values of $S_{\omega_N}\hat{f}$, which means that uniqueness is lost, but this difficulty can be overcome if the Fourier extrapolation process is sufficiently stable. We shall discuss three such methods which exploit the non-negativity property $f \geq 0$.

### 4.1 Maximum Determinant Method (MD) [10].

Suppose that the first N + 1 Fourier coefficients of f, namely, $\hat{f}_q$, q = 0, ..., N, are known or estimated. (N no longer corresponds to the number of real data points in §§2 and 3). Consider the (N+1)×(N+1) hermitian Toeplitz matrix

$$T_N = \begin{pmatrix} \hat{f}_0 & \hat{f}_1 & \cdots & \hat{f}_N \\ \hat{f}_{-1} & \hat{f}_0 & \cdots & \hat{f}_{N-1} \\ \vdots & \vdots & & \vdots \\ \hat{f}_{-N} & \hat{f}_{-N+1} & \cdots & \hat{f}_0 \end{pmatrix}, \quad (4.1)$$

and the (N+2)×(N+2) augmented matrix $T_{N+1}(\alpha)$ obtained from (4.1) when $f_{N+1}$ assumes the value $\alpha \in \mathbb{C}$. The following two lemmas are easily proved:

Lemma 1. $f \geq 0$ if and only if each matrix $T_M$, $M = 0, 1, \ldots,$ generated from its Fourier coefficients, is non-negative definite.

Lemma 2. Let $\hat{\underline{f}}_M \equiv (\hat{f}_{M+1}, \ldots, \hat{f}_1)^T \in \mathbb{C}^{M+1}$. Then

$$\det T_{M+1} = \begin{cases} (\det T_M)(\hat{f}_0 - \hat{\underline{f}}_M^H T_M^{-1} \hat{\underline{f}}_M), & \det T_M \neq 0, \\ 0, & \text{otherwise.} \end{cases}$$

From these lemmas we deduce that the unknown Fourier coefficient $\alpha \in \mathbb{C}$ must satisfy

$$\det T_{N+1}(\alpha) = (\det T_N)[\hat{f}_0 - \hat{\underline{f}}_N(\alpha)^H T_N^{-1} \hat{\underline{f}}_N(\alpha)] \geq 0, \tag{4.2}$$

where $\hat{\underline{f}}_N(\alpha) = (\alpha, \hat{f}_N, \ldots, \hat{f}_1)^T$. It can be shown [10] that the allowable values of $\alpha$ in general lie in a disc in the complex plane, the radius $\rho_{N+1}$ and centre $\alpha_{N+1}$ of which depend on the first N+1 values of $\hat{f}_q$ only. Moreover, the value of the determinant $\det T_{N+1}(\alpha)$ is maximized at the centre of the disc, $\alpha = \alpha_{N+1}$.

The MD method generates a sequence of extrapolated Fourier coefficients $\alpha_{N+1}, \alpha_{N+2}, \ldots$, by adopting the maximizing value of $\alpha$ for $\det T_{N+m}(\alpha)$, for each value of $m = 1,2,\ldots$. The coefficient $\alpha_{N+m}$ may be found very simply by a partial Cholesky decomposition, and a detailed analysis of the method may be found in [10]. It is possible to give a statistical interpretation of the method in terms of entropy [1].

4.2 Biraud's Method (BM) [3].

This is a method which stabilizes the Fourier extrapolation by "functional regularization". Since $f \geq 0$, a real valued function c exists such that

$$f(x) = [c(x)]^2 \tag{4.3}$$

which in Fourier space corresponds to the auto-convolution

$$\hat{f}(\omega) = \hat{c}(\omega) * \hat{c}(\omega) \tag{4.4}$$

where $\hat{c}$ is hermitian. The restriction $S_{\omega_M} \hat{c}$ of $\hat{c}$ to an interval $[-\omega_M, \omega_M]$, where $\frac{1}{2}\omega_N \leqslant \omega_M \leqslant \omega_N$, may be approximately constructed by minimizing the nonlinear functional

$$\| S_{\omega_N}\hat{f} - (S_{\omega_M}\hat{c}) * (S_{\omega_M}\hat{c}) \|^2_{L_2(0,\omega_N)} \tag{4.5}$$

Let $\hat{f}_{2M} \equiv (S_{\omega_M}\hat{c}) * (S_{\omega_M}\hat{c})$. Then it follows that
(i) supp $\hat{f}_{2M} = [-2\omega_M, 2\omega_M] \supseteq [-\omega_N, \omega_N]$, and
(ii) $f_{2M} = (\hat{f}_{2M})^{\vee} \geqslant 0$.
Thus $f_{2M}$ is a non-negative function with an FT which is an approximate extrapolation of $S_{\omega_N}\hat{f}$. Numerical experiments have shown that the choice $\omega_M = \omega_N$ would seem to be optimal, thereby doubling the support of $S_{\omega_N}\hat{f}$.

If we take equally spaced Fourier points $\omega_q = qH$, then provided supp $f \subset [0, (2\pi/MH)]$, we may use the approximate representation

$$f_{2M}(x) = \left[ \sum_{q=-M}^{M} \hat{c}_q \exp(i\omega_q x) \right]^2, \quad \hat{c}_{-q} = \overline{\hat{c}_q}, \tag{4.6}$$

where the coefficients $\hat{c}_q$ are found from the nonlinear least-squares problem

$$\min_{\underline{\hat{c}} \in \mathbb{C}^{M+1}} \sum_{r=0}^{N} \left| (S_{\omega_N}\hat{f})(\omega_r) - \sum_{q=-M}^{M} \hat{c}_q \hat{c}_{r-q} \right|^2. \tag{4.7}$$

Further practical and theoretical details may be found in [6,10].

4.3 <u>Lent and Tuy Iteration</u> (LT) [9].

This method executes repeated projection onto the (convex) positive cone of a suitable subspace of $L_2(0,1)$, using a form of Brègman iteration [4]. A simplified form of the algorithm is as follows:

0. Let $r = 0$. Set $f_0(x) = (S_{\omega_N}\hat{f})^{\vee}$.
1. Set $f_r^{[1]}(x) = \operatorname{Re} f_r(x)$.
2. Set $f_r^{[2]}(x) = S_1^+ f_r^{[1]}(x)$, where $S_1^+$ denotes restriction to $[0,1)$.
3. Set $f_r^{[3]}(x) = \max\{f_r^{[2]}(x), 0\}$.
4. Set $\hat{f}_{r+1}(\omega) = \begin{cases} \hat{f}_r^{[3]}(\omega), & |\omega| > \omega_N \\ \hat{f}(\omega), & |\omega| \leqslant \omega_N \end{cases}$
5. Set $f_{r+1} = (\hat{f}_{r+1})^{\vee}$.

6. Let $r \to r+1$. If the sequence $\{f_r\}$ has converged, stop; otherwise repeat steps 2 - 6.

Under ideal conditions it may be proved that the algorithm converges weakly to the exact solution f [4].

4.4 An Optimal Rectangular Filter

Prior to Fourier extrapolation, an optimal value of $\omega_N$ for the filter (1.3) may be determined by cross-validation. The optimal N, $\omega_N = NH$, minimizes the expression [6]

$$V(N) = \frac{\frac{1}{N_0} \sum_{|q| > N} |\hat{g}_q|^2}{\left( 1 - \frac{2N+1}{N_0} \right)^2} , \qquad (4.8)$$

where $\{\hat{g}_q\}$ are the discrete Fourier coefficients of the data function obtained from $N_0$ real data values $\{g_n\}$.

## 5. A Comparison of Methods

The above statistical regularization and Fourier extrapolation methods have been compared on a range of deconvolution problems, varying from mildly ill-posed problems to severely ill-posed problems as characterized by the decay rate of $\hat{k}$ and the smoothness of f. In each case random noise was added to the data function g in the form of pseudo-random numbers normally distributed with mean zero and variance $\sigma^2$. The standard deviation $\sigma$ was taken to be proportional to $g_{max}$, the largest data value, and in what follows, the notation "$\sigma$ = X%" means that $\sigma = (0.01X)g_{max}$. The actual random error added was then not likely to exceed 3X%. For each problem, X was varied in the range 0 to 10.

Full details of these numerical experiments may be found in [7] and [10], but the main conclusions may be demonstrated by considering a bi-modal problem in which the width of the kernel k is altered to change the degree of ill-posedness of the problem. In Figure 1(a) the function f (a sum of two gaussians) upon convolution with the triangular kernel k gives rise to the broadened data function g. In this case the deconvolution problem is only mildly ill-posed since k is relatively sharp and narrow so that its transform $\hat{k}$ decays relatively slowly. The bi-modality of f is also apparent in g. In Figure 1(b) the width of the

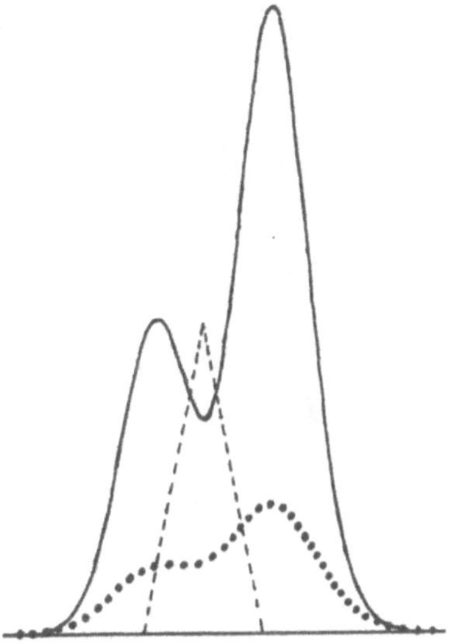

Figure 1(a). Mildly ill-posed.

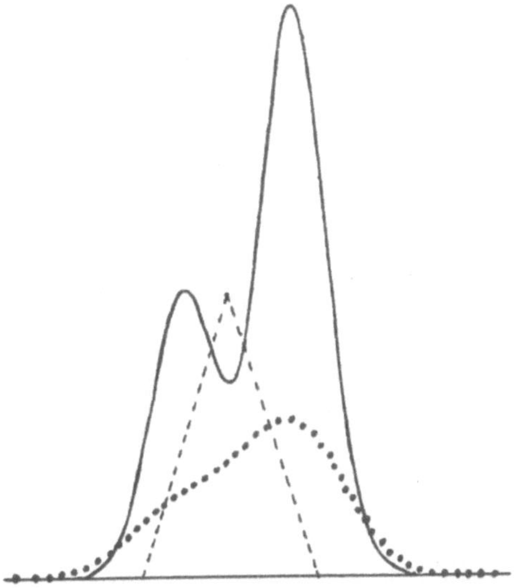

Figure 1(b). Moderately ill-posed.

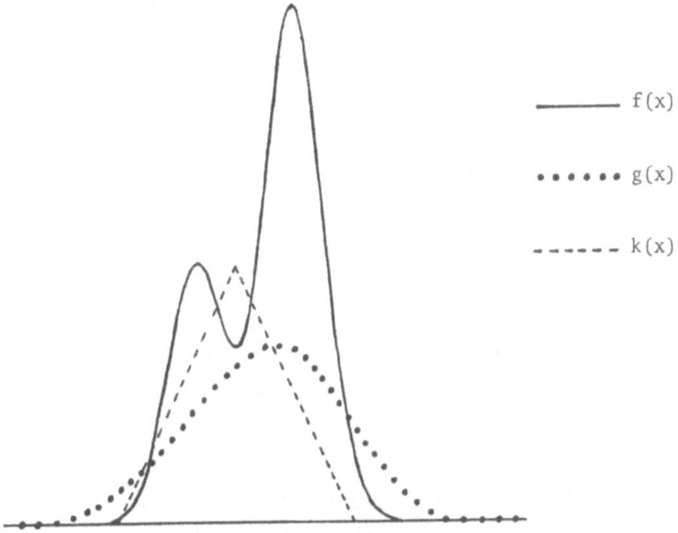


Figure 1(c). Severely ill-posed.

triangular kernel k has been increased to give a moderately ill-posed problem; $\hat{k}$ decays more rapidly, and there is only a faint trace of bi-modality in g. Figure 1(c) shows a severely ill-posed problem for which $\hat{k}$ decays very rapidly, so that the ratio $\hat{g}(\omega)/\hat{k}(\omega)$ is completely dominated by noise amplification except for very low frequencies.

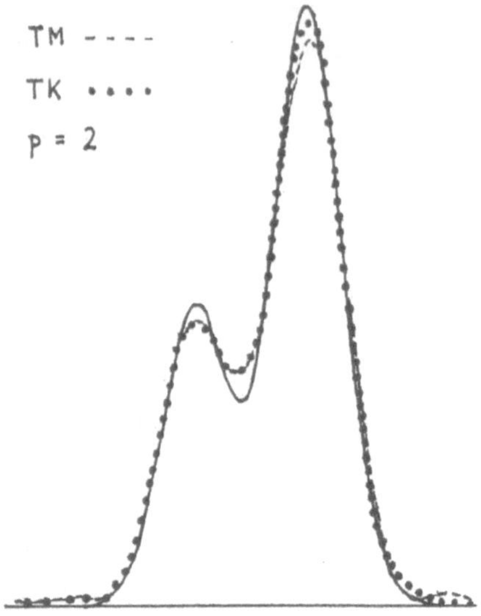


Figure 2(i). TM and TK results for case (a), $\sigma$ = 1.7%.

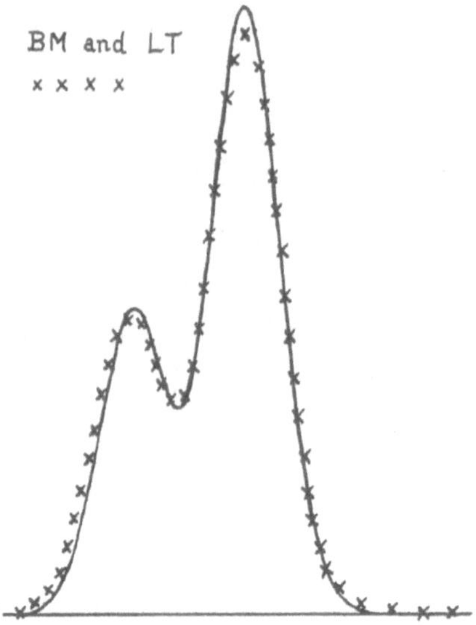


Figure 2(ii). BM and LT results for case (a), $\sigma$ = 1.7%.

In each of the cases (a)-(c), 64 sample points were used to calculate the discrete Fourier coefficients of g and k ((2.4)).

5.1 Case (a).

For this mildly ill-posed problem statistical regularization yielded successful deconvolution for noise levels in g from $\sigma$ = 0% to $\sigma$ = 10%. It was found that in this range, when $\sigma$= X%, the maximum error in the regularized solution $f_{N;\lambda}$ did not exceed 3X% of $f_{max}$. Solutions for $\sigma$ = 1.7% and regularization order p = 2 are shown in Figure 2(i). TM and TK worked well at this noise level provided an accurate estimate of $\sigma^2$ was supplied. ML and TK gave marginally better solutions than TM and CV for $\sigma$ = 1.7% and p = 2. When the order p was increased, the solutions yielded by TM and TK changed insignificantly, whereas the CV solution improved slightly (best at p = 4) while the ML solution deteriorated.

With $N_0$ = 64 and $\sigma$ = 1.7% the optimal rectangular filter for this problem had a cut-off after the 8th Fourier coefficient $\hat{g}_8/\hat{k}_8$. Fourier extrapolation beyond $\omega_8$ using MD failed to achieve any improvement to the truncated solution with N = 8. Both BM and LT, however, produced excellent extrapolations to $\omega_{16}$ as shown in Figure 2(ii). BM, in fact, was able to cope successfully with noise levels up to $\sigma$ = 10%.

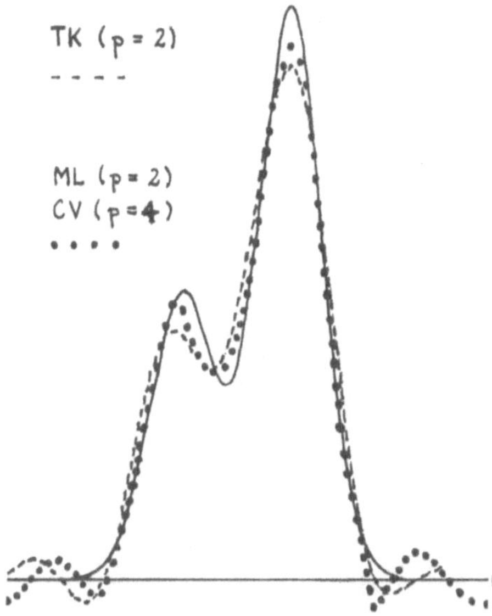


Figure 3(i). TK, ML and CV results for case (b), $\sigma$ = 1.7%.

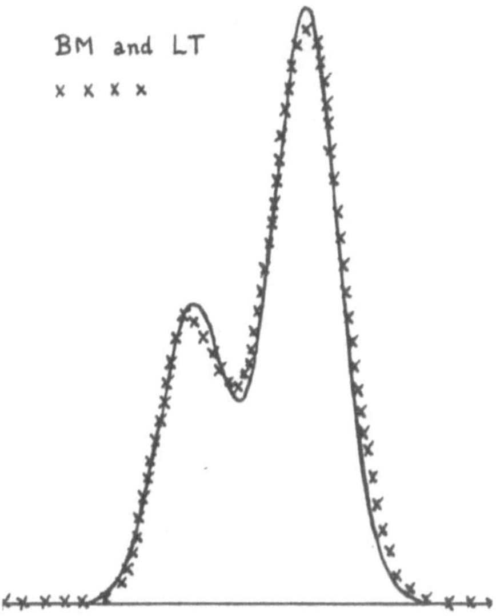


Figure 3(ii). BM and LT results for case (b), $\sigma$ = 1.7%.

5.2 Case (b).

This moderately ill-posed problem required stronger filters for the deconvolution than in case (a). At all noise levels it was found that TM consistently gave too large a value of $\lambda$, thereby yielding very strong filters which failed to render the necessary resolution. For $\sigma$ = 1.7% and p = 2, TK gave a satisfactory solution as shown in Figure 3(i), which also shows the ML (p = 2) solution and CV (p = 4) which are indistinguishable. Again the CV solution had been improved by increasing p, while ML gave its best solution for p = 2. Both ML and CV gave satisfactory solutions up to a noise level of $\sigma$ = 5%.

With $\sigma$ = 1.7%, the optimal rectangular filter had a cut-off at $\omega_5$. MD failed to improve this truncated series, but both BM and LT gave satisfactory extrapolations as shown in Figure 3(ii).

5.3 Case (c).

This difficult problem could not be satisfactorily treated using unconstrained regularization. At the low noise level $\sigma \doteq$ 0.7%, neither TM nor TK could cope even when the variance $\sigma^2$ was given exactly. ML (p = 4) and CV (p = 6) gave almost identical solutions as shown in Figure 4(i), where for comparison the inferior solution obtained from CV (p =2) is also shown. At best, large negative lobes were always present, and the size of these lobes increased with noise level.

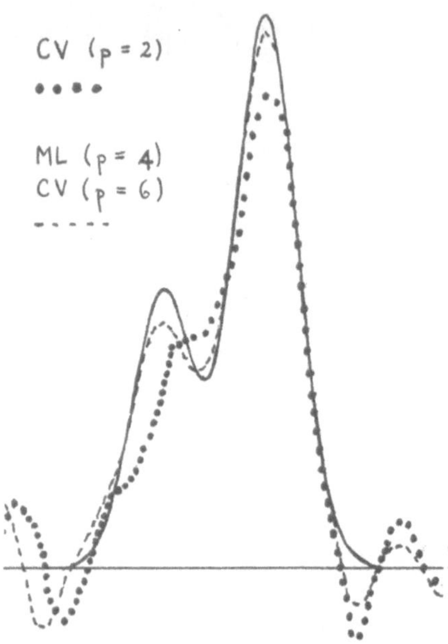


Figure 4(i). ML and CV results for case (c), $\sigma$ = 0.7%.

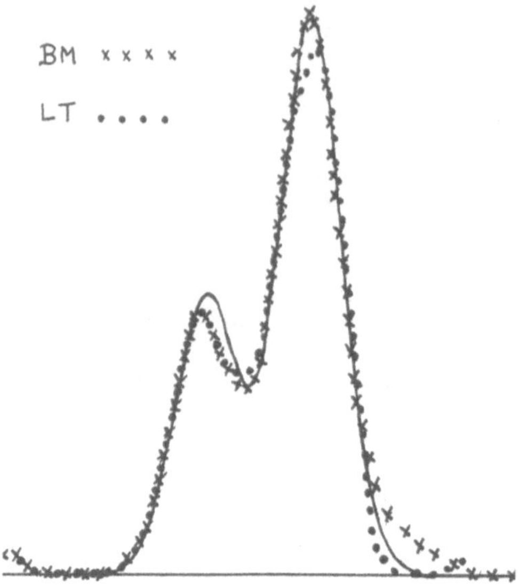


Figure 4(ii). BM and LT results for case (c), $\sigma$ = 0.7%.

With $\sigma$ =0.7% only the first 4 Fourier coefficients $\hat{g}_q/\hat{k}_q$ could be trusted (N = 3), but again BM and LT both gave very satisfactory extrapolations (Fig. 4(ii)). In fact, for this case BM could cope with noise levels above $\sigma$ = 3%.

5.4 Concluding Remarks.

The main conclusions to be drawn from the above examples and also from the results of other test problems are as follows.

The Class 1 methods TM and TK for estimating an optimal $\lambda$ only worked satisfactorily when the problems were mildly ill-posed and the variance $\sigma^2$ of the noise was given accurately. TK was generally more powerful than TM.

The Class 2 methods CV and ML were on the whole dramatically superior. Usually the function $V_{CV}(\lambda)$ could be minimized with respect to the order of regularization p as well as $\lambda$, to give improved solutions. Both CV and ML could handle severely ill-posed problems provided the noise level was low, and there was little to choose between them in performance. When the noise level was high, oscillations appeared in the functions $V_{CV}(\lambda)$ and $V_{ML}(\lambda)$ ((3.4) and (3.5)), and it could be difficult to locate the correct minimum. On the whole, for noisy, severely ill-posed problems, extra information was needed, such as non-negativity.

Of the Fourier extrapolation methods, MD generally behaved poorly in the presence of noise; the determinants became negative, and simple techniques to correct for this rarely succeeded in improving the solution. MD is not to be recommended for restoring smooth functions.

BM performed reasonably well in all cases, and of all the methods was least sensitive to noise. It had the disadvantage, however, of being costly to run for large data sets. LT also worked well in most cases and was very economical when FFTs were used. The rate of convergence might well be slow, however, for higher dimensional problems [9]. Both BM and LT seemed at least as powerful as CV and ML in the presence of noise, and these extrapolation methods were superior when constraint information was needed.

No more general approximating spaces than trigonometric polynomials were used in this investigation, and our conclusions must be interpreted in this light.

## Acknowledgements

Three of us were enabled to take part in this work by research scholarships from the following bodies: (M.I.) University of Punjab, Lahore; (K.M.) Iran University of Science and Technology, Tehran; (T.C.R.) Science and Engineering Research Council, United Kingdom.

## References

[1] Ables, J.G. Maximum entropy spectral analysis. Astron. Astrophys. Suppl. 15, 383-393 (1974).

[2] Bertero, M., de Mol, C. and Viano, G.A. The stability of inverse problems. In "Inverse Scattering Problems in Optics", ed. H.P. Baltes. Springer-Verlag, Berlin (1980). pp161-214.

[3] Biraud, Y. A new approach for increasing the resolving power by data processing. Astron. Astrophys. 1, 124-127 (1969).

[4] Bregman, L.M. The method of successive projection for finding a common point of convex sets. Sov. Math. Dokl. 6, 688-692 (1965).

[5] Davies, A.R. On the maximum likelihood regularization of Fredholm convolution equations of the first kind. In "Proceedings of the Durham Symposium on the Numerical Treatment of Integral Equations", eds. C.T.H. Baker and G.F. Miller. Academic Press (1983).

[6] Davies, A.R. On a constrained Fourier extrapolation method for numerical deconvolution. In "Proceedings of the Oberwolfach Meeting on Improperly Posed Problems and their Numerical Treatment", eds. G. Hämmerlin and K.-H. Hoffmann. Birkhäuser-Verlag (1983).

[7] Davies, A.R. et al. On the statistical regularization of Fredholm convolution equations of the first kind. To be published.

[8] Klein, G. On spline functions and statistical regularization of ill-posed problems. J. Comp. Appl. Math. 5, 259-263 (1979).

[9] Lent, A. and Tuy, H. An iterative method for the extrapolation of band-limited functions. J. Math. Anal. and Appl. 83, 554-565 (1981).

[10] Redshaw, T.C. The Numerical Inversion and Extrapolation of Integral Transforms. Ph.D. Thesis, Univ. of Wales (1982).

[11] Tikhonov, A.N. and Arsenin, V.Y. Solutions of Ill-Posed Problems. Wiley/Winston, London (1977).

[12] Turchin, V.F. Solution of the Fredholm equation of the first kind in a statistical ensemble of smooth functions. USSR Comp. Math. Math. Phys. 7, 79- 96 (1967).

[13] Wahba, G. Practical approximate solutions to linear operator equations when the data are noisy. SIAM J. Numer. Anal. 14,651 -667 (1977).

# DECONVOLUTION OF GAUSSIAN AND OTHER KERNELS

Johan Philip

Abstract: We consider the numerical solution of the convolution equation $h * f = d$ for various nonnegative kernels $h$. Obviously, the Dirac measure is the easiest kernel to deconvolve. In a certain sense, there also exists a unique most difficult kernel to deconvolve, namely the Gaussian $\exp(-x^2)$. A method for deconvolution of the Gaussian is suggested.

## 1. Introduction

We shall study the common linear model of an imaging system with error

$$h * f + e = d , \tag{1.1}$$

where $f$ is sought, $d$ is measured, $e$ is random errors and $h$ is a nonnegative kernel. Equation (1.1) models spectrographs, cameras and other instruments measuring intensity. After solution of (1.1), the obtained $f^{\wedge}$ will have a component which is an amplified rearrangement of $e$.

In this paper, we shall try to make the following theorem plausible and refer the reader to ref. [8] for the proofs.

Theorem 1.1 In numerical deconvolution, there exists in a certain sense a unique nonnegative kernel, which causes maximal error amplification. This kernel is the Gaussian $\exp(-x^2)$.

Having realized that the Dirac measure is the easiest kernel to deconvolve, a first guess for the most difficult one could be "the flattest function", i.e. $h = \text{const.}$. However, this is not what you actually mean by a deconvolution problem and it is ruled out by the normalization below.

It is well known that the deconvolution problem is ill-posed and several methods for its solution have been derived which all try to find a solution by supplementing the calculations with some extra information. One can mention generalized inverses, by Nashed [5] and regularization by Elden [4] and Tikhonov [9]. Andrews and Hunt [1] describe application of Generalized Wiener Filtering and other methods to images. Other methods are given by Philip [6] and Carasso et. al. [2]. The extreme ill-posedness of the Gaussian seems not to have been observed before but [2] gives a method for deconvolution of the Gaussian.

## 2. Normalizations

We let all functions be defined on $(0,2\pi)$, which is practical in connection with the Discrete Fourier Transformation (DFT) to be used. Any problem supported by another interval shall first be mapped on $(0,2\pi)$. When discretizing for the numerical calculations, we divide this interval into $N$ equal subintervals and denote the discretization of $f$ by $f_N$.

Comparison of the shape of different kernels $h \geq 0$ is only meaningful if they are normalized as to "size" and "width", so we let all $h$ have total mass one

$$\int h(x)\,dx = 1 . \tag{2.1}$$

Then, we can think of $h$ as a probability density function and use the standard deviation $\sigma$ as measure of its "width".

## 3. Method of Solution

We shall use Discrete Fourier Transformation (DFT), which is the natural tool for handling convolution equations. The DFT of a function $f_N$ will be denoted by the corresponding capital letter $F_N$

$$F_N(j) = \sum_{k=0}^{N-1} w^{jk} f_N(k) \quad , 0 \leq j \leq N-1, \quad w = \exp(2\pi i/N) \; .$$

The DFT of (1.1) is

$$H_N(j)\cdot F_N(j) + E_N(j) = D_N(j) \quad , 0 \leq j \leq N-1 \; . \qquad (3.1)$$

Using simple assumptions about the error $e$, the solving of (3.1) can by the maximum likelihood method be reformulated as the following least squares problem (see [6])

$$\min_{F_N(j)} \sum_{j=0}^{N-1} \left| D_N(j) - H_N(j)\cdot F_N(j)\right|^2 \; . \qquad (3.2)$$

The solution is

$$F_N(j) = \begin{cases} D_N(j)/H_N(j) & \text{if} \quad H_N(j) \neq 0 \\ \text{arbitrary} & \text{if} \quad H_N(j) = 0 \end{cases} \qquad (3.3)$$

If any $H_N(j) = 0$, the solution is not unique and a small $H_N(j)$ will blow up $D_N(j)$, which has a part stemming from the error.

To analyse the propagation of the error, we assume that the data $d = h * f_0 + e$ for some $f_0$, so that the solution $\hat{f}$ satisfies $h * \hat{f} = h * f_0 + e$. We get

$$h * (\hat{f} - f_0) = e \; . \qquad (3.4)$$

Since $e$ is a random variable, $\hat{f}$ becomes a random variable. We take as measure of ill-posedness of $h$ a discrete analogue of

$$R = \frac{\text{variance of } (\hat{f} - f_0)}{\text{variance of } e} \qquad (3.5)$$

This $R$ is the square of the error amplification of the deconvolution process. Of course, the value of $R$ depends on the method of calculation of $\hat{f}$, in particular on the size of $N$.

If all $H_N(j) \neq 0$, one can show (see formula (16) of [7]) that solution (3.3) has the following ill-posedness

$$R_N(h_N) = N^{-1} \sum_{j=0}^{N-1} \left| H_N(j) \right|^{-2} . \tag{3.6}$$

One can show that a translation of $h$ leaves $R_N$ invariant and that the Dirac measure has $R_N = 1$ . If just one $H_N(j) = 0$, this method breaks down and $R_N(h_N)$ becomes infinite.

The "correction method" of [6] is designed for the case with one or more $H_N(k) = 0$. In this method, the $F_N(k)$ for $H_N(k) = 0$, are determined with the aid of the extra condition that the support of $f_N$ shall satisfy

$$\text{supp}(f_N) + \text{supp}(h_N) < N.$$

Denoting the set of indices for which $H_N(k) = 0$ by $K$ , the ill-posedness of the correction method is by formula (21) of [7]

$$R_N(h_N) = N^{-1} \sum_{j \notin K} \left(1 + \sum_{k \in K} | u_{kj} |^2\right) \left| H_N(j) \right|^{-2} , \tag{3.7}$$

where

$$| u_{kj} | = \prod_{k \neq p \in K} \left| \frac{w^j - w^p}{w^k - w^p} \right| , \qquad k \in K, \; j \notin K . \tag{3.8}$$

Notice that the $u_{kj}$, and so $R_N$ become big if $K$ containes indices close together.

The "derivative method" of [6] , which also can handle the degenerate case, has an $R_N$ of the same character. For comparison, we give the expression for $R_N$ of Generalized Wiener Filtering in the simplest case when all involved random variables are independent

$$R_N(h_N) = N^{-1} \sum_{j=0}^{N-1} \frac{\left| H_N(j) \right|^2 + 1}{\left(\left| H_N(j) \right|^2 + s^{-2}\right)^2} ,$$

where $s$ is the signal to noise ratio.

## 4. Gausslike Functions

A discretization of the Gaussian $g(x) = (\sqrt{2\pi}\,\sigma)^{-1}\cdot \exp(-x^2/2\sigma^2)$ will be called a Gausslike function and denoted by $gl_\sigma$. Our standard way of discretization gives the following nonzero values

$$gl_{\sigma N}(j) = A \cdot \int_{(j-.5)c}^{(j+.5)c} \exp(-x^2/2\sigma^2)\,dx \quad , \quad 0 \leq |j| \leq m \ , \qquad (4.1)$$

where $c = 2\pi/N$ , $m$ is chosen so that the wanted accuracy is achieved and $A$ is a normalizing constant making the $gl_{\sigma N}(j)$ sum to 1.

We obtain a second discretization of $g$ with the aid of the following relation of De Moivre from 1733 in which he describes the convergence of sums of binomial probabilities to integrals of $g$

$$(\tfrac{1}{2})^n \sum_{\lambda_1 \frac{\sqrt{n}}{2} < k - \frac{n}{2} \leq \lambda_2 \frac{\sqrt{n}}{2}} \binom{n}{k} \longrightarrow \frac{1}{\sqrt{2\pi}} \int_{\lambda_1}^{\lambda_2} \exp(-t^2/2)\,dt \ . \qquad (4.2)$$

$$\text{as } n \longrightarrow \infty$$

We shall use this formula in the opposite direction for estimating the integral in (4.1) by a sum of binomial probabilities. The sum will only have one term if

$$\sigma = \frac{c\sqrt{n}}{2} = \frac{\pi\sqrt{n}}{N} \qquad \text{for some integer } n < N. \qquad (4.3)$$

For the values of $\sigma$ given in (4.3), we get the binomial Gausslike function

$$\begin{aligned} bi_{\sigma N}(j) &= (\tfrac{1}{2})^n \binom{n}{j} && 0 \leq j \leq n \\ bi_{\sigma N}(j) &= 0 && n+1 \leq j \leq N-1 \ . \end{aligned} \qquad (4.4)$$

It is the binomial discretization of $g$ that will be shown to have the extremal property mentioned in theorem 1.1.

## 5. The $^{4*s}$-operation

We shall define an operator $^{4*s}$, which transforms a kernel $h$ to another kernel $h^{4*s}$ with the same $\sigma$ as $h$. First, let $^r$ denote reversion of the x-axis so that $h^r(x) = h(-x)$. Next, let the $^{4*}$-operation be defined by

$$h^{4*} = h * h^r * h * h^r \quad . \qquad (5.1)$$

If $h$ has standard deviation $\sigma$ i.e. variance $\sigma^2$, the variance of $h^{4*}$ is $4\sigma^2$, so that the standard deviation of $h^{4*}$ is $2\sigma$.

For functions symmetric around the origin, we define the shrink operation $^s$ by

$$f^s(x) = 2 \cdot f(2x) \quad . \qquad (5.2)$$

The shrink reduces the standard deviation to a half, so we have

Lemma 5.1 $h^{4*s}$ has the same $\sigma$ as $h$ .

For discrete functions, the shrink is performed by doubling of $N$

$$\begin{aligned} f^s_{2N}(j) &= f_N(j) & |j| < N/2 \\ f^s_{2N}(N/2) &= \tfrac{1}{2} f_N(N/2) & \\ f^s_{2N}(j) &= 0 & |j| > N/2 \quad . \end{aligned} \qquad (5.3)$$

## 6. Binomial Kernels Are Worst

The idea of the proof of theorem 1.1, is as follows for functions defined on the whole real axis.

1) start with an arbitrary kernel $h$ and form the sequence $h^{(4*s)^k}$
2) show that this sequence tends to the Gaussian
3) show that the sequence $R(h^{(4*s)^k})$ is increasing
4) conclude that the arbitrary kernel $h$ has a smaller $R$ than the Gaussian.

Point 2) is a direct consequence of the central limit theorem of probability theory (see e.g. [3]), which implies

Theorem 6.1 For every normalized kernel $h$ with $\sigma > 0$, the sequence $h^{(4*s)^k}$, $k = 1,2,3,\ldots$, converges to the Gaussian $g_\sigma$.

The points 1) - 4) and theorem 6.1 are formulated for functions of a continuous variable defined on the whole real axis, while we want a proof for the discretization of a function with bounded support. For functions defined on $(0,2\pi)$, theorem 6.1 is approximately true if $\sigma \ll \pi$ and the functions are cut off outside $(0,2\pi)$ and renormalized. For discretized functions, the cutoff and renormalization is done when the support extends beyond $N$.

Theorem 6.2 For the binomial discretization of the Gaussian, with $n < N/2$ in (4.4), we have

$$(bi_{\sigma N})^{4*s} = bi_{\sigma\, 2N} \quad . \tag{6.1}$$

By this theorem, the family of binomial kernels is mapped on itself by $4*s$, so if $h$ in 1) is a binomial discretization of $g$, all elements in the sequence will be binomial discretizations of $g$. The approximation of $g$ is improved as the sequence proceeds.

In the discrete case, the following theorem replaces point 3).

Theorem 6.3. Let $h(x)$ be normalized and have $\sigma > 0$. If the discretizations $h_N$, $N = 2,3,4,\ldots$ not are binomial kernels, there exists a kernel $v$ (close to $h^{4*s}$) with the same $\sigma$ and a number $N_0$ such that

$$R_{2N}(v_{2N}) > R_{2N}(h_{2N}) \qquad \text{for} \quad N > N_0 \; . \tag{6.2}$$

The following corollary is a more precise statement of theorem 1.1 and point 4).

Corollary 6.4. All kernels but the binomial ones are superseded in ill-posedness, so they are the most ill-posed kernels.

Outline of proof of theorem 6.3. First by (6.1), (6.2) holds with equality if h is binomial. Since convolution of kernels corresponds to multiplication of their FT:s, we have

$$H_N^{4*}(j) = [H_N(j)]^4 \quad . \tag{6.3}$$

Substituting (6.3) in the expression (3.7) for R and using Hölders inequality, we obtain

$$R_N(h_N^{4*}) \geq U_N \cdot [R_N(h_N)]^4 \quad , \tag{6.4}$$

where $U_N$ stems from the $u_{kj}$.

The shrink changes R only a little. We have after some computation

$$\begin{aligned} R_{2N}(h_N^{4*s}) &= \frac{1}{2N}\left(\sum_{m \text{ even}} |H_N(m)|^{-8} + \sum_{m \text{ odd}} |H_N(m)|^{-8}\right) = \\ &= \frac{1}{2} \cdot R_N(h_N^{4*}) + \frac{1}{2N} \sum_{m \text{ odd}} |H_N(m)|^{-8} \; . \end{aligned} \tag{6.5}$$

Skipping the last term, we get

$$R_{2N}(h_N^{4*s}) \geq \frac{U_N}{2} [R_N(h_N)]^4 \; . \tag{6.6}$$

There certainly exists a function v (close to $h^{4*s}$) such that $v_{2N} = h_N^{4*s}$. Since it is not neccesary, we will not try to show that v can be taken as $h^{4*s}$. We shall compare the functions v and h discretized with the same number of points, which by the construction must be 2N.

The binomial kernels have an exponential increase of $R_N$, making the following arguements invalid for them. For other kernels the rate of increase of $R_N$ is, both empirically and theoretically, of the order $N^q$, where q is about 3 or 4.

$$\frac{R_{2N}(h_{2N})}{R_N(h_N)} \approx \frac{(2N)^q}{N^q} = 2^q \; . \tag{6.7}$$

Combining (6.6) and (6.7) we have

$$\frac{R_{2N}(v_{2N})}{R_{2N}(h_{2N})} \gtrsim \frac{U_N}{2^{q+1}}(R_N(h_N))^3 \approx \text{const.}\ U_N \cdot N^{3q} \quad . \tag{6.8}$$

In general, $U_N$ decreases with $N$, but much slower than $N^{-3q}$, so the RHS of (6.8) will exceed 1 for an $N$ of moderate size. For the binomial discretizations of the Gaussian, however, $U_N$ decreases so fast that the above arguement fails.

For the binomial kernels, one can show that

$$R_N(bi_{\sigma N}) \approx \frac{4}{N}\left(\frac{N}{\pi}\right)^{2\cdot(N\sigma/\pi)^2} \tag{6.9}$$

The immense rate of increase of this ill-posedness can be seen by taking e.g. $\sigma = \pi/8$, $N = 16$ and $N = 32$

$$R_{16} = 1.2\cdot 10^5 \qquad R_{32} = 2\cdot 10^{31} \tag{6.10}$$

As a comparison, the Cauchy distribution kernel $h(k) = \text{const.}\cdot(1 + k^2)^{-1}$, normalized to the same $\sigma$ as above, has for $N = 32$ the ill-posedness $R_{32} = 88$ .

## 7. A Method for Deconvolution of Gaussian and Other Kernels

The factor $N$ in the exponent of (6.9) can by (4.3) be expressed in the size $n$ of the support of the kernel

$$R_N(bi_{\sigma N}) \approx \frac{4}{N}\left(\frac{N}{\pi}\right)^{2\cdot n} \tag{7.1}$$

Obviously, it is the size of the support of the kernel that must be kept small. In our experience, problems with $n = 10$ can be solved if some extra stabilizing conditions are used. When $n = 15$ , however, our Dec-10 computer does only produce garbage.

A closer study of the expression for R shows that it is not exactly the size of the support that determines the ill-posedness but rather the number of points in which the kernel is strictly positive. Therefore, $R_N$ can be reduced by use of e.g. the following roughcast discretization of g , which only is positive in every second point

$$gr_{\sigma N}(j) = \begin{cases} gl_{\sigma N/2}(j/2) & \text{if } j \text{ is even} \\ 0 & \text{if } j \text{ is odd} \end{cases} \tag{7.2}$$

With this roughening of the kernel, the ill-posedness is reduced from $R_N$ to $R_{N/2}$ , which can be from $2 \cdot 10^{31}$ to $1.2 \cdot 10^5$ as in (6.10).

## References


[1] Andrews, H.C., and Hunt, B.R., "Digital Image Restoration", Prentice Hall Englewood Cliffs, 1977

[2] Carasso, A.S., Sanderson, J.G. and Hyman, J.M., "Digital Removal of Random Media Image Degradations by Solving the Diffusion Equation Backwards in Time", SIAM J. Numer. Anal. vol.15, no.2, April 1978

[3] Cramer, H., "Mathematical Methods of Statistics" Princeton University Press 1946

[4] Elden, L., "A Program for Interactive Regularization, Part I", Report LiTH-Mat-R-79-25, Linköping University, Sweden

[5] Nashed, M.Z. "Generalized Inverses and Applications", Academic Press 1976

[6] Philip, J., "Digital Image and Spectrum Restoration by Quadratic Programming and by Modified Fourier Transformation", IEEE Trans. PAMI vol. 1, no.4, Oct. 1979

[7] Philip, J., "Error Analysis of the Derivative and Correction Methods for Digital Image and Spectrum Restoration", TRITA-MAT-1981-16, Mathematics, Royal Institute of Technology, Stockholm, Sweden

[8] Philip, J., "Deconvolution of Gaussian Kernels", TRITA-MAT-1982-4, Mathematics, Royal Institute of Technology, 10044-Stockholm, Sweden. (Submitted to SIAM J. Numerical Analysis)

[9] Tikhonov, A.N., "Solution of Incorrectly Formulated Problems and the Regularization Method", Dokl.Akad.Nauk, SSSR 151 (1963), pp.501-504 = Soviet Math. Dokl.4 (1963) pp. 1035-1038

# REGULARIZATION BY LEAST-SQUARES COLLOCATION

Heinz W. Engl

The term "least-squares collocation" (abbreviated "LSC" in the sequel) appears in many different contexts, where functions have to be approximated by terms resulting from finitely many measurements.

Example 1: Let $k \in L^2[0,1]$. We consider the integral equation of the first kind

$$\int_0^1 k(t,s)x(s)ds = y(t) \quad (t \in [0,1]) \tag{1}$$

for $x \in L^2[0,1]$, where $y$ is given (in the range of the integral operator). For $T_n := \{t_1,\dots,t_n\} \subseteq [0,1]$ ("set of collocation points"), $x_n$ should denote the element in $L^2[0,1]$ with minimal norm satisfying

$$\int_0^1 k(t_i,s)x(s)ds = y(t_i) \ (t_i \in T_n). \tag{2}$$

This method of determining approximations for the (minimal norm) solution of (1) is called "LSC" or "moment discretization" and has been investigated in [7], where convergence (with rates) has been studied for the case that $y$ is exactly known.

Example 2: In [8], two-point boundary value problems of the form

$$L_m f = g, \quad f \in B \tag{3}$$

are studied, where $L_m$ is an m-th order linear differential operator and $B$ incorporates linear boundary conditions; results about convergence of LSC and about convergence rates are given. With $T_n$ as in Example 1, the approximation to (3) determined by LSC is defined as the minimal norm solution of

$$(L_m f)(t_i) = g(t_i) \quad (t_i \in T_n).$$

The spaces used are reproducing kernel Hilbert spaces determined by assumed smoothness properties of the solution, by the differential operator, and by the boundary conditions. As in Example 1, the data $g$ are assumed to be known exactly in [8].

Example 3: In physical geodesy, something a little different is called LSC: Assume that $\psi$ is a function of interest to the geodesist (usually, $\psi$ is the anomalous potential of the earth), and that finitely many discrete linear measurements $\ell_1\psi,\dots,\ell_n\psi$ are available. Typically, the functionals $\ell_i$ do not all represent the same physical quantity, but are different in nature. In what geodesists call LSC, one looks for a function of minimal norm in some reproducing kernel Hilbert space that satisfies the given measurements. The choice of the "right" reproducing kernel is thoroughly discussed in the geodetic literature. We refer to [5],[4],[3] for a discussion of LSC in geodesy. The question of convergence of the solutions obtained by LSC and of the effect of measurement errors is not discussed in the literature available to the author.

We now present a common (abstract) framework for LSC that incorporates the special cases described above and that permits us to study convergence and the effect of data perturbations in a unified way. Since the problem of Example 1 is nearly always ill-posed, it is no surprise that a regularization parameter which is hidden in the LSC-process will emerge.

From now on, let $H$ be an arbitrary real Hilbert space, $H_Q$ a reproducing kernel Hilbert space ("RKHS" in the sequel) with continuous reproducing kernel $Q$. Let $K : H \to H_Q$ be linear, bounded, and surjective, and $(m_1,m_2,m_3,\dots)$ be a sequence of linearly independent bounded linear functionals on $H_Q$ (that represent our observations). Let $(d_1,d_2,\dots)$ be a sequence in $\mathbb{R}$ (representing the exact data). The process of LSC is contained in the following setting:

"Problem $(P_n)$": Find $x \in H$ such that

$$m_i(Kx) = d_i \quad (1 \leq i \leq n) \tag{4}$$

and $\|x\|$ is minimal under all solutions of (4).

Remark 4: Since the $(m_i)$ are assumed to be linearly independent and $K$ is surjective, the elements $\{m_i \circ K \mid i \in \mathbb{N}\} \subseteq H^*$ are linearly independent; thus (4) is always solvable, so that the solution of $(P_n)$ exists and is of course unique. Note that our assumptions imply that $R(K) = H_Q$ is infinite-dimensional.

By $\langle\ ,\ \rangle$, $\langle\ ,\ \rangle_Q$, $\langle\ ,\ \rangle_n$ we denote the inner products of $H$, $H_Q$, and the Euclidian inner product in $\mathbb{R}^n$, respectively (similarly for the norms). We assume from now on that the structures of $H$ and $H_Q$ are related in such a way that

$$\langle y_1, y_2 \rangle_Q = \langle K^\dagger y_1, K^\dagger y_2 \rangle \tag{5}$$

holds for all $y_1, y_2 \in H_Q$. Here $K^\dagger$ is the Moore-Penrose inverse of $K$ (see [6],[2]).

Remark 5: The special cases of LSC treated in Examples 1-3 can be brought into the framework of our general setting. For the context of Example 1, it has been shown in [7] that the range of the integral operator is indeed a RKHS with kernel

$$Q(t,s) := \int_0^1 k(t,\tau)k(s,\tau)d\tau. \tag{6}$$

The validity of (5) follows from Proposition 2.1 in [7].
In the context of Example 2 or in the more general setting of [8], it follows from (2.21) of [8] that the surjectivity assumption about $K$ is fulfilled; the functionals $m_i$ in [8] are point-evaluation functionals, the space $H$ also is a RKHS in [8]. To show that (5) holds, it suffices (because of (2.18) in [8]) to show that in the notation of [8], $K^\dagger Q_t = \eta_t$ holds for all $t \in T$, which can be done easily by showing that $K\eta_t = Q_t$ and $\eta_t \in N(K)^\perp$. Finally, the geodetic LSC-problem of Example 3 is a special case of our setting for $H = H_Q$, $K = I$.

In the special cases of Examples 1-3, explicit representations for the solution of $(P_n)$ are well-known. Our first objective is to give such a representation in our framework. To this end, we need the following notation, which is used at least in the geodetic literature.

Definition 6: Let $Q$ be the reproducing kernel of $H_Q$, $m, \ell \in H_Q^*$.

a) $Q(\ell,.)$ is defined to be the element of $H_Q$ for which $Q(\ell,.)(s) = \ell(Q(.,s))$ holds for all $s$ in the interval where the functions of $H_Q$ are defined.

b) $Q(\ell,m) := m(Q(\ell,.)) \in \mathbb{R}$.

Remark 7: It follows from standard RKHS-theory that $Q(\ell,.)$ is the representor of $\ell$ according to the Riesz representation theorem, i.e., $\ell(y) = \langle y, Q(\ell,.)\rangle_Q$ holds for all $y \in H_Q$. Especially, $\langle Q(\ell,.), Q(m,.)\rangle_Q = Q(\ell,m)$, which is a generalized reproducing property. Note that for evaluation functionals $\ell(y) := y(t)$, $m(y) := y(s)$ (for all $y \in H_Q$) we have $Q(\ell,.) = Q(t,.)$ and $Q(\ell,m) = Q(t,s)$, so that our notation makes sense.

Proposition 8: Let $x_n$ be the solution of $(P_n)$. Then we have

$$x_n = (K^\dagger Q(m_1,.),\dots,K^\dagger Q(m_n,.)).Q_{n,(m)}^{-1}\cdot\begin{pmatrix} d_1 \\ \vdots \\ d_n \end{pmatrix}, \tag{7}$$

where

$$Q_{n,(m)} := (Q(m_i,m_j))_{1\le i,j\le n} \; . \tag{8}$$

Proof: Since $\{m_1,\dots,m_n\}$ is linearly independent in $H_Q^*$ by assumption and since $Q(m_i,.)$ is the Riesz representor of $m_i$, $\{Q(m_1,.),\dots,Q(m_n,.)\}$ is linearly independent. Thus $Q_{n,(m)}$ is regular as the Gramian of these elements. Since (because of the surjectivity of $K$) $N(K^\dagger) = \{0\}$, also $\{K^\dagger Q(m_1,.),\dots,K^\dagger Q(m_n,.)\}$ is linearly independent; let $V_n$ be the subspace of $H$ spanned by these elements. It is easily seen that the element in (7) solves (4) and is the only solution of (4) in $V_n$. Now, let $x_n$ be given by (7) and assume that $h \in H$ is another solution of (4), $w := h - x_n$. With $z := Q_{n,(m)}^{-1}\cdot(d_1,\dots,d_n)^T$ we have $\langle w,x_n\rangle = \sum_{i=1}^{n} \langle w,K^\dagger Q(m_i,.)\rangle.z_i = \sum_{i=1}^{n} z_i\langle K^\dagger Kw,K^\dagger Q(m_i,.)\rangle$, since $R(K^\dagger) = N(K)^\perp$ and $(I-K^\dagger K)w \in N(K)$. Because of (5), we obtain $\langle w,x_n\rangle = \sum_{i=1}^{n} z_i\langle Kw,Q(m_i,.)\rangle_Q = \sum_{i=1}^{n} z_i m_i(Kw) = 0$ because of the definition of $w$. Therefore, $\|h\|^2 = \|x_n\|^2+\|w\|^2>\|x_n\|^2$ as soon as $h \ne x_n$. This concludes the proof. □

We are now in the position to prove the following convergence result:

Theorem 9: Assume that $\text{lin}\,\{m_1,m_2,\dots\}$ is dense in $H_Q^*$ and there is a $y \in H_Q$ such that

$$m_i(y) = d_i \quad \text{for all} \quad i \in \mathbb{N}\,. \tag{9}$$

Then this $y \in H_Q$ is unique. Let for all $n \in \mathbb{N}$, $x_n$ be the unique solution of $(P_n)$. Then $(x_n)$ converges (in the norm) to the solution $\bar{x}$ of

$$Kx = y \tag{10}$$

with minimal norm.

Proof: The uniqueness of $y$ is an easy consequence of the Hahn-Banach theorem. Since for $y = 0$ the result obviously holds, we assume $y \ne 0$ and show first that

$$(Kx_n) \rightharpoonup y, \tag{11}$$

where "$\rightharpoonup$" denotes weak convergence.
Let $v = \sum_{j=1}^{i} \lambda_j m_j \in \text{lin}\,\{m_1,m_2,\ldots\}$ be arbitrary. Then, for $n \geq i$, $v(Kx_n) = v(y)$. Thus, $(v(Kx_n)) \to v(y)$ for all $v$ in the dense set $\text{lin}\,\{m_1,m_2,\ldots\}$. To show (11), it thus suffices (by the Banach-Steinhaus theorem) to show that $(\|Kx_n\|_Q)$ is uniformly bounded. By definition of $x_n$, we have for all $n \in \mathbb{N}$

$$\|x_n\| \leq \|K^\dagger y\|, \tag{12}$$

since $y = KK^\dagger y$ and $K^\dagger y$ solves (4). Because of Proposition 8, $x_n \in R(K^\dagger) = N(K)^\perp$, so that $K^\dagger K x_n = x_n$. Thus $\|x_n\|^2 = \langle K^\dagger K x_n, K^\dagger K x_n\rangle = \|Kx_n\|_Q^2$ by (5). Similarly, $\|K^\dagger y\|^2 = \|y\|_Q^2$. It follows now from (12) that for all $n \in \mathbb{N}$, $\|Kx_n\|_Q \leq \|y\|_Q$. Thus (11) holds.
We now show that

$$(\|Kx_n\|_Q) \to \|y\|_Q. \tag{13}$$

Because of the density assumption, we have with $V := \text{lin}\,\{m_1,m_2,\ldots\}$

$$\|y\|_Q = \sup\,\{v(y)/v \in V,\ \|v\| = 1\} \tag{14}$$

(note the symmetry of $V$ and the Hahn-Banach theorem). Let $0<\varepsilon<\|y\|_Q$ be arbitrary, but fixed. Because of (14), there is a $v \in V$ with $\|v\| = 1$ and $\|y\|_Q \leq v(y)+\varepsilon$. Let $i \in \mathbb{N}$ be such that $v \in \text{lin}\,\{m_j/1 \leq j \leq i\}$. Then for all $n \geq i$ we have $v(Kx_n) = v(y)(>0)$ because of (4) and (9). Thus, we have for all $n \geq i$, $\|Kx_n\|_Q \geq |v(Kx_n)| = v(y) \geq \|y\|_Q - \varepsilon$. Together with $\|Kx_n\|_Q \leq \|y\|_Q$ (see above), this implies (13).
By well-known properties of Hilbert spaces, (11) and (13) imply

$$(Kx_n) \to y. \tag{15}$$

Since $x_n - K^\dagger y \in N(K)^\perp$, we have because of (5): $\|x_n - K^\dagger y\|^2 = \langle K^\dagger K(x_n - K^\dagger y), K^\dagger K(x_n - K^\dagger y)\rangle = \langle Kx_n - y, Kx_n - y\rangle_Q = \|Kx_n - y\|_Q^2$. Together with (15) this implies the assertion. □

Remark 10: As special cases, we obtain the convergence results of [7] and [8]. Moreover, for the case of geodetic LSC, we obtain the result that if the span of the measurement functionals is dense and the data come from an actual function in the space considered, then the approximations obtained by LSC converge to that function in the norm of the underlying RKHS, which implies uniform convergence, if the reproducing kernel is continuous and the interval of definition of the functions considered is bounded.

So far, we considered only the case that the data are known exactly. We now assume that for computing the solution of $(P_n)$, we have perturbed data $(\tilde{d}_1,\dots,\tilde{d}_n)$ and know that

$$\|(d_1,\dots,d_n) - (\tilde{d}_1,\dots,\tilde{d}_n)\|_n \le \delta_n. \tag{16}$$

We do not assume that the $\tilde{d}_i$ are the same for $(P_n)$ and $(P_{n+1})$, but avoid using a second index.
Instead of the solution of $(P_n)$, we look at the solution of

"Problem $(\tilde{P}_n)$": Find $x \in H$ such that

$$m_i(Kx) = \tilde{d}_i \quad (1 \le i \le n) \tag{17}$$

and $\|x\|$ is minimal under all solutions of (17); here $(\tilde{d}_1,\dots,\tilde{d}_n)$ is assumed to fulfill (16).

As $(P_n)$, $(\tilde{P}_n)$ has a unique solution which we will denote by $\tilde{x}_n$ and which has the representation (7) with $d_i$ replaced by $\tilde{d}_i$.
Our aim is to obtain necessary and sufficient conditions for both strong and weak convergence of $(\tilde{x}_n)$ to $K^\dagger y$, where $y$ is defined by (9). For the special case of Example 1 and strong convergence, this has been done in [1]. Since integral equations of the first kind fit in our framework, one expects that some kind of regularization has to take place.

Notation 11: By $\lambda_n$ we denote the smallest eigenvalue of $Q_{n,(m)}$, where $Q_{n,(m)}$ is defined as in (8).

Lemma 12: Let $x_n$ and $\tilde{x}_n$ be the solutions of $(P_n)$ and $(\tilde{P}_n)$, respectively. Then

$$\|x_n-\tilde{x}_n\|^2 = \|Kx_n-K\tilde{x}_n\|_Q^2 = \langle Q_{n,(m)}^{-1}(d^n-\tilde{d}^n), (d^n-\tilde{d}^n)\rangle_n,$$

where $d^n = (d_1,\dots,d_n)$, $\tilde{d}^n = (\tilde{d}_1,\dots,\tilde{d}_n)$.

Proof: The first equality follows from (5), since because of Proposition 8, $x_n-\tilde{x}_n \in N(K)^\perp$. We put $v_n := x_n-\tilde{x}_n$, $c := d_n-\tilde{d}_n$. Because of Proposition 8 and the surjectivity of $K$, we have $Kv_n = \sum_{i=1}^{n} w_i \cdot c_i$ with $w_i = \sum_{k=1}^{n} Q(m_k,.)r_{ki}$, where $r_{ki}$ is the $(k,i)$-th entry of $Q^{-1}_{n,(m)}$. We have for all $i,j \in \{1,\dots,n\}$: $\langle w_i,w_j\rangle_Q = \sum_{s,t=1}^{n} r_{si}\cdot r_{tj}\cdot\langle Q(m_s,.),Q(m_t,.)\rangle_Q = \sum_{s,t=1}^{n} r_{si}\cdot r_{tj}\cdot Q(m_s,m_t) = \sum_{s=1}^{n} r_{si}\cdot\delta_{js} = r_{ij}$, since $Q(m_s,m_t)$ is the $(s,t)$-th entry of $Q_{n,(m)}$. Thus, $\langle Kv_n,Kv_n\rangle_Q = \sum_{i,j=1}^{n} c_i\cdot c_j\cdot\langle w_i,w_j\rangle_Q = \sum_{i,j=1}^{n} r_{ij}c_j\cdot c_i = \langle Q^{-1}_{n,(m)}c,c\rangle_n$. This proves the assertion. □

Lemma 13: Let $x_n$ solve $(P_n)$.

a) If $\tilde{x}_n$ solves $(\tilde{P}_n)$ and $\delta_n$ is as in (16), then $\|x_n-\tilde{x}_n\|^2 = \|Kx_n-K\tilde{x}_n\|_Q^2 \leq \delta_n^2\cdot\lambda_n^{-1}$.

b) There is a vector $(\tilde{d}_1,\dots,\tilde{d}_n)$ fulfilling (16) such that for the solution $\tilde{x}_n$ of the corresponding problem $(\tilde{P}_n)$, $\|x_n-\tilde{x}_n\|^2 = \|Kx_n-K\tilde{x}_n\|^2 = \delta_n^2\cdot\lambda_n^{-1}$.

Proof: a) By a usual variational formula, $\lambda_n^{-1} = \sup\{\langle Q^{-1}_{n,(m)}z,z\rangle_n \mid \|z\|_n = 1\}$. Thus, the result follows from Lemma 12.

b) Let $z \in \mathbb{R}^n$ with $\|z\|_n = 1$ be such that $\langle Q^{-1}_{n,(m)}z,z\rangle_n = \lambda_n^{-1}$; such a $z$ exists. With $(\tilde{d}_1,\dots,\tilde{d}_n) := (d_1,\dots,d_n) - \delta_n z$, the assertion holds. □

As an immediate consequence, we obtain the desired convergence result.

Theorem 14: Let the assumptions of Theorem 9 hold and assume $\lim_{n\to\infty}\delta_n = 0$.

a) For each $n \in \mathbb{N}$, let $\tilde{x}_n$ be the solution of $(\tilde{P}_n)$. If $\lim_{n\to\infty}(\delta_n^2\cdot\lambda_n^{-1}) = 0$, then $(\tilde{x}_n) \to K^{\dagger}y$ and $(K\tilde{x}_n) \to y$.

b) Let $\tilde{x}_n$ be as in a). If $\limsup_{n\to\infty}(\delta_n^2\cdot\lambda_n^{-1}) < \infty$, then $(\tilde{x}_n) \rightharpoonup K^{\dagger}y$ and $(K\tilde{x}_n) \rightharpoonup y$.

c) If $\liminf_{n\to\infty}(\delta_n^2\cdot\lambda_n^{-1}) > 0$, then for each $n \in \mathbb{N}$, there exist $(\tilde{d}_1,\dots,\tilde{d}_n)$ fulfilling (16) such that for the solutions $\tilde{x}_n$ of the corresponding problems $(\tilde{P}_n)$, we have $(\tilde{x}_n) \not\to K^{\dagger}y$ and $(K\tilde{x}_n) \not\to y$.

d) If $\lim_{n\to\infty}(\delta_n^2\cdot\lambda_n^{-1}) = \infty$, then for each $n \in \mathbb{N}$, there exist $(\tilde{d}_1,\dots,\tilde{d}_n)$ fulfilling (16) such that for the solutions $\tilde{x}_n$ of the corresponding problems $(\tilde{P}_n)$, we have $(\tilde{x}_n) \not\rightharpoonup K^{\dagger}y$ and $(K\tilde{x}_n) \not\rightharpoonup y$.

Proof: a) follows from Theorem 9 and Lemma 13.

b) Let $v \in \text{lin}\{m_1,m_2,\dots\}$ be arbitrary, but fixed and $i \in \mathbb{N}$ be such that $v = \sum_{j=1}^{n}\alpha_j m_j$ with $\alpha_1,\dots,\alpha_i$ suitable. Then we have for $n \geq i$ with $d^n := (d_1,\dots,d_n)$, $\tilde{d}^n := (\tilde{d}_1,\dots,\tilde{d}_n)$ (here we have to take into

account that the $\tilde{d}_i$ are not the same for $(\tilde{P}_n)$ and $(\tilde{P}_{n+1})$):
$|v(x_n-\tilde{x}_n)|^2 = |\sum_{j=1}^{i}\alpha_j(d^n-\tilde{d}^n)_j|^2 \leq (\sum_{j=1}^{i}\alpha_j^2)\cdot\delta_n^2$, so that $\lim_{n\to\infty} v(x_n-\tilde{x}_n)\to 0$.
Since $\mathrm{lin}\,\{m_1,m_2,\dots\}$ is dense in $H_Q^*$ and $(\|x_n-\tilde{x}_n\|)$ is uniformly bounded by Lemma 13, we obtain from the Banach-Steinhaus theorem that $(x_n-\tilde{x}_n)\rightharpoonup 0$.

Together with Theorem 9, this implies the assertion.
c,d) Let $(\tilde{d}_1,\dots,\tilde{d}_n)$ and $\tilde{x}_n$ be constructed according to Lemma 13b. Since by Theorem 9, $(x_n)\to K^{\dagger}y$, it is necessary for strong (weak) consequence of $(\tilde{x}_n)$ to $K^{\dagger}y$ that $(x_n-\tilde{x}_n)$ converges strongly (weakly) to 0. But since $\|x_n-\tilde{x}_n\|^2 = \delta_n^2\cdot\lambda_n^{-1}$, this contradicts the assumptions in both cases.
We considered only convergence in $H$. The analogous results in $H_Q$ claimed in the Theorem follow from the (weak) continuity of $K$ and from (5). □

<u>Remark 15:</u> Note that in the situation of Theorem 14b, we obtain pointwise convergence of $(K\tilde{x}_n)$ to $y$, while we obtain uniform convergence under the assumptions of Theorem 14a (for bounded reproducing kernel $Q$). Theorem 14 generalizes the main results of [1] and settles the conjecture mentioned there about weak convergence.
Theorem 14 identifies $\lambda_n$ as "regularization parameter": The error level $\delta_n$ has to be decreased in such a way that $\delta_n^2\cdot\lambda_n^{-1}$ goes to 0 or remains bounded in order to ensure convergence. Note that typically $\lambda_n\to 0$ as $n\to\infty$. For concrete situations, where $Q$ is given, such as in geodesy or for a concrete integral equation , estimates of $\lambda_n$ in dependence on $Q$ and the measurement functionals $m_i$ would be of interest in order to find some optimal choice of number and location of the measurement functionals in dependence of the error level.

In [1], we illustrated our results for the problem of numerical differentiation. Here, we treat a simple boundary value problem:

<u>Example 16:</u> We treat the problem considered as an example in [8]. Since this problem can be solved by integration, it is only a model problem here.
Let $H = \{x:[0,1]\to\mathbb{R}\,|\,x''$ absolutely continuous, $x'''\in L^2[0,1]$, $x(0)=x(1)=0\}$ with inner product $\langle x_1,x_2\rangle := \int_0^1 x_1'''(s)x_2'''(s)ds + x_1''(0)x_2''(0)$ and let $H_Q := \{y:[0,1]\to\mathbb{R}\,|\,y$ absolutely continuous, $y'\in L^2[0,1]\}$

with inner product $\langle y_1,y_2\rangle_Q := \int_0^1 y_1'(s)y_2'(s)ds + y_1(0)y_2(0)$. Let $K : H \to H_Q$ be defined by $Kx := x''$. All our general assumptions are fulfilled, $H_Q$ has the reproducing kernel $Q(t,s) = 1 + \min\{s,t\}$.

For each $n \in \mathbb{N}$, let $T_n = \{2^{-n}, 2.2^{-n}, 3.2^{-n}, \ldots, 1\}$ be a set of equally spaced collocation points. We want to solve the problem $x'' = y$, $x(0) = x(1) = 0$ with $y \in H_Q$ given by LSC. The solution of the corresponding problem $(P_n)$, which we call $x_n$, is determined by $x_n''(t) = y(t)$ $(t \in T_n)$, $x_n(0) = x_n(1) = 1$, $\|x_n\| \to \min$. Let $\tilde{y}_n$ be a vector with $\|(y(t)|t \in T_n) - \tilde{y}_n\| \leq \delta_n$. The solution of the corresponding problem $(\tilde{P}_n)$ will be denoted by $\tilde{x}_n$. In order to apply our results, we have to estimate $\lambda_n$.

Since the $m_i$ are point evaluation functionals at $t_i$, the matrix $Q_{n,(m)}$ is given by

$$Q_{n,(m)} = 2^{-n}.\begin{pmatrix} 1+2^n & 1+2^n & \ldots\ldots & 1+2^n \\ 1+2^n & 2+2^n & \ldots\ldots & 2+2^n \\ \cdots & \cdots & \cdots & \cdots \\ 1+2^n & 2+2^n & \ldots\ldots & 2^n+2^n \end{pmatrix} .$$

The inverse can be directly calculated:

$$Q^{-1}_{n,(m)} = 2^n.\begin{pmatrix} \frac{2+2^n}{1+2^n} & -1 & 0 & 0 \ldots 0 & 0 & 0 \\ -1 & 2 & -1 & 0 \ldots 0 & 0 & 0 \\ 0 & -1 & 2 & -1 \ldots 0 & 0 & 0 \\ \cdots & \cdots & \cdots & \cdots & \cdots & \cdots \\ 0 & 0 & 0 & 0 \ldots -1 & 2 & -1 \\ 0 & 0 & 0 & 0 \ldots 0 & -1 & 1 \end{pmatrix}$$

(the format of this matrix being $2^n$). Since the row-sum norm of $Q^{-1}_{n,(m)}$ is bounded by $4.2^n$, we have $\lambda_n \geq 2^{-n-2}$. On the other hand, trace $(Q^{-1}_{n,(m)}) \geq 2^n.(2^{n+1}-2)$, so that the largest eigenvalue of $Q^{-1}_{n,(m)}$ is at least $2^{n+1}-2$. Thus, $\lambda_n \leq 2^{-n}$. By Theorem 14, necessary and sufficient (in the precise sense stated there) for strong (weak) convergence of $(\tilde{x}_n)$ to the solution of the boundary value problem is that $\delta_n^2.2^n \to 0$ $(\delta_n^2.2^n$ bounded, respectively). From the optimal convergence rates for the error free case given in [8] and our sharp estimates for the effect of error it appears that the best possible convergence rate in dependence of $\delta_n$ is $O(\delta_n^{2/3})$, which is achieved by a choice of $n \sim \frac{2}{3}\log(\delta_n)$ (which actually determines the necessary accuracy for the n-th step). However, this has not been thoroughly investigated, since the error bounds given in [8] are pointwise.

## References

[1] H.W. Engl, On least-squares collocation for solving linear integral equations on the first kind with noisy right-hand side, Boll. di Geodesia e Sc. Affini 41, 3 (1982), to appear.

[2] C.W. Groetsch, Generalized Inverses of Linear Operators, Dekker, New York 1977.

[3] T. Krarup, Some remarks about collocation, in: H.Moritz-H.Sünkel (eds.), Approximation Methods in Geodesy, Wichmann, Karlsruhe 1978, 193-209.

[4] P. Meissl, Hilbert spaces and their application to geodetic least squares problems, Boll. di Geodesia e Sc. Affini 35, 1 (1976), 49-80.

[5] H. Moritz, Advanced Physical Geodesy, Wichmann, Karlsruhe 1980.

[6] M.Z. Nashed (ed.), Generalized Inverses and Applications, Academic Press, New York 1976.

[7] M.Z. Nashed - G. Wahba, Convergence rates of approximate least squares solutions of linear integral and operator equations of the first kind, Math. Comp. 28 (1974), 69-80.

[8] G. Wahba, A class of approximate solutions to linear operator equations, J. Approx. Theory 9 (1973), 61-77.

# CONTRIBUTORS

H.G. Bock
Institut für Angew. Mathematik
Universität Bonn
Wegelerstr. 10
D-5300 Bonn
Fed.Rep.Germany

Prof.G. Dahlquist
Department of Computer Science
Royal Institute of Technology
S-10044 Stockholm 70
Sweden

Prof. A.R. Davies
Department of Applied Mathematics
Physical Sciences Building
The University College of Wales
Penglais
Aberystwyth
Dyfed SY23 3BZ
UK

Prof.P. Deuflhard
Institut für Angewandte Mathematik
der Universität Heidelberg
Im Neuenheimer Feld 293
D-6900 Heidelberg
Fed.Rep.Germany

Prof.L. Eldén
Department of Mathematics
Linkoeping University
S-58183 Linkoeping
Sweden

Prof.H. Engl
Johannes Kepler Universität
Institut für Mathematik
Altenberger Str. 69
A-4040 Linz
Austria

Prof.R. England
Dept.of Numerical Analysis
Universidad Nacional Autonoma de Mexico
Apartado Postal 20-726
Mexico 20, D.F.
548-54-65
Mexico

Prof.R.E. Ewing
Department of Mathematics
University of Wyoming
Laramie, Wyoming 82071, and

Mobil Research and Development Corp.
Field Research Laboratory
P.O.Box 900
Dallas, Texas 75221
USA

Prof.L.R. Fletcher
Department of Mathematics
University of Salford
Salford
Lancashire M5 4WT
UK

Prof.P.C. Franzone
Istituto di Analisi Numerica
Corso Carlo Alberto 5
27100 Pavia
Italy

Prof.C.W. Gear
University of Illinois at Urbana-Champaign
Department of Computer Science
1304 West Springfield Avenue
Urbana, Illinois 61801-2987
USA

Prof.O.H. Hald
University of California
Department of Mathematics
Berkeley, California 94720
USA

Prof.A.J. Hermans
Technische Hogeschool Delft
Afdeling der Algemene Wetenschappen
Onderafdeling der Wiskunde
Julianalaan 132
2628 BL Delft
The Netherlands

Prof.U. Hornung
Institut für Numerische und Instrumentelle Mathematik
Universität Münster
Einsteinstr. 62
D-4400 Münster
Fed.Rep.Germany

Dr.H. Lachmann
Institut für Physikalische Chemie
der Universität Würzburg
Marcusstr. 9-11
D-8700 Würzburg
Fed.Rep.Germany

Prof.R. Lipperheide
Hahn-Meitner-Institut
Glienicker Str. 100
D-1000 Berlin 39
Fed.Rep.Germany

Dr.E. Mundry
Niedersächsisches Landesamt
für Bodenforschung
Stilleweg 2
D-3000 Hannover 51
Fed.Rep.Germany

Prof.F. Natterer
Institut für Numerische und Instrumentelle Mathematik
Universität Münster
D-4400 Münster
Fed.Rep.Germany

Prof.N.K. Nichols
Department of Mathematics
University of Reading
Whiteknights
Reading, Berks.
UK

U. Nowak
Institut für Angewandte Mathematik
der Universität Heidelberg
Im Neuenheimer Feld 293
D-6900 Heidelberg
Fed.Rep.Germany

Dr.J. Philip
Department of Mathematics
Royal Institute of Technology
S-10044 Stockholm 70
Sweden

Dr.S.W. Provencher
European Molecular Biology Laboratory
Meyerhofstr. 1
D-6900 Heidelberg
Fed.Rep.Germany

J. Schloeder
Institut für Angewandte Mathematik und Informatik
der Iniversität Bonn, SFB 72
Wegelerstr. 6
D-5300 Bonn
Fed.Rep.Germany

Prof.F.F. Seelig
Lehrstuhl für Theoretische Chemie
Universität Tübingen
Auf der Morgenstelle 8
D-7400 Tübingen
Fed.Rep.Germany

Prof.P.E. Zadunaisky
Centro Espacial San Miguel
Av.Mitre 3100 - (1663) San Miguel (B.A.)
Argentina